WEB PROCESSING AND CONVERTING TECHNOLOGY AND EQUIPMENT

Edited by

Donatas Satas

VNR VAN NOSTRAND REINHOLD COMPANY
NEW YORK CINCINNATI TORONTO LONDON MELBOURNE

Copyright © 1984 by Van Nostrand Reinhold Company Inc.

Library of Congress Catalog Card Number: 83-12417
ISBN: 0-442-28177-3

All rights reserved. No part of this work covered by the copyright hereon may be reproduced or used in any form or by any means—graphic, electronic, or mechanical, including photocopying, recording, taping, or information storage and retrieval systems—without permission of the publisher.

Manufactured in the United States of America

Published by Van Nostrand Reinhold Company Inc.
135 West 50th Street, New York, N.Y. 10020

Van Nostrand Reinhold
480 Latrobe Street
Melbourne, Victoria 3000, Australia

Van Nostrand Reinhold Company Limited
Molly Millars Lane
Wokingham, Berkshire, England

Macmillan of Canada
Division of Gage Publishing Limited
164 Commander Boulevard
Agincourt, Ontario MIS 3C7, Canada

15 14 13 12 11 10 9 8 7 6 5 4 3 2 1

Library of Congress Cataloging in Publication Data
Main entry under title:

Web processing and converting technology and equipment.

   Includes index.
   1. Paper coatings. 2. Paper converting machinery.
I. Satas, Donatas.
TS1118.F5W4 1984   667'.9   83-12417
ISBN 0-442-28177-3

Contributors

William G. Baird, Jr. is Vice President, Science & Technology, Cryovac Division, W. R. Grace & Co. He holds B.S. (1941) and M.S. (1943) degrees in chemical engineering from Washington University, St. Louis, Mo.; advanced management program, Harvard Business School. Mr. Baird pioneered the development of shrink films and the use of electron beam crosslinking for packaging materials and he has many patents and publications in this field. He was winner of the Packaging Institute Professional Award (1977). In 1980 he received the principal award at the Third International Meeting of Irradiation Processing, Tokyo.

John Eric Bell has studied mechanical engineering at Bradford Technical College, England and has over 20 years experience in the coating and converting industry. Mr. Bell was the product manager for rotary screen process machinery, Stork Screens America, Inc. Currently he is associated with J. Josephson, Inc., So. Hackensack, N.J.

Wayne M. Collins is Sales Manager, Enercon Industries Corporation, Menomonee Falls, Wisc. Mr. Collins is a graduate of the Cleveland Institute of Electronics and has close to 20 years experience in the converting industry, mainly in various applications of electronic and electrical equipment.

Dan Eklund, since 1974, has been professor in paper chemistry and head of the laboratory of paper chemistry at the University of Åbo Akademi, Finland. Dr. Eklund holds an M.Sc. in chemical engineering from Chalmers University of Technology in Sweden and a Ph.D. from Åbo Akademi, Finland. He has authored and co-authored over 50 papers, mainly in the pigment coating field.

Orland W. Grant is Regional Manager, Ross-Waldron Division of Midland-Ross Corporation, New Brunswick, N.J. He holds a B.M.E. degree in mechanical engineering from George Washington University, Washington, D.C. and has pursued advanced engineering management courses. Mr.

Grant has over 30 years experience in the application of coating, drying and converting equipment.

Thomas S. Greiner is Management Consultant for the Jagenberg group of companies and was President of Jagenberg USA, Inc. (1961-1979). A graduate of the Elizabeth Institute, Vienna, he also studied at the Polytechnic Institute, London.

David R. Hardt is President, American Tool & Machine Company, Fitchburg, Mass. His educational background is in engineering and business administration. He has been associated with a number of converting equipment manufacturers, and he holds patents in printing, laminating and drying technology.

Ernst K. Hartwig is associated with Leybold-Heraeus GmbH, Hanau, West Germany. Mr. Hartwig is a graduate with a degree in physics from the Free University of Berlin. He is the author of numerous technical publications.

Frank Jacobi is Sales Manager for Brueckner Maschinenbau, Siegsdorf, West Germany. Mr. Jacobi is an engineer with a degree in process engineering. His main field is film orientation equipment.

Joseph J. Keers is Market Manager, Industrial Markets, for the Static Controls Systems Division, 3M Company, St. Paul, Minn. Mr. Keers has a B.S. degree in physics from Penn State University and an M.B.A. from Mankato State University. His experience is mainly in the field of static elimination.

Michael J. Larkin attended Rutgers University and was associated with a number of leading equipment producers in various capacities. Mr. Larkin is a resident of Mechanicsville, Va.

Kenneth A. Mainstone, Manager of Plastics Operations with Black Clawson Company, Fulton, N.Y., was educated in England, majoring in mechanical and marine engineering. He is the author of two U.S. patents related to extrusion converting.

Lee A. Mushel is Manager of Process Development with Faustel, Inc., Butler, Wisc. Mr. Mushel has a B.S. degree in chemistry from the University of Wisconsin (1965). His experience is mainly in converting and the converting equipment industry.

John A. Pasquale III is Vice President of Liberty Machine Co., Inc., Paterson, N.J. Mr. Pasquale is a graduate of the Stevens Institute of Technology, with B.S. and M.S. degrees in mechanical engineering. He has been involved with various aspects of web processing equipment and has contributed several articles to the *Modern Plastics Encyclopedia*.

Joseph E. Radomski, Product Manager, Sheeting Systems, Lenox Machine Co., Lenox, Mass., has a B.A. degree from Beloit College and an M.B.A. from Northwestern University. He has spent over 30 years in the paper manufacturing and paper processing equipment industries. Mr. Radomski is a charter member of TAPPI, Finishing Division, and he is presently the chairman of its Packaging Committee.

Donatas Satas is an independent consultant specializing in adhesives, coating and laminating technology (Satas & Associates, 99 Shenandoah Rd., Warwick, R.I.). Mr. Satas has a B.S. degree in chemical engineering from the Illinois Institute of Technology (1953) and has done graduate work at the same school. He conducts technical seminars and has authored some 12 patents and 20 technical papers. He edited the book *Handbook of Pressure-Sensitive Adhesive Technology*.

E. Raymond Schaffer is an Applications Engineer and Acting Field Service Supervisor for Bobst Champlain, Inc., Roseland, N.J.

Preface

Processing of continuous webs is an old technology which started with textile treatment and paper coating for wallcovering application. These materials were joined by plastic films and by metal foils, all requiring surface treatment for one or another purpose. In addition, combining of various substrates by lamination has become an important technology.

The equipment used for these web converting and treatment applications has evolved gradually. Invention of continuous paper manufacturing also created a need for continuous paper treatment, such as coating and printing. Equipment developed for one specific purpose found applications in other fields. Materials which were introduced later (e.g., plastic films) benefited by the web handling technology developed for paper and textiles.

While a part of web coating technology is old, new equipment development continues. The coating and laminating industry is changing, because new products are constantly developed and new techniques introduced. Increased use of hot melt coatings created a need for different application equipment. Introduction of radiation curable coatings based on reactive monomers and oligomers created a need for new application and curing equipment.

A large number of technical personnel are involved in this industry. Some are engaged in the equipment design and construction, more in product development and manufacturing. The person in the product development area might not be too familiar with various equipment and processes available to him and choosing of the proper hardware might be a difficult task. The equipment designer often does not have much background in the processing area. This gap between the user and the manufacturer of web processing equipment can cause expensive mistakes and interfere with a successful product development effort. One of the purposes of this book is to help to bridge this gap.

The emphasis in preparing this book has been on the equipment, rather than process. The equipment has been described from the point of view of the user, not the designer. The focus is on web converting, rather than web manufacturing, although some equipment is interchangeably used for both.

I would like to thank the contributing authors as well as the many individuals whose work has been used in reviewing this field. My special gratitude is to Mr. G. W. Eighmy, Jr., of Farrel, and to Messrs. R. E. Carlson and L. W. Sundberg, of Hydralign, Inc., for their help in preparing some of the chapters. I would also like to thank the editorial staff of Van Nostrand Reinhold Company, especially Ms. Susan Munger and Ms. Alberta Gordon for their help and guidance in putting this book together. Finally, I extend thanks to my son Paul, who found time to prepare many drawings, my daughter Audrone, for typing help, and my wife Saule, for her help with index preparation.

D. Satas

Warwick, R.I.

Contents

Contributors/v
Preface/ix

1. **REVERSE AND TRANSFER ROLL COATING, Donatas Satas/1**

    Reverse roll coaters/3   Transfer roll coaters/10   Second generation transfer coaters/12

2. **GRAVURE COATERS, Michael J. Larkin/15**

    Direct gravure coating/16   Elements of a gravure coater/18   Hybrid gravure coaters/24   Applications/27   Gravure in laminating/28   Engraving methods/31

3. **BLADE AND AIR-KNIFE COATING, Dan Eklund/34**

    Pigment coatings/34   Blade coating/35   Air-knife coating/54

4. **OTHER KNIFE AND ROLL COATERS, Orland W. Grant and Donatas Satas/60**

    Knife coaters/60   Levelon coaters/67   Bar coaters/68   Squeeze roll coaters/72   Kiss roll coaters/74   Cast coaters/76   Meniscus coaters/78   Brush coaters/78

5. **ROTARY SCREEN COATING, John Eric Bell/81**

    Screens/81   Rotary screen system/87   Machine mechanics/89   Screen choice/90   Knife/screen mechanics/91   Rheology/93

xii CONTENTS

6. SPRAY COATING, Donatas Satas/97

Atomization/97 Air spraying/98 Hydraulic spraying/100 Electrostatic spray guns/100 Rotary disks/102 Roll flinger type coaters/103 Auxiliary equipment/104

7. CALENDERING, Donatas Satas/105

Calender types/107 Laminating/108 Pressure-sensitive tapes/110 Friction calendering/112 Rolls/114 Drives/119 Accessories/120

8. EXTRUSION, Kenneth A. Mainstone/122

Materials/123 Adhesion/124 Primer coaters/125 Extruders/126 Dies and Adapters/129 Coextruders/132 Combining adapter/133 Dual slot die/137 Multi-manifold die/138 Web handling/140 Safety/145

9. HOT MELT COATERS, Donatas Satas/147

Melters/148 Curtain coaters/150 Slot orifice coaters/156 Roll coaters/158 Gravure coaters/161 High-viscosity melt coaters/162 Wax coaters/168 Finishing/170 Foaming/171

10. POWDER COATING, Donatas Satas/173

Scatter coating/174 Powder spot coating/175 Rotary screen printing/176 Fluidized bed coating/178 Spray coating/179 Knife coating/181

11. HIGH-VACUUM ROLL COATING, Ernst K. Hartwig/182

Thin film technology/183 Thin film properties/183 Coating materials and substrates/184 Process and equipment/184 Standard types of roll coating equipment/203 Product application/207

12. SATURATORS, David R. Hardt/213

Pre-wet section/213 Immersion section/214 Metering systems/218

13. LAMINATING, Lee A. Mushel/224

Web path control/228 Wet bond laminating/229 Wet bond problem analysis/230 Dry bond laminating/231 Dry bond problem analysis/232 Thermal laminating/233 Thermal laminating problem analysis/235 Ancillary equipment/236

14. SURFACE TREATMENT, Wayne M. Collins/241

Methods of corona generation/242 Corona treating applications/246 Corona treater selection/247

15. DRYING, Donatas Satas/250

Drying process/250 Convection driers/255 Through drying/268 Conduction drying/268 Infrared radiation drying/271 Dielectric driers/285 Construction of driers/288 Air handling/301 Solvent recovery and incineration/308 Inert gas drying/310 Curling/313

16. ULTRAVIOLET IRRADIATION, Donatas Satas/319

Radiation sources/321 Reflectors/324 Cooling/326 Application of UV-curable coatings/329

17. ELECTRON BEAM IRRADIATION, William G. Baird, Jr./331

Electron beam generator system components/333 Electron beam system parameters/342 Electron beam power output and production capacity/345

xiv CONTENTS

18. **FILM ORIENTATION, Frank Jacobi/352**

Sequential film orienting equipment/352 Simultaneous orienting equipment/358 Tubular film orienting equipment/358 Other methods/362

19. **WINDING, SLITTING AND SPLICING, Thomas S. Greiner/364**

Primary winders/364 Secondary winders/369 Slitting systems/377 Spreaders/381 Surface winders/382 Center winders/389 Splicing/390

20. **WEB HANDLING, Donatas Satas/394**

Tension/394 Web guiding/401 Squaring rolls/413 Spreader rolls/414 Slitting and slicing/418 Accumulators/425 Thickness gauging/428

21. **SHEETING, Joseph E. Radomski/434**

Equipment design and selection/438 Roll unwind equipment/440 Slitting/446 Metering section/448 Cutting section/451 Delivery system/453 Collection section/456 Static removal/458

22. **DIE CUTTING, E. Raymond Schaffer/461**

Folding cartons/463 Reciprocal die cutters/464 Feeder/467 Cutter creaser/472 Stripper/476 Delivery/478 Dies/481

23. **EMBOSSING AND RELATED PROCESSES, John A. Pasquale III/485**

Thermoplastic webs/485 Products/487 Embossing machines for thermoplastic webs/489 Non-thermoplastic embossing/495 Vacuum embossing/496 Valley printing/497 Polishing/501 Buffing and sanding/502 Perforation/502 Transfer printing/505

24. STATIC ELECTRICITY, Joseph J. Keers/511

Conductive versus nonconductive/513 Static elimination/514
Induction neutralizers/514 Electrically-powered neutralizers/515
Nuclear-powered neutralizers/517 Combination bar/518
Static-caused problems/519 Fire and explosion problems/520
Dust and lint problems/521 Degraded-destroyed products/522
Handling Problems/522 Personnel Shocks/523

Subject Index/525

WEB PROCESSING AND CONVERTING TECHNOLOGY AND EQUIPMENT

1
Reverse and Transfer Roll Coating

Donatas Satas
Satas & Associates
Warwick, Rhode Island

Coating operations which employ rolls to deposit, meter or level polymeric coatings over a running web are the most often used methods of web coating. Coaters can be subdivided into two categories on the basis of whether the equipment applies a premetered amount of coating, resulting in a uniform coating thickness regardless of the irregularities in the web thickness, or applies the coating to a constant total thickness by postmetering the coating after its application. This is illustrated in Fig. 1-1. Roll coaters may belong to either of the two categories, but most of them (reverse roll, gravure, calender) apply premetered coatings.

Roll coating can be subdivided into three categories on the basis of the forces acting on the coating film:

- Film peeling
- Film wiping
- Film splitting.

Film, after it is formed on the roll, may be peeled away cleanly and transferred to the substrate. Calender coating belongs to this category. The cohesive strength of the coating used must be high in order to prevent splitting of the coating between the substrate and the roll. This requirement limits the materials that can be used for calendering to elastomers and polymer blends of substantial cohesive strength yet of sufficient thermoplasticity to allow formation into a film in the nips of the rolls. Calendering is discussed in Chapter 7.

If a film is a liquid of a low cohesive strength, such as a polymer solution

Fig. 1-1. Comparison of premetered and postmetered coatings. a) Postmetered coating resulting in a constant total thickness. b) Premetered coating resulting in a constant coating thickness.

either in an organic solvent or in water, aqueous emulsion, or a hot melt, it will split between the substrate and the roll. For this process to take place, the coating must adhere sufficiently well to both web and the roll and the roll must rotate in the same direction and at the same surface speed as the web.

The wet film thickness on rolls rotating in the same direction at equal speeds was investigated experimentally by Schneider.[1] The liquid film is compressed between the rolls and then split on the other side of the nip. A simple mathematical relationship describes the dependency of film thickness on the gap between the rolls.

$$t = K c/2 \tag{1}$$

where

t = wet film thickness
c = gap between the rolls
K = a function of all the other variables (its value varies between the following limits: $1.192 < K < 1.333$, for the conditions in practical coating).

This equation corresponded to the experimental data within 5½%. Roll speed had no effect on the wet film thickness. The increase of viscosity in the range 0.1–1 Pa·s has shown only a slight increase in the wet film thickness. Surface tension had no effect.

Meyers[2] has discussed the splitting of viscoelastic coating between the roll and the substrate. He divides the flow in the nip into four regions: laminar flow, cavitation, cavity expansion, and filamentation.[3] Figure 1-2 illustrates this behavior. The film surface is rough after splitting. It levels at a rate dependent on the rheological properties of the coating. Sufficient time must be

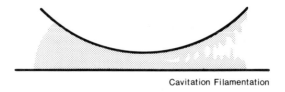

Fig. 1-2. Splitting of a fluid in a nip between roll and substrate.

allowed to elapse before the film surface is hardened in order to obtain a smooth coating surface. In some cases a leveling bar might be employed to accelerate the surface leveling. Roll coating with film splitting has also been analyzed by Middleman,[4] Greener and Middleman[7] and Tharmalingam and Wilkinson.[8]

The split film coating process takes place in direct coaters, where the roll rotates in the direction of the web travel: transfer roll coating, size press, gravure coating, direct roll hot melt coating.

REVERSE ROLL COATERS

Reverse roll coaters are the most versatile and accurate coating machines and their use is widely spread for many applications. These coaters can handle a large range of viscosities: from water thin coatings to viscosities as high as 50 Pa·s. Reverse roll coaters have also been adapted for hot melt coating and the roll temperature can be maintained at 150°C. The coating weight up to 50 g/m^2 can be deposited. This is on the lower side of the coating range, but certainly sufficiently heavy for most of decorative and adhesive coatings.

Reverse roll coaters are expensive to build. A high degree of accuracy is required in the machined parts in order to have a properly operating machine. Some of the steel rolls may be built to the TIR* of 1 μm.[5] The rolls are usually constructed from chilled cast iron and are chrome plated for corrosion resistance. The backing roll that carries the substrate is rubber covered.

The width of the machines varies from narrow pilot plant coaters to as wide as 480-cm coaters. The most popular are 137-cm (54″) and 183-cm (72″) machines. Speeds are possible up to 300 m/min on some of the designs.

The main operating feature of reverse roll coaters is the application of the coating by a roll rotating in the opposite direction to the substrate movement. The name reverse roll coater obviously comes from this feature. In addition, the coating is premetered on reverse roll coaters and the deposit thickness is constant regardless of the substrate.

There are several designs of reverse roll coaters. The main characteristics which distinguish various designs are the type of coating feed arrangement—either nip or pan, or the more recently proposed die fountain; the location of the feed—either top or bottom; and the number of rolls used—either three or four.

Nip feed requires only a small amount of coating to be exposed, and the evaporation rate in case of solvent based coatings is minimized. The foaming

---

*Total Indicator Runout (TIR) is the deviation of a roll as read from the total movement of a dial indicator. It includes deviation because of roll eccentricity as well as deviation from a straight surface.

in the coating supply bank is also maintained at a minimum. Nip feed requires well constructed dams to prevent leakage of the coating and it may be difficult to handle low-viscosity coatings. The dams can also cause a wear on the roll surface. Speeds up to 250 m/min and coating weight up to 50 g/m² are possible.

Pan feeding allows the use of lower viscosity coatings at the expense of increased splashing, solvent loss and foaming in the pan.

Die fountain feed allows elimination of end dams in a nip fed machine since the die fountain maintains the coating within the die width.

Four-roll coaters can be run faster than three-roll coaters (up to 300 m/min). The splashing and foaming in the pan are decreased in a four-roll coater, because a slower rotating fountain roller can be used.

The simplest reverse roll coater is a hybrid between an offset gravure and the reverse roll methods as shown in Fig. 1-3. The basic feature, an applicator roll rotating against the movement of the substrate, is the same as in reverse roll coaters. The metering function is done by an engraved roll. The roll accuracy is less important in these machines and they are less expensive to construct. They are widely used in the metal coil coating. These machines run at speeds up to 300 m/min and can deposit wet coating weights up to 35 g/m².[5]

A top nip fed three-roll coater is shown in Fig 1-4. This machine has the most often used roll configuration for nip fed coaters. The gap between the metering and applicator rolls can be adjusted between 0.025 and 6 mm, but the gap most frequently is set between 0.15 and 0.4 mm. The gap determines the thickness of the wet coating film deposited on the surface of the applicator roll. Depending on the rheological properties of the coating, the wet film thickness is approximately 0.6–0.8 times that of the gap width. The higher the viscosity, the heavier the film.

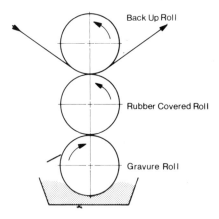

Fig. 1-3. Gravure-reverse roll coater.

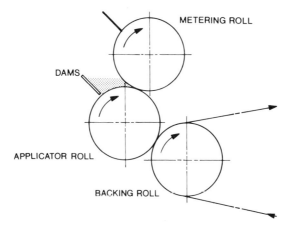

Fig. 1-4. Top nip fed three-roll reverse roll coater.

The metering roll rotates in the opposite direction of the applicator roll rotation. Its speed is 2-20 times slower than the speed of the applicator roll. Higher viscosity requires slower metering roll speed in order to maintain a smooth film surface. For some coating the metering roll might be stationary.

In addition to the gap between the applicator and metering rolls, the coating thickness is controlled by the speed ratio between the applicator and backing rolls, backing roll running at the substrate speed. This ratio is called the wipe ratio and it can be as low as 0.6, although it is usually between 1 and 2, but can be as high as 4.[6] Obviously, the higher the ratio, the faster the rotation of the applicator roll in relation to the web speed, and the higher the deposited coating weight. In order to maintain the same weight throughout the run, it is necessary to maintain the wipe ratio constant, even if the web speed might vary.

The gap between the applicator and the rubber covered backing roll is set below the web thickness so that the web is slightly compressed in the nip. This assures that the coating is completely removed from the applicator roll. Excessive compression should be avoided: it increases the power consumption, wear on the rubber surface and also the strike-through of the coating in case of porous substrate.

Nip fed reverse roll coaters are also used in other roll configurations. Figure 1-5 shows a coater where metering and applicator rolls are arranged horizontally. This eliminates the need of the coating dam and only the edge dams are required. In this configuration the web cannot be run at an angle above the horizontal; the web should remain in contact with the backing roll after it leaves the coating nip, otherwise the coating finish will be damaged by the applicator roll contacting the unsupported web. Moving the backing roll slightly up out of the vertical line helps to alleviate this situation.

6 WEB PROCESSING AND CONVERTING TECHNOLOGY AND EQUIPMENT

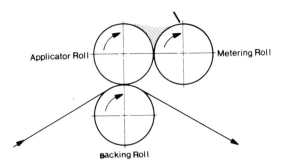

Fig. 1-5. Three-roll reverse roll coater with applicator and metering rolls on the same horizontal axis.

The nip feed may be located on the bottom, as shown in Fig. 1-6. This configuration gives a good visibility of the coated surface which is right in front of the operator.

The most often used pan fed reverse roll coater is shown in Fig. 1-7. The metering roll is smaller than the applicator roll in order to avoid the submersion of the metering roll in the coating. Maximum speed of 150 m/min with a low-viscosity coating is possible. Another variety of the same configuration has a metering roll equal in diameter to the applicator roll and submerged in the coating. This increases the agitation and foaming in the pan. Other variations of the same coater are shown in Figures 1-8 and 1-9. A photograph of reverse roll/knife over roll coater is shown in Figure 1-10.

Four-roll pan fed coaters can be run at higher speeds, because the splashing in the pan is reduced by a slower rotating fountain roll. Figure 1-11

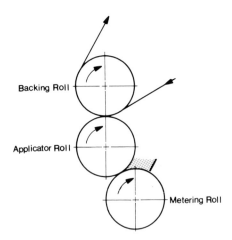

Fig. 1-6. Bottom nip fed three-roll reverse roll coater.

REVERSE AND TRANSFER ROLL COATING 7

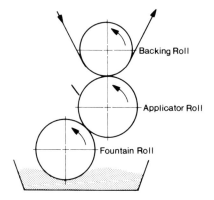

Fig. 1-7. Pan fed three-roll reverse roll coater.

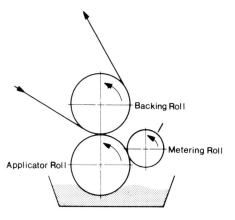

Fig. 1-8. Pan fed reverse roll coater, V-arrangement.

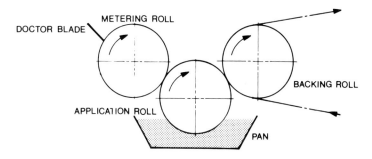

Fig. 1-9. Pan fed reverse roll coater, angular arrangement.

Fig. 1-10. Reverse roll/knife over roll combination. (*Courtesy American Tool and Machine Co.*)

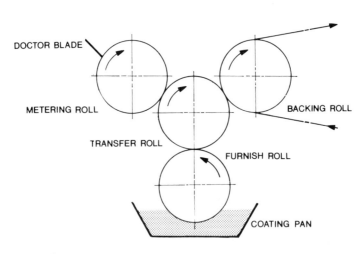

Fig. 1-11. Pan fed four-roll reverse roll coater (Contracoater).

REVERSE AND TRANSFER ROLL COATING 9

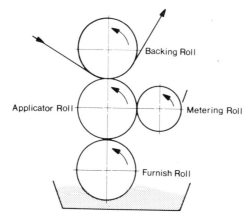

Fig. 1-12. Pan fed four-roll reverse roll coater with a vertical roll configuration.

shows a popular configuration of a four-roll coater known as the Contra-coater. A similar machine with a somewhat different roll arrangement is shown in Fig. 1-12.

Edge doctors might be used on the applicator rolls of pan fed reverse roll coaters to provide a coating-free edge, or to prevent the coating from flowing to the other side of the substrate, especially in case of low-viscosity coatings and thin substrates.

Fountain fed reverse roll coaters eliminate the use of dams. They are especially suitable for striping, because it is simple to restrict the flow of the fountain in the desired areas. A die fountain fed reverse roll coater is shown in Fig. 1-13.

It also has been proposed to use die fountain to coat the substrate directly and then remove the excess by a metering roll.[5] Such a coater as shown in

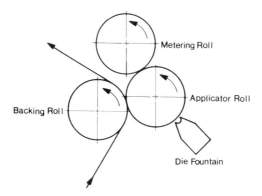

Fig. 1-13. Die fountain fed reverse roll coater.

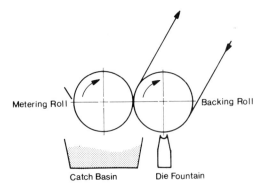

Fig. 1-14. Die fountain coater with direct application on the web.

Fig. 1-14 is not really a reverse roll coater; it is basically the same coater as the Levelon shown in Fig. 4-11.

TRANSFER ROLL COATERS

Transfer roll coaters have been developed for publication grade paper coating. These coaters apply heavily filled clay coating to papers for magazine publishing and other uses requiring a smooth ink-accepting surface. These are large machines designed for fast-speed rugged and continuous service usually in line with paper making. Introduction of air knife and later blade coaters for the same purpose has practically eliminated building of new transfer coaters for paper coating applications. Transfer coaters in existence are still being used. Their use was mainly concentrated in North America. They are being replaced by less expensive blade coaters.

Transfer coaters consist of a series of horizontally mounted rolls, some chrome plated, some rubber covered. The original transfer coater was developed by Consolidated Papers, Inc. and Peter J. Massey, and its schematic diagram is shown in Fig. 1-15. The coating is introduced into the nip of gate rolls which might be as large as 40 cm in diameter. The gap setting, speed of the gate rolls, and especially the solid content of the coating determine the amount of material deposited onto the paper surface. These machines have a range of 3-20 g/m^2 of dry weight coating. The transfer rolls are driven at progressively faster speeds, but below the web speed, except the applicator rolls which run at the web speed. The transfer rolls distribute the coating uniformly across the roll width, disperse any agglomerates and generally smooth out the coating. The applicator rolls might be 120 cm or larger in diameter. The coating is transferred from the applicator rolls to the web. Most of these coaters are designed to apply the coating on both sides of the

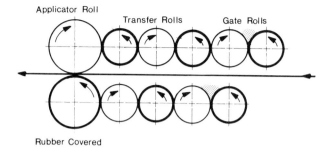

Fig. 1-15. Consolidated-Massey transfer coater.

paper: the demand for two-side coated paper is much larger than for one-side coated.

The surface speeds of various rolls for a typical Consolidated-Massey coater are given below.[6]

	UPPER COATING TRAIN	LOWER COATING TRAIN
	(SPEED, M/MIN)	
Gate roll	67	68
Gate roll	67	68
Transfer roll	238	—
Transfer roll	301	297
Transfer roll	301	298
Applicator roll	314	308

The applicator rolls are run at slightly different surface speeds in order to prevent paper fluttering and resultant cross machine chatter marks in the coating. The surface speed of the lower roll is that of the web. The top roll is run at slightly higher speed so that the paper would follow it after leaving the nip. The transfer coaters are generally run at speeds of 300–950 m/min.

Slight pressure is applied between the transfer rolls. It should not be as high as to cause forming a coating bead at the nip. The applicator rolls operate at a pressure of 3.5–8.5 kg/cm^2.[9] This assures a complete contact with the web and the coating transfer uniformly over the surface.

In addition to the Consolidated-Massey coater, several other coaters have been designed by various paper manufacturers. The Westvaco coater is similar to the Consolidated-Massey, except smoothing rolls are added. The Kimberly-Clark-Mead coater is a two-side coater consisting of large-diameter applicator rolls, metering and a transfer roll. The Combined Locks coater consists of several stations for paper processing in line with paper making. The coater designed by St. Regis-Faeber is similar to the

Consolidated-Massey machine, except the coating is applied on two sides in tandem, rather than simultaneously. The Champion-Hamilton coater meters the coating not by gate rolls, but by a reversely rotating small-diameter rod mounted against the applicator roll.

SECOND GENERATION TRANSFER COATERS

New transfer coaters were not constructed for some time, because of their complete replacement in the paper manufacturing industry by air-knife and by blade coaters. More recently the use of various 100% or high-solid coatings has increased for applications such as silicone release coatings, UV radiation curable coatings and film laminating adhesives. These coatings require an accurate deposition of a low-weight coating and many of the existing coaters have been found deficient for this application. This has revived the interest in the transfer coating process. The process is suitable for light weight applications, because of the speed difference between the rolls. It also is suitable for a complete transfer of the coating, including surface irregularities that might be present in the substrate, because of the deformation of the backing in the coating nip.

Figure 1-16 shows a transfer coater designed for applying 100% solids silicone release coating.[10] Coating may be applied as thin as 5 µm. The metering and applicator rolls are machined to the accuracy of 0.0025 mm TIR. This assures that the metering and applicator rolls rotate to a surface accuracy within 0.005 mm and consequently that the coating thickness deposited on

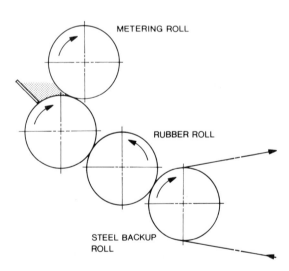

Fig. 1-16. Transfer roll coater. (Microtransfer by Cameron Waldron.[10])

the applicator roll is within these tolerances. The coating thickness deposited is slightly less than the gap between the metering and applicator rolls. For example, a typical gap of 0.25 mm might give a coating 0.23 mm thick.

The metering roll is run countercurrently to the applicator roll at a relatively slow speed. Typical surface speed might be 10 m/min. The applicator roll is run faster, perhaps at a surface speed of 50 m/min. These rolls must be run sufficiently slowly to avoid introduction of air bubbles into the coating.

The coating roll is run 5-25 times faster than the applicator roll. Thus the coating film thickness transferred to the rubber covered coating roll is decreased by this ratio. The coating roll is running at the same speed as the web and the coating is transferred to the web at the nip between the coating and backing rolls.

A very similar transfer coater is shown in Fig. 1-17.[9] This coater has been used for such applications as coating moisture curable polyurethane adhesive for flexible packaging laminates. The machine is designed to apply coating at an elevated temperature of about 95-105°C. Higher temperature decreases the viscosity of the coating and makes it easier to apply in low coating weights.

The adhesive is preheated and pumped into the gate roller nip. The metering roll is heated and the rubber covered transfer roll is cooled, although the rubber surface remains at the coating temperature. The cooling is to protect the rubber cover from deterioration and especially to prevent the rubber-steel bond failure. All three coating cylinders have independently adjustable speeds. The applicator roll is run at the web speed, the transfer roll at about 25% and the metering roll at about 10% of the web speed. The coating width can be controlled by inserting edge dams, using edge doctors or undercutting either the rubber covered transfer roll or the backing roll. The machine is run at speeds up to 250 m/min.

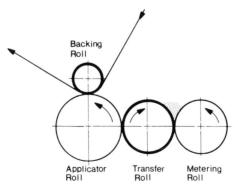

Fig. 1-17. Heated transfer roll coater.

REFERENCES

1. Schneider, G. B. *Trans, Soc. Rheology* **VI**:209-221 (1962).
2. Meyers, R. R. *J. Polym. Sci.* Part C, **35**:3-21 (1971).
3. Miller, J. C. and Meyers, R. R. *Trans. Soc. Rheology* **II**:77-93 (1958).
4. Middleman, S. *Fundamentals of Polymer Processing*. New York: McGraw-Hill, 1977.
5. Zink, S. C. Coating Processes. *Kirk Othmer Encyclopedia of Chemical Technology,* Third Edition. New York: Wiley Interscience, 1979, Vol. 6, 386-426.
6. Booth, G. L. *Coating Equipment and Processes*. New York: Lockwood, 1970.
7. Greener, J. and Middleman, S. *Ind. Eng. Chem. Fundam.* **18**(1): 35-41 (1979).
8. Tharmalingam, S. and Wilkinson, W. L. *Polym. Eng. Sci.* **18**(15): 1155-1159 (1978).
9. Weiss, H. L. *Coating and Laminating Machines*. Converting Technology Co., Milwaukee, Wisc., 1977.
10. Kosta, G. U.S. Patent 4,029, 833 (1977) (assigned to Midland Ross Corp.).

2
Gravure Coaters

Michael J. Larkin
Mechanicsville, Virginia

A series of events and inventions during the 1800's and the early part of the 20th century spurred the development of today's web fed coating equipment. Two of these inventions, namely the Fourdrinier paper making machine, c. 1803, and the first rotating cylinder flat bed letterpress, c. 1814, were significant. Then, in the 1860's, the first web fed letterpress came on stream and in the 1870's the gravure (intaglio) printing process was developed.

After the turn of the century, the trend toward higher speed and the development of new techniques and materials began to meet the demands of an expanding population. A new world opened up for converters, once rolls of paper, paperboard, films and foils could be put in a machine and transformed quickly and economically into a semi-finished or finished product. There was a gradual transformation from labor and scrap intensive sheet fed converting to web fed lines.

It was a transition wherein the gravure coating method emerged as a substitute for what had been slow-speed, plain roll coaters. Since gravure coating was preceded by the gravure printing method, it was probably a printer using a sheet or web fed press who realized it was adaptable for coating as well as applying inks. Today the use of gravure coating is still expanding as a result of not only its accuracy, but its ability to apply the new generation of high-solids and 100% solid liquids. Converters of packaging materials, wallcovering, upholstery, magnetic tape and pressure-sensitive products, to name a few, use gravure coating in their operations.

In this chapter we will address gravure primarily in the sense of it being used for coating only. However, out of necessity some suggestions for its use will be in the discipline of printing (one color only).

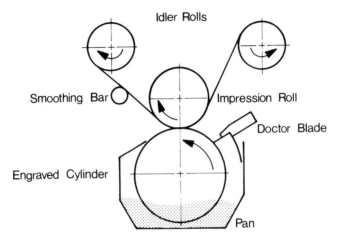

Fig. 2-1. Gravure coating unit.

DIRECT GRAVURE COATING

Direct gravure coating is a process utilizing a driven engraved cylinder, impression roll, web transport rolls, doctor blade, pan and/or applicator to apply a liquid to a web (see Fig. 2-1).

The process operates on the principle of pressing the web onto the engraved cylinder to remove liquid from the engraved cells or lines by capillary action, vacuum or a combination of those two phenomena. Thus the web can be porous or impervious, smooth or rough, but it must be easy to wet. If it is not, it has to be primed or treated.

Advantages and Limits

Gravure coating provides the converter with one of the few methods that will apply a very accurate, predetermined volume of liquid, because it uses a cylinder engraved below its surface to meter the coating before it is applied. Since the engraved area's cells or lines are all uniform in size, except when a moire is desired, the engraved cylinder cannot apply more than the desired volume of coating. Therefore, it will produce consistent run-to-run results over the life of the cylinder.

The key element of a gravure coater, namely the engraved cylinder, can be engraved over its entire surface, or with a single discrete design, stripe, dot or multiples thereof.

This coating method is generally considered as best suited for long run jobs that are repeated often and require very thin (to 25 micron-thick) wet film thicknesses. Theoretically, the thinnest or thickest coating that can be applied is limited only to the size of the cells or lines that can be engraved in a cylinder. But the holding ability of the engraving is dependent on the viscosity of the fluid. Thus, there is a variety of engraved cell designs available (see *Engraving Methods*). There have been occasions when a very viscous coating, with very little lateral flow, has been applied from lines actually machined or hand-engraved in a cylinder. Currency and special note intaglio printing is accomplished with this method.

Some thick coatings are applied by a sand blasted roll, which gives more control than a plain roll and flooded nip.

A typical gravure coating is generally water-like, or as high as 3–5 Pa·s in viscosity. In some cases, viscosities as high as 10 Pa·s have been used in thermoplastic coatings.

The coatings can be at room temperature, hot or cooled. The solids can be in a water or solvent vehicle, or they can be a hot wax or melt. Furthermore, they can be low or high in solids or a 100% solid liquid. Those in the latter category are cured by electron beam or ultraviolet light irradiation or application of heat. The prime criterion for all of these methods for gravure coating is their ability to flow, or the possibility to induce flow by viscosity reduction or pressure.

An engraved cylinder can be small in diameter, or as large as one that would produce a five-foot repeat. The diameter is a function of web width, impression pressure, speed of the process and the repeat required if a registered coat to print job is run.

A standard direct gravure coater has a minimum of parts, with only one driven cylinder, and it can be changed over quickly. Most units are designed to allow manual or cart removal of the cylinder, etc. from the side. Others have a cartridge or trolley type design, wherein all the elements, except the impression roll, are in a mobile module that is inserted from the front or side of the unit. The intent of this configuration is to provide very rapid changeover by having spare modules on hand and ready to be inserted into the coater as soon as the one in the unit is removed.

The accuracy of a gravure coater is not speed-sensitive, so long as each element, which is minimal, is properly adjusted. Therefore, a gravure unit can be the major processing unit with an appropriate drier or curing and web handling systems. Or it can be integrated with other converting equipment, such as a laminator. In either case, the gravure coater should be preceded and followed by a unit that will maintain constant web tension in and out of the unit, to prevent pre- or post-coating web wrap on the engraved cylinder.

There are gravure coaters to coat on the face or back of a web, alternately, without the need of a web turning bar, but with a bi-directional dryer or curing unit. This style of coater is supplied with a doctor blade that can be put on either side of the engraved cylinder, and the cylinder drive is reversible.

Gravure coating also has an advantage for porous webs, because the short dwell time between the coating nip and the dryer or curing unit minimizes penetration of the liquid into the pores and fibers. Extremely porous or rough webs are excluded as suitable for direct gravure coating. The porosity may be so great that the liquid strikes through the web and onto the impression roll. A rough though dense web does not pull liquid from the cells of an engraving that line up with a valley in the web. If either of these conditions prevail, you can investigate using one of the hybrid gravure coating methods. Some converters employ, for rough webs, an electrostatic assist unit. It is being used in most cases for printing, but also deserves consideration for coating.

As one would expect, there are limits to this system that are somewhat constraining, especially coating thickness. In some cases, what may seem to be a major obstacle may be surmounted by reformulating the coating or changing the elements in the coater.

A coating should not be too acidic or abrasive because it will act as a deplater, or lapping compound, and wear the doctor blade, chrome and engraving inordinately. There are substitutes for chrome, such as ceramic, that can be plasma-sprayed on a cylinder. When acidic liquids are used, appropriate base, engraving media and plating are required.

Film splitting with some coatings, especially hot melts and other thermoplastics might take place. They have a tendency to split such that part adheres to the web and part to the engraved cylinder, resulting in cobwebbing around the coater.

Coating weight changes in a direct or offset gravure coater can only be accomplished by changing the engraved cylinder. When web widths change, a series of impression rolls cut back to the required widths are also required, unless the web is very thick. If they are not cut back, the rubber on each side of the web will pick up coating from the engraved cylinder and transfer it to the edges and back of the web (see Fig. 2-2).

Bear in mind, however, that the cost of the base of an engraved cylinder and impression roll can be spread over a long period of time because each of them can be restored to its original condition many times.

ELEMENTS OF A GRAVURE COATER

The Coating Cylinder

The coating (engraved) cylinder is generally the integral type (see Fig. 2-3). It consists of a steel tube into which shafts and gudgeons are shrunk and welded to create a strong one-piece member. Following the turning of

GRAVURE COATERS 19

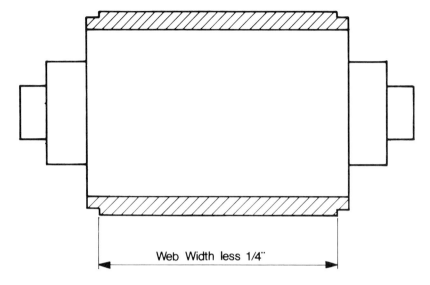

Fig. 2-2. Impression roll with cutbacks for web width

the shafts and journals for the bearings, and machining for the drive gear, the outside of the cylinder is ground and finished with a surface favoring the eventual copper plating and engraving. In some cases, engraving is done on the steel base.

The cylinder base should be hot or cold drawn tubing with a diameter, wall thickness and shafts to withstand the impression pressure. Strive for a maximum deflection of 0.38 mm and a total indicator runout (TIR) of not more than ± 0.025 mm. It should be statically and/or dynamically balanced, depending on the speed. Each base should have an identification number and its wall thickness permanently stamped on it.

A good specification with a drawing, similar to Fig. 2-3, sent to the vendor with the order, is a good practice to assure receipt of a good base cylinder. If

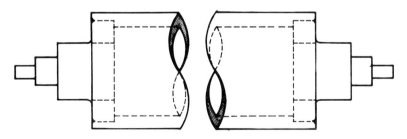

Fig. 2-3. Engraved cylinder base drawing.

20 WEB PROCESSING AND CONVERTING TECHNOLOGY AND EQUIPMENT

the vendor also does the engraving and chrome plating, there is double insurance because he will have central responsibility.

The Doctor Blade

The purpose of the doctor blade is to remove the liquid from the unengraved portions of the engraved roll. Unlike doctor blades used in other systems, such as a plain roll coater, where the blade is used just to clean, a gravure doctor blade is operating on an expensive roll. Needless to say, the doctor must be given careful attention to make sure it is smooth and free of nicks or burrs that can ruin a cylinder in one revolution.

There are two types of doctor blades in common use: a conventional wiping blade and the reverse blade angle or shearing blade (see Fig. 2-4). The conventional blade is generally wiping at an angle of between 17° and 27° to

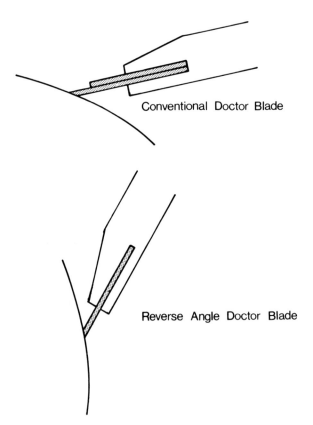

Fig. 2-4. Doctor blade types.

the cylinder axis, a reverse angle at 30°–35°. The reverse angle blade is purported to be less problematical to adjust for pressure, because hydraulic pressure of the liquid keeps it on the cylinder. But alignment and setting of the reverse angle blade requires very careful attention to prevent it from digging into the cylinder.

A conventional doctor blade is usually 0.15-mm blue spring steel, backed up by a 0.25-mm blade to reinforce the doctor and prevent too much flexure under pressure. The combined set is held in a clamp secured to a plate or bar with a back shower deflector, and the assembly is equipped with a series of adjustments. They are: pressure, induced mechanically or pneumatically; parallelism; height; and angle. Most often the assembly is also equipped with an oscillating drive. It induces an even wear pattern on the blade and cylinder and assists to displace foreign matter from behind the blade.

When an engraving consists of a line or lines parallel to the cylinder axis, a doctor blade has a tendency to jump in and out and accelerate wear. Sometimes, a slight skew of the blade will help to alleviate the condition.

Impression Roll

This roll presses the web onto the engraved cylinder and will be found in two designs: an integral type with external bearing, or a tube with internal bearings on a stationary shaft.

The type and thickness of the rubber covering depends on the diameter, impression pressure, web and coating characteristics. The roll covering vendor is the best judge of what elastomer and hardness should be used. Some rules of thumb are: 2-cm minimum thickness; hardness of 60 Shore Gage A for films, 70 for paper and 90 for paperboard.

An impression mechanism will have a pneumatic or hydraulic pressure system, with an off impression circuit energized at machine stop, to allow the unit to go into idle mode, and to prevent a flat from forming on the rubber. It should also have a paralleling adjustment and a calibrated wedge or some type of rigid stop. The latter provides the operator with a repeatable adjustment that is easy to duplicate.

There are occasions when a back-up roll is used on the impression roll to prevent deflection of the roll when it is small in diameter. This design is generally used when a range of web widths is run, and the low weight of the small impression roll hastens changeover. But each of them is then under the influence of the energy produced from two pressure points and will require more frequent regrinds or recoverings.

Some designs have a turret that has two or more different width impression rolls around the perimeter of the back roll to promote quick changeover, without the necessity of removing a roll.

Smoothing Bar

This appliance is used as a post-coating technique to smooth liquids that do not flow well. As it implies, it smooths or levels an irregular coating that would be undesirable. It should be used judiciously, because it does induce pressure and hence tension on the web.

A smoothing bar is usually a driven integral steel roll with a polished chrome, or other smooth surface or covering. The drive should be variable speed and reversible. The assembly will have a pressure system, with an off circuit energized at machine stop, and a pivoting and paralleling adjustment. The bar should be placed as close as possible to the coating nip between two rolls to provide good web support.

The advantages of a smoothing bar lend themselves to being used in other coating methods, such as squeeze roll metering, when the coating has striations or an orange peel effect.

A smoothing bar must be given close attention, to prevent any damage on it from affecting the web and coating.

Coating Pan and Enclosures

The engraved roll is supplied with liquid from a pan with enclosures that are designed to not only act as a reservoir, but also as a container to keep splashing, slinging and pre-coating evaporation of volatiles to a minimum. In some units the pan is equipped with an inner pan that can be adjusted to provide a range of bath depths, as well as induce some hydraulic pressure on the fluid and force it into the engraving.

An ideal pan will be ported, supplied with an overflow weir, and designed to be attached to hoses or rigid piping with quick disconnect couplings. Non-ported pans must be replenished periodically to restore the proper bath depth and are normally used for short runs only.

Materials of construction are preferably stainless steel with polymer end bearers for the coating roll journals. Some pans for wide webs are made of fiberglass. It is also possible to find disposable pan liners to reduce wash-up time and prevent cross-contamination.

Most units support the pan on a pneumatic or mechanical elevator with height adjustment. Simple versions or cartridge-type units will have a stationary support plate and alignment pins.

Coating Applicators, Pumps and Circulating Systems

These accessories make an important contribution to the overall performance of a gravure unit and help to conserve on the ink. Over the years, applicators have been supplied in a number of configurations. They are designed to apply the coating directly onto an engraved cylinder and decrease the size of the pan, if not to eliminate it entirely and substitute just a drip tray.

They will be found in as simple a design as an overflow trough or perforated pipe, to sophisticated units that completely encapsulate a quadrant of the engraved cylinder and have integral doctor blades. They also serve to scuff air (which at high speeds can be an impediment to the coating) out of the engraving. In some cases, the applicator is mounted at a point immediately after the coating nip, to wet the cylinder rapidly and prevent dry-in of any remaining coating. Others may be put under the doctor blade. Those with integral doctor blades are put as close to the coating nip as possible.

An applicator can be cooled or heated and operate at or above atmospheric pressure. Since they are used with a pump and circulating system, they have a built-in capability to prevent settling of solids.

Few gravure coaters will be found with a non-ported pan. Thus they will have a pump, tank and circulating hose or pipe to deliver coating into the pan or applicator, receive the overflow and put it back into circulation. This type of system not only protects the coating, but it also lends itself to incorporation of controls that constantly monitor the viscosity and automatically add vehicle as needed. The tank has a hinged lid to permit replenishment and viscosity sampling, which is a much safer procedure than attempting to do it in the pan near an expensive engraved cylinder.

A filter can also be included in the piping system, to remove foreign matter or over-sized solids that might create scrap or clog an engraving.

Each coating will have specific properties and constituents that must be considered before selecting a pump and circulating system. Discretion must be used in the choice by giving attention to the system's ability to resist acids or solvent, ease of clean-up and, of course, economy. Pumps that create heat should be avoided since they unduly raise the temperature of a volatile coating.

Precautions must be taken with foam-prone materials. In this case, there is an option of using a gravity feed, with a pump for the overflow to return it to a settling and defoaming tank with enough residence time to allow defoaming and eventual return to circulation.

Web Transport Rolls

Conveying of the web in and out of a gravure unit is performed by a series of idler or driven rolls, depending on the web tension.

The rolls should be a minimum of an inch or two wider than the maximum web width, balanced and strong enough to withstand the tension and not deflect. The surface should be smooth, or for films or foils, textured with a herringbone or helical pattern to help break up the air layer they carry with them. If not expelled, the air creates a cushion and prevents the web from touching the rolls.

The roll just prior to the coating nip should have an adjustment to correct for a baggy web edge. If wrinkle-prone webs are run, a bowed roll before the nip will flatten the web before it enters the coating nip.

Driven rolls are used when the web is under very low tension, or when the roll has a high inherent friction. A heated or water-cooled roll would fall into the latter category. Two types of drive are used: a direct 1:1 to web speed and tendency drive. The direct drive is used for integral rolls with external bearings, and the belts and pulleys are attached to the shaft. A tendency drive is used on rolls with internal bearings on a shaft that is driven at a constant speed, and prevents a roll from stalling if a web's tension is not high enough

HYBRID GRAVURE COATERS

There are a few hybrid gravure coaters that can be used to circumvent the shortcomings of a direct gravure coater for some processes. Each of them retains an engraved cylinder to provide accurate coating weight control.

Offset Gravure

This method's name is derived from the fact that the coating is removed from the engraved roll by the offset roll, and not the web (see Fig. 2-5). The rubber-covered offset roll is driven 1:1 with the engraved cylinder and pressed onto it to remove the fluid, and it applies the coating to the web as it is pressed against it by a steel roll.

Offset gravure can be used if the coating does not have good flow characteristics or the web is highly irregular. The dwell time of the coating on the offset roll allows for some lateral flow, and along with the pressure in the nip, they mutually contribute to smoothing of the coating.

Another application for this method is when the web should not be subjected to high impression nip pressure that may adversely affect the web's surface, or something on it. It can also be used to apply a stripe or image to a web without a special engraved roll. The covering on the roll can be sculpted to suit, or covered with a flexible plate. Then, only the raised areas of the offset roll pick up coating from the engraved roll and transfer it to the web.

A version of this coater (see Fig. 2-6) can be used to apply a coating to both sides of a web simultaneously. Notice that the offset roll does not have to be under the web as in a conventional unit. It can, in fact, be arranged to coat a web in a horizontal, vertical or angular plane.

GRAVURE COATERS 25

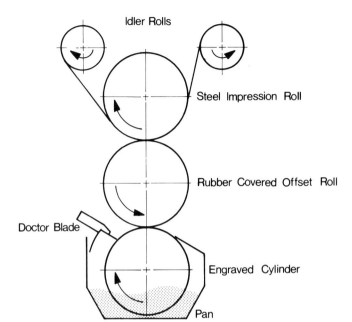

Fig. 2-5. Offset gravure coating unit.

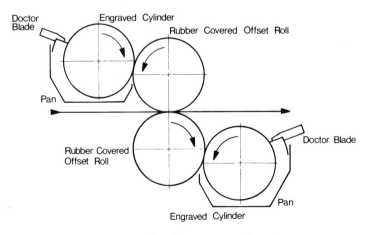

Fig. 2-6. Double offset gravure coating unit.

26 WEB PROCESSING AND CONVERTING TECHNOLOGY AND EQUIPMENT

There are also special direct and offset gravure coaters that have been incorporated in web and sheet fed offset printing lines.

Reverse Offset Gravure

This coating method can be used to prevent high-transfer pressure in the offset coating nip, or when the coating weight needed is above or below the one obtainable from the engraved cylinder (Fig. 2-7).

All of the offset gravure parts are used. The engraved cylinder and offset roll are driven 1:1 and are equipped with a reversible, variable speed drive, to allow running them opposite to and above or below the web speed. The steel impression roll is driven at line speed and equipped with precision gapping devices. The purpose of this design is to literally wipe the coating off the offset roll onto the web. Therefore, it is possible by running the offset roll opposite to and faster than web speed, to apply a thicker coating, or a thinner coating if the offset roll is run slower.

Reverse Gravure

This is a unique, although seldom used, method, but deserving of consideration to apply abrasive or foam-prone coatings, or to coat gage banded or irregular webs that cannot be put under a great deal of pressure or tension (see

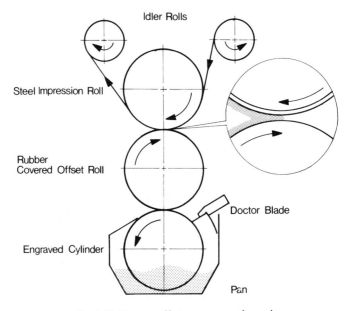

Fig. 2-7. Reverse offset gravure coating unit.

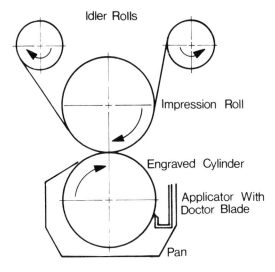

Fig. 2-8. Reverse gravure coating unit.

Fig. 2-8). It consists of the engraved cylinder with a reversible drive, an impression roll with gapping devices or a web lay on roll and an integral doctor blade and applicator.

The principle of the operation is to apply coating into and onto the unengraved portion of the engraved roll by using light doctor blade pressure, run the engraved roll opposite to web direction and in the nip wipe the coating from the surface of the roll and some from the cells or lines, onto the web. A tri-helical or other type of open engraved pattern that can be easily rinsed is recommended for this coating method.

APPLICATIONS

Over the years two schools of thought have emerged for using web processing equipment. One teaches to do everything in-line. The other instructs separating and segregating the processing steps enroute to a final product. The first is based on the theory that one rotary process should be compatible with another. But it follows that any weak link in a line will be then seriously hinder the overall performance of the others.

Judging which school of thought to follow depends on the coating that is to be applied and on the final product. The decision to apply them in-line or off-line will have to be made after close scrutiny of the sequence required to obtain the final product, available existing equipment or the magnitude of investment that can be made for new equipment.

Some guidelines, based on having the coater in operation as many hours as possible are listed below.

a) The release or slip agents are best applied in a machine dedicated to it only. The residue left after a job, even after the ultimate in clean-up and purging, can seriously affect another coating that must adhere tenaciously to the web.
b) Barrier coatings may have to be applied in two or more laydowns to assure pinhole-free results. An off-line single or multiple unit coater is best suited for this type of work.

Obviously one would not put a gravure coater in-line with another rotary unit that has to be changed over every two hours and where each changeover takes an hour. Likewise, using a gravure coater that can run 150 m/min with a system that cannot exceed 50 m/min is not an efficient solution.

Gravure coating is suitable with any web fed press. One in front of a press, for example, could apply a hold-out coat to a porous paper. A unit at the end of a press can be used to apply a top coating over the printing, or heat sealing stripes on the back of the web. Some flexographic presses are equipped with a gravure unit at the end to apply a thermoplastic adhesive, followed by a laminating station. Most laminators for aluminum foil and paper have a gravure coater to apply an overall coating to the foil, to prevent corrosion from moisture and salts in the paper as well as to prime it for printing.

Some hints to use in trouble-shooting are mentioned below. Many times a gravure coater will apply an off-target weight, purely because of improper set-up or housekeeping. Causes of low coating weight are: a) insufficient impression pressure; b) clogged cells in the engraving, or a worn wear surface; c) change in the coating's viscosity or settling of solids from vehicle; d) damaged impression roll; and e) a change in the surface tension, porosity, roughness or cleanliness of the web.

GRAVURE IN LAMINATING

Gravure is used for adhesive application in the production of many laminations. The two laminating methods used are given below.

Wet Laminating (see Fig. 2-9)

This method has traditionally been used to combine one imprevious and one porous, or two porous webs, with an aqueous, solvent, wax or hot melt adhesive. Now it is coming into wider use because it can also be used for 100% solid adhesives.

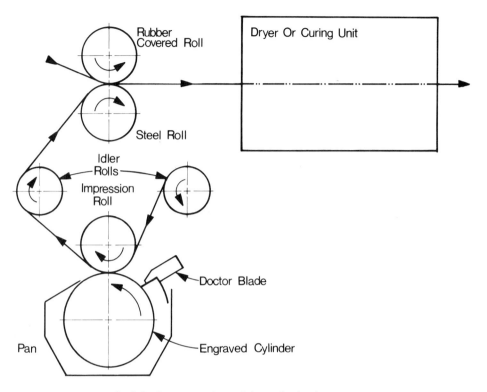

Fig. 2-9. Gravure coating unit in wet laminating system.

The adhesive is applied to the impervious web in order to minimize the adhesive usage and combined with the second web as soon as possible thereafter, followed by drying, cooling or curing. The practice of applying the adhesive to the impervious web does not necessarily have to be done with 100% solid adhesives, because little if any penetration takes place on a porous web before the curing process. Electron beam curable adhesives can be used with transparent or opaque webs. Ultraviolet curables can only be used if one web is transparent.

Some of the 100% solid adhesives are supplied as two component systems, and one has to be applied to each web, and then combined. Therefore, two gravure coaters are required.

Dry Laminating (see Fig. 2-10)

In contrast to wet laminating, this method is used to combine two impervious webs. One web, preferably the easiest to handle and pass through a

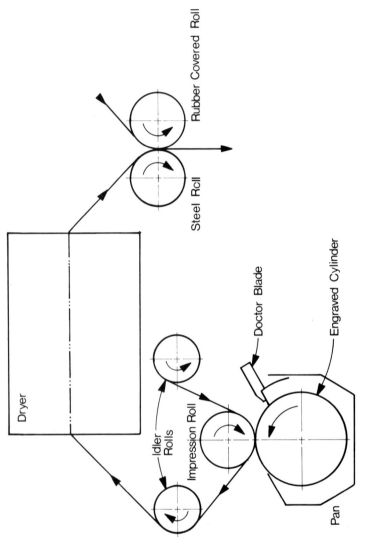

Fig. 2-10. Gravure coating unit in dry laminating system.

drying unit, is coated, the volatiles removed, and then combined with the second web in a heated and pressurized nip.

Some examples of the laminations are below.

WET LAMINATIONS	DRY LAMINATIONS
Foil to paper or paperboard	Film to film
Film to paper or paperboard	Film to foil
Film to fabrics	Film to panels

ENGRAVING METHODS

There are three methods used to engrave a gravure cylinder. They are:

- Mechanical (knurling)
- Acid etching
- Electromechanical.

The one used is dependent on the thickness of the wet laydown; the coating or adhesive; the gravure coating method; and whether or not the coating is to be overall or a pattern.

Mechanical Engraving

This method uses a lathe-like machine that runs a tool into a smooth cylinder, to displace metal and leave a replication of the tool's design on the steel or copper plating. After engraving, the cylinder is chrome plated or sprayed with an abrasion-resistant coating.

Three common mechanical engravings are used (see Fig. 2–11): quadrangular, pyramid, and tri-helical. A quadrangular cell engraving is recommended for direct gravure coating, pyramid for offset gravure and tri-helical for either when heavy adhesives or coatings are used or when the formula requires quick flushing and rinsing.

Commercial engravers have an extensive library of tools, to produce the volumetric capacities required for most jobs. Charts showing various patterns and what volume is deposited from each of them are available.

Acid Etching

This method is used to engrave a copper plated cylinder using a screened resist and an acid. It is suggested for lightweight coatings or ink when they are applied in other than an overall laydown. The process is quite involved and time-consuming.

32 WEB PROCESSING AND CONVERTING TECHNOLOGY AND EQUIPMENT

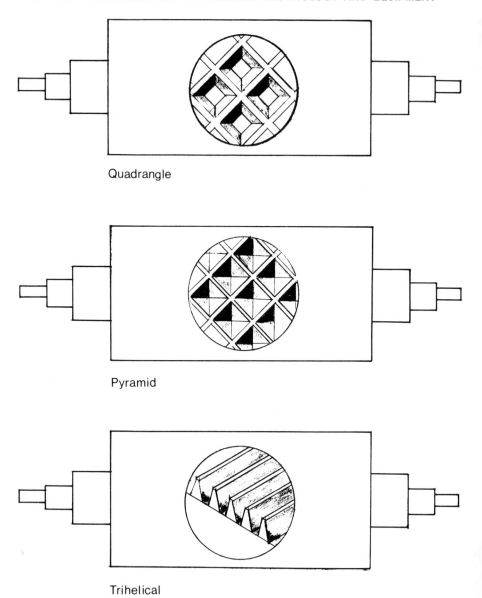

Quadrangle

Pyramid

Trihelical

Fig. 2-11. Mechanical engravings.

Electromechanical Engraving

This is a relatively new method of engraving, developed to eliminate the multitude of steps in acid etching. It consists of a special lathe like machine with two pair of head and tailstocks, an electromechanically pulsed diamond stylus, a photo optical sensor and a computer. One pair of head and tailstocks support and rotate the copper plated cylinder in front of the stylus. The second pair support and rotate a tube covered with a reflective separation in front of the optical scanner. The values of gray on the separation are scanned by the optical system, analyzed by the computer which sends impulses to the stylus, driving it in and out of the copper to remove cells.

3
Blade and Air-Knife Coating

Dan Eklund
University of Åbo Akademi
Turku, Finland

PIGMENT COATING

Pigment coating is one of the most important means for the paper maker to increase the quality of the paper or board. The goal is primarily an enhanced printability, but the optical properties of the paper, as, for example, opacity, brightness and gloss, which have importance for the general appearance of the paper, are also of importance and can be a reason for pigment coating.

In pigment coating, a coating color is applied onto the paper or board surface. The amount might vary from a few g/m^2 to an amount in excess of 30 g/m^2, depending on quality and end use. The coating is a water dispersion of a pigment to which binders and additives have been added. The amount and kind of binder and additives vary depending on the requirements on the paper from a printing point of view. Of course, the requirements set by the base paper and the process must also be fulfilled.

The pigment coating can be considered to consist of the following operations.

- Mechanical processes for web handling, as, for example, unwinding, reeling, web guidance, web threading, drives and web tension control.
- Coating unit.
- Coating color preparation and handling.
- Drying of the coated web.

The development of pigment coaters have gone from brush coaters and different roll coaters to the situation prevailing today, where blade coaters and air-knife coaters are the dominating processes in use.

BLADE COATING

The general principle in blade coating is that the applied coating is leveled with a thin steel blade of a 0.2–0.5-mm thickness. By varying the pressure of the blade against the paper, the final coat weight is adjusted.

The forces which act on the blade are in equilibrium with one another. The forces can be divided into two groups: the mechanical, which are dependent on the construction and which can be varied by the operator; and the dynamic, which originate from the coating. The difference between the mechanic and dynamic forces defines the pressure of the blade against the paper and thereby also the final coat weight. Even if the blade coaters are founded on the same basic principle, there are many different kinds of constructions which differ in essential ways. The difference can be in the applicator or in the blade unit, or in both. Schematically, the situation can be described as follows.

Inverted blade	Roll application	Bevelled blade
	Slot- orifice application	Low angle blade
	Jet fountain application	Rod-blade
Puddle coaters		Blade-blade
Simultaneous two-sided coating		Roll-blade

Inverted Blade Coaters

Most of the blade coaters today are designed according to the inverted blade principle (Fig. 3-1). First, an excess of coating color is applied with some kind of an applicator. Most common are roll applicators, as shown Fig. 3-1, but slot- orifice (fountain) and jet applicators are also used. The excess of coating is doctored away, and the coat weight regulated with a flexible steel blade in an inverted position, with 95–98% of the coating coming to the blade doctored away. The process consists of two distinct parts: the application and the coat weight regulation with the blade.

Application. As mentioned above, the application can be done with a roll, a slot orifice or a jet. Roll applicators are the most common, and the basic principle is shown in Fig. 3-2. A rubber-covered roll is rotating in a pan and an excess of coating is fed to the nip between the applicator roll and the paper supported by the rubber-covered backing roll. The distance between the applicator roll and the backing roll is usually 0.3–1 mm. The applicator roll rotates in the same direction as the web and with a speed which normally is 10–20% that of the machine speed. The velocity must be sufficiently high in order to lift enough coating to maintain a flooded nip. If the amount is in-

36 WEB PROCESSING AND CONVERTING TECHNOLOGY AND EQUIPMENT

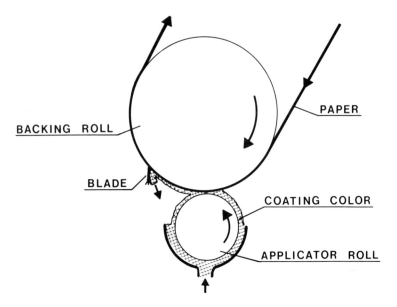

Fig. 3-1. The blade coating process.

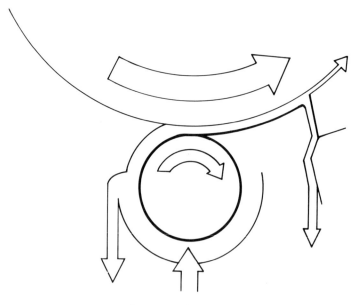

Fig. 3-2. Roll applicator.

sufficient, uncoated spots in the paper will occur, causing so-called skip coating. The amount which is put on by the applicator is of importance for the operation of the blade and is mainly dependent on the distance between the applicator roll and the paper (d), the machine speed (u) and the viscosity of the coating (η). The mass flow (m) from the applicator roll can then be written:

$$m = k \cdot d \cdot \eta \cdot u \tag{1}$$

where k is a constant depending, among other factors, on the roll characteristics, its material and its deformation.

The pressure which develops in the nip between the applicator roll and the paper is relatively small, 100–250 kPa, depending on the elasticity and deformation of the rolls. Roll applicators have been proven to work also on high-speed commercial coaters. Trials with speeds up to 25 m/s show that such speeds are not the limit for roll applicators. Roll applicators are, however, relatively complicated; they take up much space and they need their own drive.

In order to overcome many of these problems, the slot orifice applicators were developed. The principle is shown in Fig. 3-3. The coating is fed through distributor pipes into the applicator and pressed through a narrow

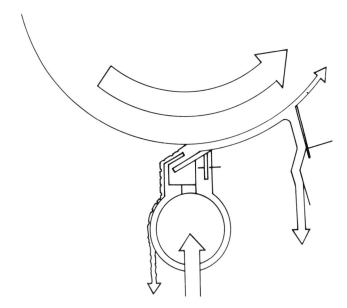

Fig. 3-3. Fountain applicator.

slot onto the paper. The distance between the applicator and the paper is of the magnitude 0.5-1 mm and adjusted in such a way that the distance on the incoming side (for the paper) is somewhat larger than on the outgoing. A certain amount of coating is pouring over the incoming lip to prevent the trapping of air and to assure that a sufficient amount of coating is applied onto the paper.

The pressure in the fountain is usually of the same magnitude as the pressures which develop in a roll applicator. As the distance between the paper and the slit is larger than the distances in roll application, the amount of coating that must be fed is larger, increasing the demand for pumping capacity. The main advantage of slot orifice applicators is a rigid construction without moving parts. A rigid element very close to the web can, however, also be a drawback. The biggest difficulty with this type of applicator is problems with entrapped air, and in most constructions there is a separate air removal system placed in the coating circulation system.

The jet applicator can be regarded as a further development of the slot orifice applicator. In the jet fountain, the coating emits from a slot orifice located about 20-40 mm from the web. This application does not require coating "back flow" at a point of application, which allows reducing the coating supply. Also, deflection compensation at the applicator point is no longer of great concern.

Blade Systems. The inverted blade coaters can be divided into two categories:

- Bevelled blades (or stiff or straight blades)
- Low-angle blades (or bent blades).

These two main types of coaters do not only differ in respect to their construction; the physical laws on which their operation is founded are not quite similar.

Theory of Bevelled Blade Coaters. Characteristic for this type of coater is that the blade forms an angle with the paper of the magnitude $40°-55°$. The configuration around the blade tip can be seen in Fig. 3-4. An excess of coating follows the paper to the blade, which is influenced by a dynamic force, and 90-99% is doctored by the blade. The blade is also influenced by a mechanical force imposed by the operator. The blade is balanced by three forces: the mechanical force (or blade pressure), the dynamic force and a supporting force from the paper to the blade. These three forces are always present and in equilibrium in stable operations.

The force on which the coating weight is dependent is the supporting force

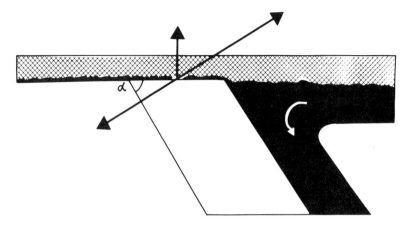

Fig. 3-4. Configuration around the blade.

from the paper to the blade. When this force is increased, the coating weight decreases, and when it is decreased, the coating weight increases. The situation can be regarded as follows: assume a coater running with a constant coating weight and with constant speed; i.e., with constant dynamic forces. The blade pressure (mechanical force) is higher than the counteracting dynamic forces; i.e., the supporting force is positive. The blade is in contact with the top of the fibers, which transmit the supporting force from the paper to the blade. The coating which passes the blade and forms the deposit is situated in the surface volume between the fiber tops in contact with the blade. The amount of coating is therefore dependent on the surface volume: the larger the volume, the higher the coat weight. If, the blade pressure is increased, the pressure against the paper is also increased, and the paper is compressed. The compression diminishes the surface volume and by that also the coating weight. If the blade pressure is decreased, the paper expands and the surface volume and coating weight are increased. This happens until the supporting force becomes zero and the paper is in its most expanded state. This is also the point where the process becomes unstable; a further small decrease of the blade pressure will make the supporting force negative (i.e., the blade is bent out from the paper). This bending is pronounced even for a very diminutive force. The bending can be mechanically restricted and high coat weights can be obtained in a relatively controlled way.

What factors, then, have an influence on the different forces? The mechanical forces, which can be regulated, are developed in different ways depending on the coater construction. The dynamic forces can, as a first assumption, be regarded to have an influence partly before the coating passage under the blade, partly under the tip of the blade. It can, however, be shown that the tip of the blade is parallel with the paper during stable running conditions, which means that dynamic forces will not develop. If parallelity does not exist, which can occur if the blade geometry is altered, dynamic forces can occur. These, however, are working only during a short time because the paper hones the blade to parallelity.

The blade tip transfers the supporting forces between the paper and the blade. This implies that at a given blade pressure and dynamic forces, a thick blade will give a lower specific pressure than a thin blade and therefore will cause less compression. A thick blade, hence, will give higher coat weights at the same conditions as a thin one (see Fig. 3-5).

Dynamic forces before the blade can be divided into two parts:

- Impulse force
- Hydrodynamic force.

When the excess coating color strikes the blade and its direction alters, the change of momentum will induce impulse-type forces. The impulse force component, F_a, that acts perpendicularly to the blade can be calculated from the following equation:

$$F_a = \frac{mu(1 + \cos\alpha)}{\sin\alpha} \quad (2)$$

where m is the mass flow doctored by the blade, u the velocity of the coating (which equals the paper speed) and α the blade angle.

The equation shows that the impulse force is increased with increasing ma-

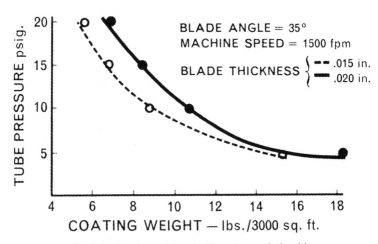

Fig. 3-5. Coating weight—blade pressure relationship.

chine speed and increasing mass flow. The mass flow up to the blade is dependent on the working conditions of the applicator, and can be altered by changing the characteristics of the fluid coating. An increased blade angle diminishes the impulse force.

The effect of frictional flow may also become important when shear stresses significantly affect liquid equilibrium. For example, this situation arises when liquid gets into the wedge-shaped space, where the relative velocity differences between the wall and liquid phase are large. If such is the case, the viscosity of the liquid becomes an influence factor. The blade geometry, the angle between the blade and base, may offer the conditions for the development of such a viscosity-dependent hydrodynamic pressure in coating. However, the direction of the flow must, in this case, be toward the narrow edge of the wedge, and hence it is clear that this pressure can develop only in the close proximity to the blade edge where the coating no longer flows down along the blade. The force F_h that causes a perpendicular pressure against the blade can be calculated from Eq.3.

$$F_h = \frac{6\eta u}{\tan^{2\alpha}} \left(\ln(1 + \nu) - \frac{2\nu}{2 + u} \right) \quad (3)$$

where η = viscosity
 $\nu = h_l/h_o - 1$
 h_l = distance between blade and base at the point origin of pressure
 h_o = distance between blade and base at the blade tip.

It can, however, be stated that the hydrodynamic forces are insignificant for blade angles below 20°, if the viscosity is in the range commonly used in blade coating (30-60 mPa·s at high shear rates). For bevelled blades, hence, where the blade angle is in the range 40°-55°, the only dynamic forces of importance are the impulse forces. Another characteristic feature is that the coating weight decreases when the pressure increases (Fig. 3-5.)

Theory for Low-Angle Coaters. Characteristic for the low-angle coaters is that the blade angle α usually is lower than 30°. The blade extension is also rather high, in the magnitude of 50-80 mm compared with 20-40 mm for bevelled blades (see Fig. 3-6). Unlike the bevelled blades, however, an increase in the blade pressure gives an increase in coat weight as well.

As in the bevelled blade coater, the main forces working on the blade are the dynamic and the mechanic forces. In most constructions the mechanic force acts directly on the blade at a distance of 15-25 mm from the blade tip.

When the blade pressure is increased, the blade is bent (bent blade coater is another name for this type of equipment) and the actual angle between the web and blade, α, is diminished. This diminution not only increases the im-

42 WEB PROCESSING AND CONVERTING TECHNOLOGY AND EQUIPMENT

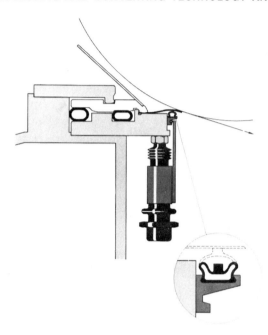

Fig. 3-6. Low-angle coating head.

pulse force substantially, as can be seen from Eq. 2, but when the angle is sufficiently small, the hydrodynamic force becomes more and more important; for very small angles ($\alpha < 10°$) the hydrodynamic force can be predominant. Hence, when the blade pressure is increased and the angle α decreased, the increase in the dynamic forces are larger than the increase in the mechanical force and the blade is bent out from the paper until a new equilibrium is obtained. The bending of the blade is thus governed by two large counteracting forces, making the system stable.

In this type of process, the blade is not in contact with the fibers, and high coating weights of 15–35 g/m² can be obtained, also on a smooth base sheet.

Different Constructions of the Blade Unit. All manufacturers of blade coaters have developed machines which in their details are different. These details, even if important, will not be dealt with in this context as they are continuously changed. The different coaters can, however, be classified according to the way the blade pressure is applied:

- Regulation of blade pressure with a pneumatic rubber tube
- Regulation with constant angle of the blade tip
- Rod-blade
- Vacuum regulation (Vacply).

Pneumatic Loading System. This type of coater has been the workhorse for many years and coaters of this type are today manufactured, especially for low-weight coating, but also for other coating operations. The best known manufacturers are Beloit (U.S.), Black-Clawson, KMW and Wärtsilä. The details of the constructions differ, but the principle is the same: the blade pressure is obtained by means of a pneumatically loaded, heavy-duty extruded tube operating directly against the blade, as shown in Fig. 3-7.

The blade is champed into the blade holder. This can be done mechanically, as in Fig. 3-7, or pneumatically. The blades used are normally 0.3-0.5 mm in thickness and have a width of about 75 mm. The free extension for the blade is usually 35-50 mm. The coating head is moved on to the paper pneumatically to a fixed position, and the blade pressure is obtained by inflating the rubber tube. To adjust for irregularities there are adjusting screws on the coating head along the width of the machine.

Most of the recent coaters of this type also have a possibility to change the blade angle during the run of the machine. Normally this is done by pivoting the coating head around the tip of the blade. This change of angle is of less importance for bevelled blades, as can be deduced from the theory presented above, but is of utmost importance for low-angle coaters. The change of blade angle, hence, makes it possible to run the coater as a low-angle coater.

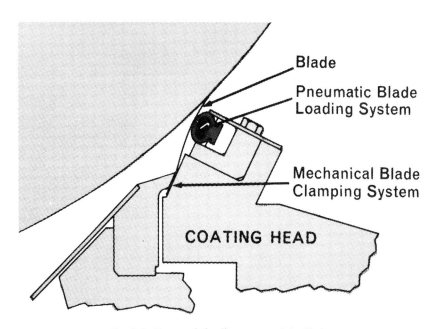

Fig. 3-7. Pneumatic loading system of the blade.

44 WEB PROCESSING AND CONVERTING TECHNOLOGY AND EQUIPMENT

In the newest Wärtsilä construction, the possibility of changing the blade is also utilized for bevelled blades in a computerized compensation for the change in the tip angle of the blade with changing blade pressure. As stated above, at stable running conditions the blade tip is parallel to the paper. However, as is evident from Fig. 3-7, if the blade pressure is altered, the angle of the tip will also somewhat alter. If the pressure is increased, the blade will run on the heel for a short moment; if the pressure is decreased, the blade runs on the tip. Neither of these cases are stable and can cause an inferior surface. In a short period (10–60 sec) the blade will hone into the new angle and the process will be stable again. However, even if the time period is short, it can mean a certain period of inferior paper quality. The computerized compensation of the change in tip angle is one way to overcome these problems. Another way is by construction of the blade unit in such a way that the tip angle is constant despite changes in blade angle.

Coaters with Constant Blade Tip Angle. This type of coating head is made by several of the leading coating equipment manufacturers, all of them working according to slightly different mechanisms. In this context, four of them will be discussed: the S-matic (Beloit-Italia), the Combiblade (Jagenberg), the Constacoat (Voith) and the Artblade (Wärtsilä).

In the *Beloit S-matic* (or S-blade), the blade is loaded by rotation around a pivot point located on the focus of the tangents to the blade tip (point B in Fig. 3-8). The blade is bent by pivoting, and the loading alters without altering the angle between the tip and the paper.

In the *Jagenberg Combiblade* (Fig. 3-9), the blade loading backstop B and the blade beam D are fixed in their positions and hence is also the blade tip at A fixed by movement of the blade clamping beam C. The blade base is shifted along a straight line which has a certain angle to the tangent of the blade base (Fig. 3-10). The angle is so chosen that it compensates for the shortening of the blade caused by its deformation, so that there is no noticeable shifting of the blade tip.

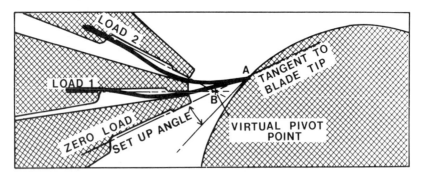

Fig. 3-8. The Beloit S-matic coating head.

BLADE AND AIR-KNIFE COATING   45

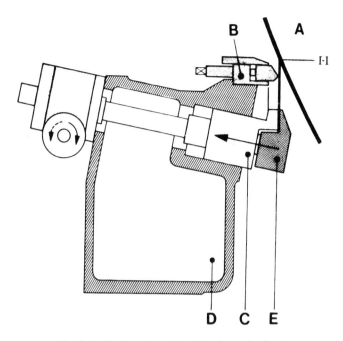

Fig. 3-9. The Jagenberg Combiblade coating head.

The principle of the *Voith Constacoat* can be seen from Fig. 3-11. The pressure inducing element is mounted on a steel blade. The blade pressure is then altered by altering the differential pressure in the two tubes. Its theory of working is in this respect very close to the coaters with pressure tube regulation. However, by pivoting the blade around a point away from the blade tip, Fig. 3-12, the blade pressure can be altered without any appreciable change in the tip angle. This mode of operation is similar to that of the Beloit S-blade.

In the *Wärtsilä Artblade* (Fig. 3-13), the initial blade pressure is set up by deforming the blade with the beam A, which is attached to the blade beam. The blade clamping beam B is mounted on a steel blade. Blade pressure alternations are induced by movement of B, thus inducing varying degrees of deformation of the blade. As the beam A is always in the same place, the tip angle remains the same irrespective of the blade load. Besides the described features, the units mentioned above usually also pivot around the tip, thus making it possible to run the units as bevelled or as low-angle blades.

Rod-Blade. One of the problems with blade coaters is the tendency to give blade streaks. In order to overcome this, and to minimize the cleaning intervals, as well as to get rid of blade changing, which usually takes place with 6-12-hour intervals, the rod-blade was developed. In this type of equipment

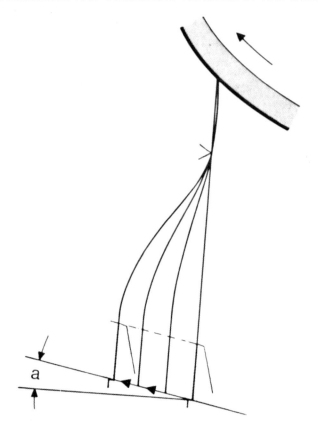

Fig. 3-10. Loading of the Combiblade.

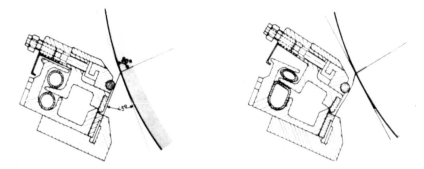

Fig. 3-11. The Voith Constacoat coating head.

BLADE AND AIR-KNIFE COATING 47

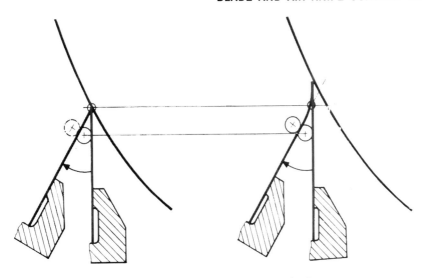

Fig. 3-12. Blade pressure alteration by pivoting.

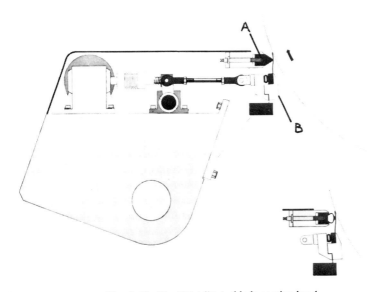

Fig. 3-13. The Wärtsilä Artblade coating head.

48 WEB PROCESSING AND CONVERTING TECHNOLOGY AND EQUIPMENT

the blade is exchanged to a rod mounted on a flexible blade. The principle of operation can be understood from Fig. 3-14. The rod-blade works according to the theory for bevelled blades presented above. However, as the blade construction has a high degree of stiffness, the possibilities to alter the coat weight by means of altering the pressure are diminished. (This type of equipment is used for board coating, mainly in pre-coating.)

Vacply. A unique way of applying blade pressure is the KMW Vacply. Its main features can be seen from Fig. 3-15. The coating or surface size is applied by a fountain or by an applicator roll and the coat weight adjusted with a flexible blade in a bevelled blade position. The system is under vacuum, which "sucks" the blade onto the paper. This results in an even pressure and a good profile of the coating. The use of vacuum also gives a controlled absorption and reduction of the air film which follows the web.

Puddle Coaters

The puddle type coater was the first type of blade coater for paper which was commercially installed, and equipment of this type is still used and built. The main features are seen in Fig. 3-16. The coating apparatus consists of a coating head or trough (C) to contain the coating, a flexible blade (E) which is held in jaws (D) at the edge of the blade bottom, dikes (F) at the end of the trough, and a backing roll (B). The paper is drawn around the backing roll, and forms one side of the trough. In operation, a pond of 10-15 cm deep is

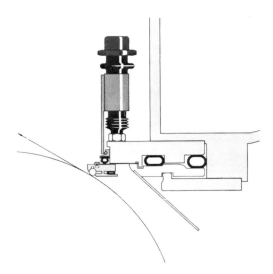

Fig. 3-14. Rod-blade coating head.

BLADE AND AIR-KNIFE COATING 49

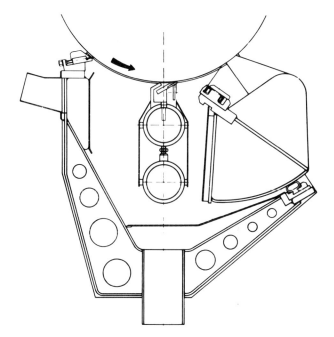

Fig. 3-15. Vacply coating head.

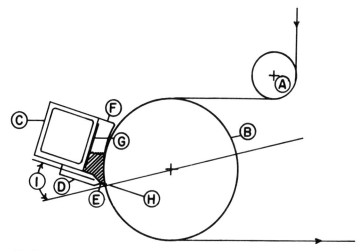

(A) Lead-in roll
(B) Backup roll
(C) Coating head
(D) Blade jaw
(E) Blade
(F) Dikes
(G) Dikes mounting rail
(H) Blade centerline
(I) Blade angle

Fig. 3-16. Puddle coater.

maintained on the paper. This is maintained in most installations by means of an overflow or recirculation of the coating back from the pond to the supply.

A thorough theoretical study of the puddle type coater has not been carried out. It is, however, probable that its mode of operation in principle is the same as for inverted blades. In this case, the dynamic forces are more difficult to calculate: the mass flow to the blade is not easy to determine, as the velocity of the coating near the paper equals the paper speed, and the speed gradually decreases away from the paper. A force originating from the hydrostatic pressure of the pond must also be added.

Puddle coaters have been successfully operated at speeds in excess of 1000 m/min commercially. High-speed operation gives rise to turbulence problems in the pond. The main feature of this coater is the short dwell time, which minimizes streaks and rheological problems as well as problems with low water retention in the coating.

Simultaneous Two-Sided Coating

Most coated paper today is coated on both sides. To do this with the coaters already mentioned, either a coater with two coating units with intermediate drying, or running the paper twice through a one-station coater, is required. Developments during the last 15 years, however, have resulted in machines where both sides can be blade coated in one pass. The best known of these processes are:

- Two-blade units
- Blade-roll units
- Separate blade units without intermediate drying.

Two-Blade Units. Best known of the two-blade units are the Sym-lam made by Allimand and the Twinblade made by Inventing. These are the only which so far have commercial units on stream. Other manufacturers, such as Beloit, Black-Clawson, Jagenberg and Wärtsilä, also have their own constructions. The operating principle of two-blade units is quite similar, and as an example the Twinblade (Fig. 3-17) unit is shown. In this coater, both sides of the web are coated simultaneously by two flexible blades. It has a vertical web-run upwards through the applicator nip, which consists of two opposed slot orifice extension dies, through which the coating is metered and deposited across the web. The two opposed flexible blades are located immediately above the applicators to minimize dwell time. The excess coating from either applicator or blade nip is directed back into the pans, located on each side of the web. The deckel width is controlled in the applicators, so

BLADE AND AIR-KNIFE COATING   51

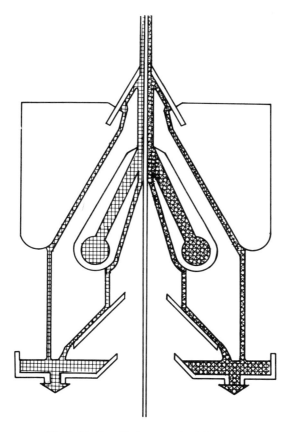

Fig. 3-17. Two-blade coater (Twinblade).

that complete separation is achieved between the two sides, thus making it possible to use different coatings on each side of the web without contamination.

The two-blade unit operates with thin blades, 0.1-0.2 mm, and with low angles; their working principle is close to the low-angle blade coating. The utilization of the two-blade principle is restricted to paper qualities which have sufficient strength to withstand the draw through the nip.

A critical point of the two-blade units is the first turn before the driers, where the web still is wet. This is usually accomplished through chilled guide rolls or by contactless turning air boxes.

Blade-Roll-Units.   The only manufacturer of this type of equipment is Inventing, with the Billblade unit and its modifications.

In the Billblade, a web runs downwards through a pond formed between the blade-holder, the bend blade and the soft backing roll, which rotates faster than the web (Fig. 3-18A). After coating, the web leaves the nip at an angle away from the backing roll to avoid film-splitting between the roll and the web. The slippage between the web and the backing roll gives a blade-like quality also on the roll side. It is also possible, by changing the guide roll to different positions, to run either one or two-sided applications (Fig. 3-18B).

If different chemicals are used on each side, an arrangement as in Fig. 3-18C, Billblade differential coater can be used. The guide roll is placed as for one-sided coating. With the metering rolls in off-position it is also possible to coat as shown in Fig. 3-18A and B.

After the passage through the coater, the web drying begins. The web is turned by means of a chilled roll or by contactless turning air boxes to the main drier of the machine.

Separate Blade Units Without Intermediate Drying (Chilply). The Chilply coating system made by KMW is based on the principle of double-sided coating in two separate stations, giving full individual control of blade

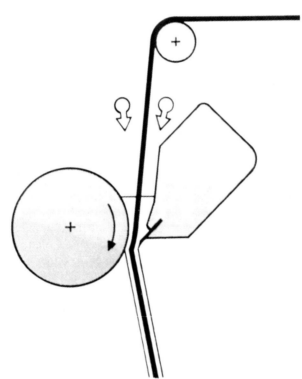

Fig. 3-18A. The Billblade coater.

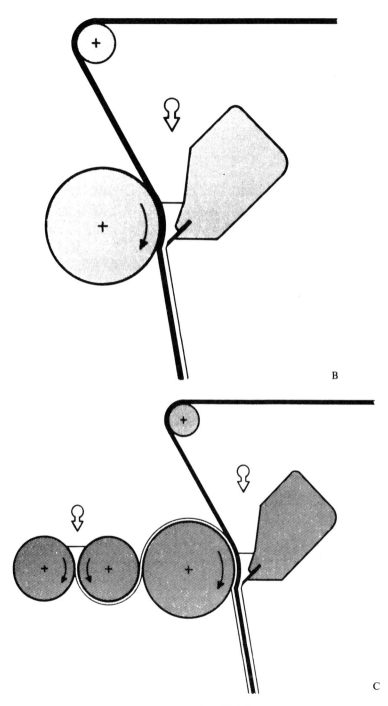

Fig. 3-18B and C. The Billblade coater.

parameters on both sides of the sheet. In Fig. 3-19, a Chilply system is shown. The first station is equipped with a conventional backing roll and the second station with a chilled rubber-faced backing roll. Between the two stations there is an infrared heater, giving the first coated side a preheating before it is turned around to the backing roll of coating station 2. Through the evaporation after the heating of the first coating layer, a condensation takes place on the backing roll of the second station. The created condensation film prevents the coating layer from sticking to the backing roll when the second side is coated.

AIR-KNIFE COATING

In the air-knife process, an excess amount of coating is applied by a single or multiple roll applicator and the excess is removed and final metering carried out with a sharp stream of air which is accelerated in a slot nozzle. Compared with blade coating, air-knife has advantages and disadvantages. The biggest advantage is the contour-coating: the air stream is removing as much coating from the hills as from the valleys in the paper, whereas the blade mainly fills the valleys of the paper. The air-knife is also a contactless coating method with the possibility to put on high coat weights, 10-35 g/m². In this respect, however, it has to compete with the new low-angle blade coaters.

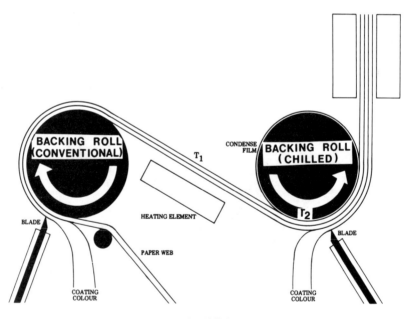

Fig. 3-19. The Chilply coater.

The main drawbacks of the air-knife is the lower speed (below 500 m/min), as compared with blade coating, where speeds up to 1500 m/min are possible. Another drawback is the low solids content, usually in the range of 40-45%, compared to blade coating, where the solids range is 55-65%. The low solids content increases the energy requirement for drying and also causes greater problems with binder migration. The air-knife equipment as such also requires energy. A further drawback is the soiling of the surroundings by the aerosol formed (misting).

Taking into account these drawbacks, it is not as astonishing that the air-knife had to give way to the blade coaters for pigment coating. However, for board coating, where the speeds are well below 500 m/min, and where the surface is rather rough and a good coverage is needed, air-knife coaters still are of importance and find their place.

The Construction of the Air-Knife

A typical air-knife can be seen in Fig. 3-20. The air is compressed to about 0.5 bar and is then forced into the air-knife head at a relatively low velocity. The low velocity assures even distribution of air in the cross-machine direction. In the air-knife head, the air is accelerated as it moves toward the slotted nozzle; i.e., the energy, which essentially is static, is converted into kinetic energy at the nozzle. In order to secure an even air stream from the nozzle, it is essential that the nozzle surfaces are finely ground and absolutely

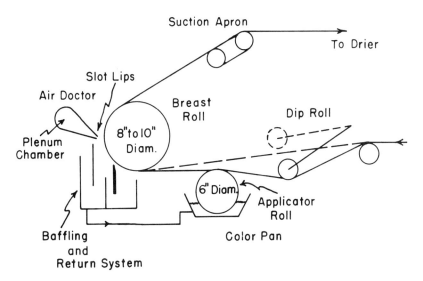

Fig. 3-20. Air-knife coater.

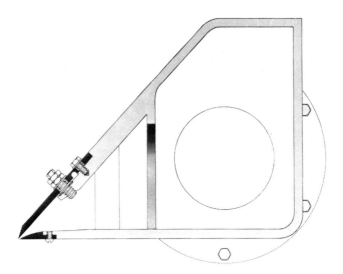

Fig. 3-21. Air-knife nozzle.

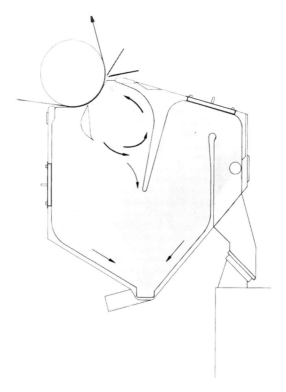

Fig. 3-22. Separator pan.

clean during operation. The narrow (usually less than 1 mm), high-velocity air stream impinges on a web to which an excess of coating has been applied. At this moment, its flow impulse transfers to the coating, and part of the fluid is carried away by the air stream. The generated aerosol is collected in a separator pan under vacuum located closely underneath the metering point, and in the pan the coating particles are separated from the air (Fig. 3-22).

It is essential that the excess amount of coating coming to the air-knife head is kept as low as possible. In practice, a most favorable situation is achieved when about half, or slightly less, of the amount meeting the air stream is blown away. However, this requires a sophisticated applicator system such as a three-roll system with metering nips, which can be set and adjusted accurately, or the use of a pre-metering unit, as for example a pre-smoothing roll. Some applicator systems for air-knife coating are shown in Fig. 3-23.

As was previously mentioned, a thin, high-velocity air stream emits from the air-knife. In order to utilize the kinetic energy of the air it is essential that the tip of the nozzle is close to the web, usually in the range of 2-5 mm. An increased distance will decrease the amount blown off; i.e., the coating weight will increase. The angle of the air jet to the web is usually about 45°. The effect of the angle on coat weight is rather small, but there is a tendency for the coating weight to decrease when the angle is increased. The usual way to control the coating weight in an air-knife coater is the air pressure. An in-

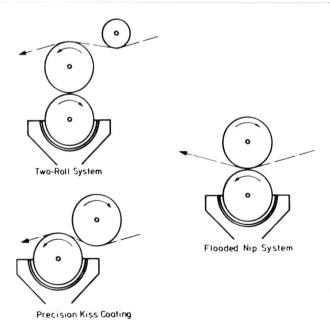

Fig. 3-23. Applicator systems for air-knife coating.

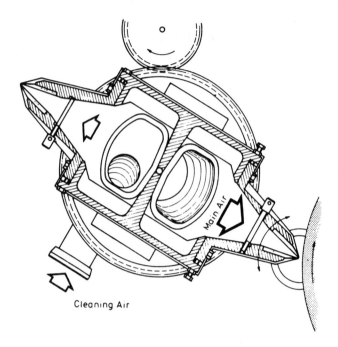

Fig. 3-24. Dual air-knife coating head.

creased pressure will result in an increased air velocity and a decreased coating weight.

One problem with air-knife coaters is the tendency of foreign matter to build up in or at the air-knife lips, causing a disruption of the air jet, which reflects this disturbance on the coated sheet. The cleaning of the lips means a production interruption for about five minutes. In order to diminish this time, the dual air-knife was developed (Fig. 3-24). In this system, the air-knife body is rotated by motor, and the soiled part exchanged for clean nozzles. The soiled nozzles are brought to a position where they are easy to clean. The installation of this type of system reduces the production interruption drastically and thereby decreases the amount of broke.

REFERENCES

1. Clark, C. Wells (ed.). *Blade Coating Technology*. TAPPI Press, 1978.
2. *1981 Air Knife Coating Seminar Notes*. TAPPI Press, 1981.
3. *1981 Blade Coating Seminar Notes*. TAPPI Press, 1981.
4. Booth, G. L. *Coating Equipment and Processes*. New York: Lockwood, 1970.
5. Waldvogel, H. *Wochenblatt für Papierfabrikation* **107**, 8:263-266 (1979).
6. Frei, H. P. *Wochenblatt für Papierfabrikation* **108**, 19:781-783 (1980).

7. Eklund, D., Kahila, S. and Obetko, D. *Wochenblatt für Papierfabrikation* **106**, 17:661–665; and **106**, 18:709–714 (1978).
8. Strenger, H. *Wochenblatt für Papierfabrikation* **106**, 7:287–290 (1978).
9. Akesson, R. Coating parameters in a KMW Fountain Blade Coater. Pilot study and mill experience. Tappi Coating Conference, May 23–26, 1982, p. 73–80.
10. Kahila, S. J. and Eklund, D. E. Factors influencing the coat weight in blade coating with bevelled blades. Theory and practice. Tappi Coating Conference, May 1–3, 1978.
11. *Pigmented coating processes. TAPPI Monograph Series No. 28.* (1964).

4
Other Knife and Roll Coaters

Orland W. Grant
Midland-Ross Corporation
New Brunswick, New Jersey
and
Donatas Satas
Satas & Associates
Warwick, Rhode Island

This chapter covers various unrelated types of coating machines, such as knife coaters, rod coaters, brush coaters, meniscus coaters and some roll coaters, such as squeeze roll application equipment. Some of the coaters are only of historical value, such as brush coaters, some like various knife coaters are widely used for many applications.

KNIFE COATERS

Knife coaters are the simplest and most direct machines for applying coatings to continuous and reasonably flat surfaces. They are simple to operate and require a minimum of maintenance. Because of the relatively simple construction, the knife coaters are generally quite inexpensive and therefore they are often the first choice for many coating applications.

Knife coaters can be used for a wide viscosity range. The limit on the low end of viscosities is the capability to keep a coating puddle without excessive leakage around the dams. It is difficult to contain coatings below 5 Pa·s in viscosity. High-viscosity coatings (above 50 Pa·s) may be coated by knife coaters, but skips, poor leveling, scratches and excessive variation in coating thickness might appear at such high viscosities, especially with dilatant coatings. Knife coating is a high-shear operation and any dilatancy of the coating might cause application problems. High-speed knife-over-roll coater might develop shear rate[1] in the range 1000–10,000 sec^{-1}, while a reverse roll coater develops shear rate in the range 100–1000 sec^{-1}.

The substrate passes between the knife and the web supporting arrangement, therefore the applied coating thickness is affected by the web thickness variation, the accuracy of the web support, and the straightness of the knife edge. The applied coating thickness can also be affected by web tension variations and line speed changes. As a result, knife coaters are generally limited to fairly low speeds.

Knife coaters have been used to make a large variety of products from filling the back side of rugs to magnetic tapes. Imitation leather, pressure-sensitive tapes, industrial belts, striped awnings and many other products are coated by knife coaters.

Knife coaters include several types of machines which employ a blade to meter the coating. The metering is accomplished by setting a gap between the supporting surface and the knife edge, or by scraping the surface of the substrate with the knife edge. These two processes are quite different: in the first case the most important variable determining the coating thickness is the gap setting, while in the second case the amount of coating deposited depends on the pressure applied by the knife.

Knife-Over-Roll Coater

This type of a machine is widely used for many applications. Two quite different variations of this method are used. The web supporting roll may be a high-precision chrome plated steel roll. The knife is machined accurately to give a constant gap width across the machine. The knife is set slightly off center, as shown in Fig. 4-1. This way a slight variation in the angle of separation of the web from the supporting roll does not affect the gap. The

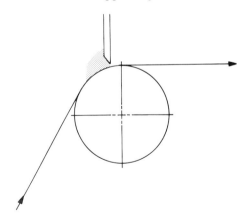

Fig. 4-1. Knife-over-roll coater with a steel roll.

amount of coating is regulated by moving the knife vertically. Some machines might also allow the change of knife angle by rotating the knife in relation to the roll surface. A typical knife-over-roll coater is shown in Fig. 4-2.

The gap width might be adjusted to the roll contour by bending the knife blade along the width of the machine. Bolts are located at regular intervals on the knife blade holder which exert the pressure onto the blade causing it to bend.

A resilient rubber covered roll might be used instead of the chrome plated steel roll. The amount of coating applied is determined by the pressure of the blade against the substrate surface. The knife is mounted at the point of web separation from the roll as shown in Fig. 4-3. The web supporting rolls are usually driven.

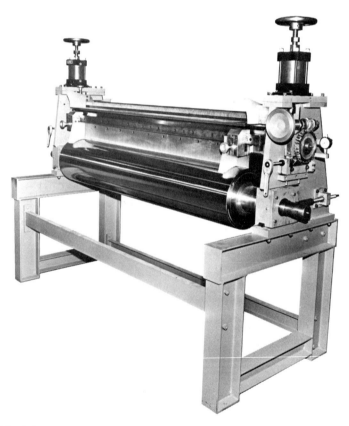

Fig. 4-2. A typical knife-over-roll coater. (*Courtesy Liberty Machine Co., Inc.*)

OTHER KNIFE AND ROLL COATERS 63

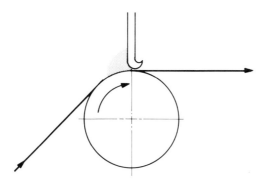

Fig. 4-3. Knife-over-roll coater with a rubber covered roll.

A coating reservoir is formed behind the knife by a pair of side plates and a backing plate. The side plates are contour-fitted to the knife and roll to contain the coating.

Because of the high-shear forces under the knife, strike-through of the coating might be a problem with such coaters if an open weave substrate is used. Double wrap is often employed to eliminate the strike through (Fig. 4-4).

Knife-Over-Blanket Coater

This coating method is similar to knife-over-roll with a resilient supporting roll. The web is supported by a resilient blanket and the amount of coating, is regulated by the pressure of the coating knife (Fig. 4-5). The coater is used

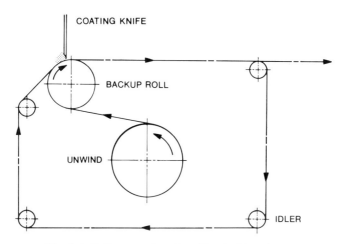

Fig. 4-4. Knife-over-roll coater with a double web wrap.

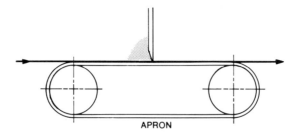

Fig. 4-5. Knife-over-blanket coater.

for coating of delicate webs which cannot be subjected to tension. Contour-fitted side plates are used to keep the coating off the blanket. One of the rolls is driven and the other is movable, allowing to change the tension on the blanket.

Floating Knife Coaters

Figure 4-6 shows a schematic diagram of a floating knife coater. The web is supported on both sides of the knife and it must be at a controlled tension. Roll supports, channels, plates or other means of web supports may be used. Side plates may be used to control the width of coating applied, or the coating may be allowed to run off at the edges into a pan from which it is recirculated. Floating knife coaters are used to fill open fabrics, back-coat of textile materials, allowing handling of open fabrics without undue penetration of the coating.

The sequential knife coater has two or more floating knife coating stations in sequence without drying between the coating. The first station may be used for a pre-coat and the second for the main coat, or both stations may apply the same coating. Sequential knife coating may be preceded by a knife-

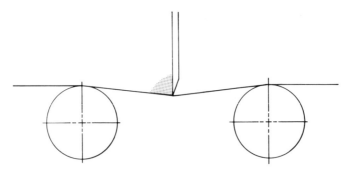

Fig. 4-6. Floating knife coater.

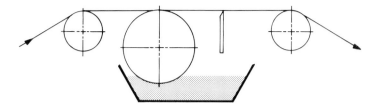

Fig. 4-7. Inverted knife coater.

over-roll coating station. Vinyl coated textiles are often made on a sequential coater.

The inverted knife coater (Fig. 4-7) is another application of the floating knife technique. Since a puddle before the knife cannot be maintained, an excess of coating is applied by an applicator roll. Dry edges may be easily produced by using edge scrapers on the applicator roll.

Blade Designs

Many different knife contours have been used. The selection of the best design is rather empirical, although there are some general rules which determine the choice of the knife design. The sharper the knife, the smaller the radius on the leading edge, and the higher the shear force exerted on the coating. Some of the most often used knife designs are shown in Fig. 4-8.

The sheeting knife (A) is often used to minimize the amount of coating applied. The straight-edged blade shears the coating exactly, while the rounded edge causes more penetration of the coating into the fabric. The tapered edge usually faces the coating puddle, but sometimes it is used in the opposite position. The land edge is narrow, about 1.5 mm wide. This type of a blade is used for plastisol coating and in floating knife coaters.

Radius knife (B and C) decreases the shear force exerted on the coating, forces more material into the backing and helps to smooth out thicker coat-

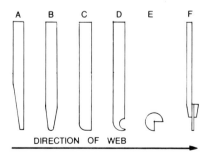

Fig. 4-8. Blade designs:
A—Sheeting knife
B, C—Radius knives
D—Hook knife
E—Bull nose
F—Spanishing knife.

ings. Radius knives in many different modifications are most often used for knife-over-roll applications.

Drawing D shows a radius knife with a groove machined into the back of the knife. This relief helps to prevent globules of coating from adhering to the back of the knife and from there dropping onto the coated surface as lumps.

A bull nosed doctor is shown in drawing E. The notch is cut from a round bar to produce a form of knife contour with a large radius. The curved side is located on the entering side. This device is often used after a reverse roll coater application of high-viscosity, pressure-sensitive adhesives to smooth out the surface of the coating.

A spanishing knife is shown in drawing F. It is a thin, flexible blade mounted in a holder similar to a cleaning doctor. It is useful to scrape the surface of an embossed material, to which a supply of fluid has been applied, in such a way as to leave the ink in the recessed sections of the embossing only. The blade is flexible, and higher pressure may be exerted, causing a high shear under the knife.

The Mechanism of Knife and Blade Coating.

The mechanism of knife coating has been discussed by several authors. Middleman[2] discusses blade coaters. Gartaganis, Cleland and Wairegi[3] has analyzed the coating process with stiff and with flexible blades. Stiff blade arrangement approximates knife-over-roll coating and the flexible blade coaters, as discussed in Chapter 3 by Eklund. The authors have shown the dependence of the volumetric flow (W) on various parameters: coating viscosity (μ), blade thickness (l), coating speed (U). The following relationships are suggested for stiff blade coaters.

$$W \alpha \mu^{1/2} \tag{1}$$
$$W \alpha l^{1/2} \tag{2}$$
$$W \alpha U^{3/2} \tag{3}$$

The authors show a good agreement with experimental results on a paper coater.

Freeston[4] has related the pressure developed in the coating during its application on a knife coater to various variables and to the knife geometry as shown in Fig. 4-9.

$$p_m = 4\mu U/l\theta^2 \tag{4}$$

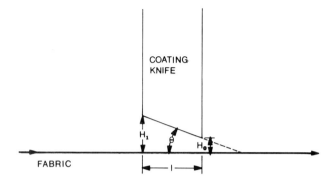

Fig. 4-9. Knife geometry (Eq. 4).

where

p_m = the difference between the total pressure and the hydrostatic pressure
U = web velocity
μ = coating viscosity
l = knife width
θ = knife angle.

The penetration of the coating into an open substrate depends upon the pressure developed under the knife.

Hwang[5] has analyzed the hydrodynamic forces involved in the knife coating process and has related the effect of various variables on the coating thickness. The gap between the knife and the substrate is the most important variable determining the coating weight. The coating thickness is usually about half of the coating gap width. The other variables account for at the most 10% of the change of film thickness. The blade thickness is the next most important variable followed by viscosity, the web speed and the liquid surface tension. The coating weight decreases with increasing viscosity and the web speed and increases with increasing surface tension.

LEVELON COATERS

This coating system in its most prevalent form consists of two rolls arranged one above the other as shown in Fig. 4-10. This coater can be equipped with one accurate metal roll and one rubber covered roll, or, for a better gap control, with two accurate metal rolls. The excess of coating in the nip is metered in a manner similar to a knife-over-roll coater. When a rubber covered backing roll is used, coating weight is controlled by means of nip pressure. When

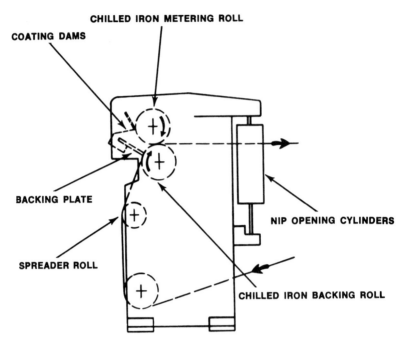

Fig. 4-10. Levelon coater vertical reservoir type (*Courtesy Midland-Ross Corporation.*)

two accurate metal rolls are used, coating weight is controlled by the gap between the rolls.

As compared to knife coating, Levelon coater generates less shear in the nip and produces smoother surface with less streaks. It has found applications in the pressure-sensitive tape coating. A horizontal and a pan fed version of the Levelon coater are shown in Figs. 4-11 and 4-12. In the pan fed type coater, the coating is applied by a kiss roll.

BAR COATERS

Wire wound rod coaters, also known as Mayer bar coaters, are used widely for metering of lightweight coatings on smooth substrates. The coater consists of a pick-up roll rotating in a pan to apply an excess of coating to the web, followed by a rotating rod to remove the excess coating, leaving a metered amount on the web surface. In Fig. 4-13 a schematic diagram of a bar coater is shown, and Fig. 4-14 shows a photograph of the wire wound rod coating head.

The coating deposit is controlled by the spaces between the wire that is spirally wound around the rod (see Fig. 4-15). The larger the wire gauge, the

OTHER KNIFE AND ROLL COATERS   69

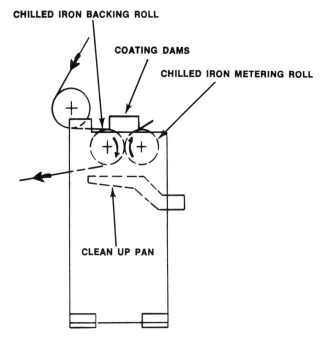

Fig. 4-11. Levelon coater, horizontal reservoir type. (*Courtesy Midland-Ross Corporation.*)

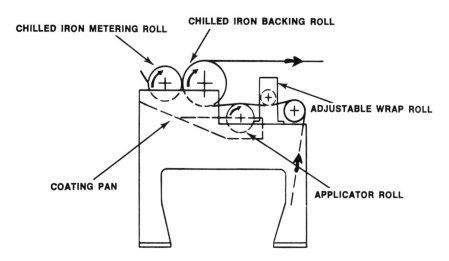

Fig. 4-12. Levelon coater pan fed type. (*Courtesy Midland-Ross Corporation.*)

70 WEB PROCESSING AND CONVERTING TECHNOLOGY AND EQUIPMENT

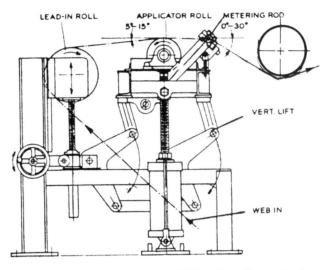

Fig. 4-13. Bar coater. (*Courtesy Midland-Ross Corporation.*)

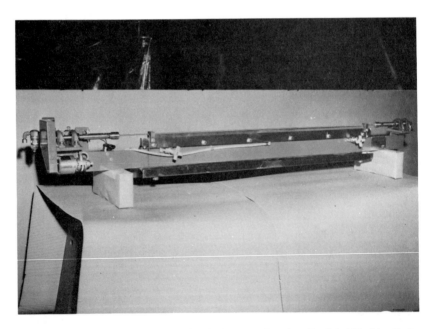

Fig. 4-14. Wire wound rod coating head (*Courtesy American Tool and Machine Co.*)

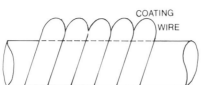

Fig. 4–15. Wire wound rod.

heavier the metered deposit. The area of one gap is equal to $2r^2 - r^2/2 = 0.429r^2$. Ideally this would give the coating thickness of $0.2145r$, if the coating would be completely removed from the spaces between the wire. In addition to the amount of the coating that can be contained by the spaces, the coating deposit also depends on web tension, wrap angle, web speed, rheological properties of the coating, bar rotating speed and other factors which might have an effect on the efficiency of wiping the coating clean from the interstices. Because of these variables, in addition to the absorbency of the substrate, the coating thickness must be determined experimentally for given coating conditions.

The wire wound rod is rotated in either direction by a separate drive in order to distribute the wear and to dislodge any particles that might cause streaks. The rod should be installed in such a way that the rotation tightens the wire to the rod.

Several different designs of rod holders are used. The holder keeps the rod in place and allows it to rotate without excessive friction. The holder could consist of simple clamps holding the rod, or it could be a magnetic holder which holds the rod firmly in place but allows it to rotate without much friction.

The coater, such as that shown in Fig. 4–13, includes an adjustable roll to control the angle of wrap over the pick-up roll, a reversible variable speed drive for the pick-up roll and the wire wound rod, and a lifting device to take the web off the pick-up roll during machine stops. The distance between the coating application and the wire wound rod should be kept at a minimum.

A smooth rod might be used to apply a small amount of the coating. This is often used to apply high-solid clay coating to the surface of the board to produce a smooth flat surface to be followed by a heavier air-knife coating of the paperboard in order to improve the ink receptivity. Wire wound rods are used to apply lightweight coatings over various films, to apply silicone release coatings, water based pressure-sensitive adhesive coatings and generally low-viscosity, low-deposit weight coatings over smooth surfaces.

Smooth bars are also used as polishing devices. Their function is not to meter the coating, but to remove the surface imperfections, such as striations, streaks, air bubbles and other surface imperfections. In case of saturation, the smoothing bars might also help to achieve a better penetration of the web.

The polishing bar may operate by simply assisting the leveling, or it may remove a substantial amount of coating and redeposit.

The bars may be stationary, or they may be rotated by an independent drive. The direction of rotation usually is opposite to the web movement. Mushel[10] describes a fast rotating polishing bar which improves the quality of the coating surface as it comes from a roll coater. Such coatings often have regularly spaced striations running in the machine direction. This is especially a problem with thickened aqueous coatings which exhibit poor leveling. A fast rotating bar does not completely eliminate these striae, but decreases their amplitude, thus improving the visual appearance.

Hot melt and other thermoplastic coatings are polished by employing heated bars.

SQUEEZE ROLL COATERS

Squeeze roll coaters consist of two rolls to form a nip through which the web passes. One roll is generally rubber covered and another is steel surfaced. Coating is applied to the web by pan feed, pipe feed or a slot die applicator. The wet coating is split between the roll and the web. The wet coating deposit depends on the roll pressure, web absorbency, coating viscosity and surface tension, the resiliency of the rubber roll and the web speed.

Since the coating splits when applied, the squeeze roll coaters produce a striated coating surface. While the appearance can be improved by low-viscosity self-leveling coating, it is often necessary to introduce a post-smoothing roll or a bar assembly to improve the surface appearance. In cases where the squeeze roll coater is used to apply a laminating adhesive, the surface appearance might not be important.

Squeeze roll coaters are simple, suitable for running at high speeds and applying low-coating weights of low-viscosity coatings. The coater can be run at 300 m/min, but the viscosity at such speeds should not exceed 0.15 Pa·s. Patterning becomes very great at higher viscosities. These coaters have been used for waxing papers, applying phenolic resin to papers for high-pressure laminates, applying clay coatings over paper and paperboard, manufacturing of gummed tape. The coater is speed-sensitive and some waste could be made while the speed is adjusted. Edge doctors are required, if coating free edges are needed.

Coating can be applied simultaneously to both sides of the substrate. It is only required that coating is fed to the both sides of the nip, or that the web is dipped into the coating.

OTHER KNIFE AND ROLL COATERS 73

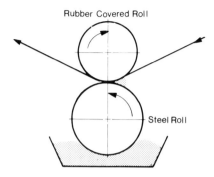

Fig. 4-16. Two-roll squeeze coater for a single-side coating.

Squeeze roll coaters consist of either two or three rolls. Several arrangements are possible with each type. Fig. 4-16 shows a two-roll, one-side squeeze roll coater. The web is passed between a chilled iron or steel pick-up roll rotating in a pan and a rubber covered top roll. It is important that a uniform squeeze nip is formed between the rolls. In some cases, crowning of the rubber roll might be required to eliminate roll bending.

Figure 4-17 shows the same type of coater adopted for two-sided coating by adding a feed pipe above the top web surface. Such a coater requires a good tension control of the web in order to get a uniform coating.

Figure 4-18 shows a horizontal roll arrangement of a two-roll coater. Such units are commonly used for size press coating of paper. It is difficult to lead such two-sided coated material into the drying oven without contacting a roll.

Figure 4-19 shows a three-roll coater. The third roll is used to premeter the coating. This arrangement allows the running of the coater at higher speeds, because the pick-up roll can be rotated at a slower speed to minimize slinging of coating from the pan.

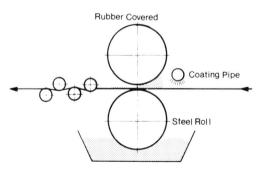

Fig. 4-17. Two-side squeeze roll coater with polishing rolls.

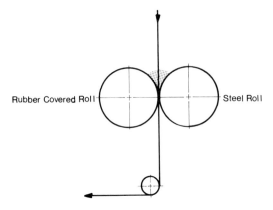

Fig. 4-18. Horizontal two-roll squeeze coater.

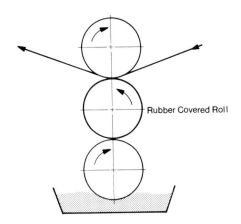

Fig. 4-19. Three-roll squeeze coater.

KISS ROLL COATERS

In kiss roll coaters the web passes over the roll without any back-up support. Idler rolls are positioned to allow the web to wrap around the applicator roll. The wrap angle is usually small, 5°–15°, and the only force keeping the web in contact with the web surface is the vertical component of the web tension. High tensions might be required for some webs to keep them flat at the point of contact. A simple one-roll kiss coater is shown in Fig. 4-20.

The applicator roll might be turning in either direction. If the roll is turning in the direction of web travel, the coating is split between the roll and the web. If the roll is rotating against the web travel, the coating is wiped clean from the roll to the web. In either case, the amount of the coating deposited

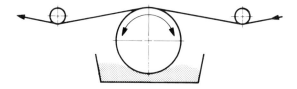

Fig. 4-20. Single-roll kiss coater.

depends on a large number of variables: web tension, relative speed, wrap angle, applicator cylinder diameter. It is difficult to control the deposit weight with kiss roll application, and such coaters are usually employed only as means of applying excess coating to the web and may be followed by a metering station consisting of wire wound rod, knife-over-roll or other equipment.

Kiss roll coaters are run at modest speeds up to 200 m/min and the coating viscosity is low at 0.3-0.4 Pa·s.[6] Kiss coaters are easily adaptable for striping. Lifting fingers or wiping straps are inserted between the web and the applicator roll preventing the coating transfer in those areas.

When the machine stops, the applicator roll remains rotating in order to avoid drying of the coating on the roll. If the web is porous, it might become quite saturated with the coating in that spot and, therefore, means of separating the web from the applicator roll are required. This is usually accomplished by lifting the web away from the applicator roll, but lowering of the roll might be also used in some designs.

While a simple single-roll kiss coater as shown in Fig. 4-20 is rarely used alone without an additional metering station, some more elaborate kiss coaters provide better means to control the coating weight. Figure 4-21 shows a two-roll coater with an external reverse metering roll. Figure 4-22 shows another design of a two-roll coater. The pick-up roll transfers the coating from the pan to the applicator roll. Usually the rolls are run at different peripheral speeds for metering and smoothing of the coating on the applicator roll.

Figure 4-23 shows a three-roll kiss coater equipped with a squeeze nip and the reverse metering roll. In all cases, the web may be run in the same or in the opposite direction of the applicator roll rotation.

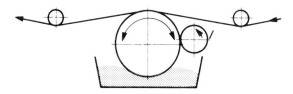

Fig. 4-21. Two-roll kiss coater with a reverse metering roll.

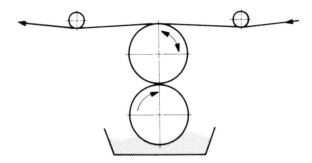

Fig. 4-22. Two-roll kiss coater with a squeeze nip.

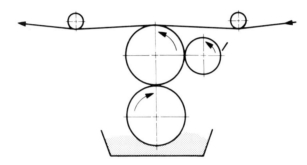

Fig. 4-23. Three-roll kiss coater.

CAST COATERS

The basic characteristic of the cast coating is that the polymeric coating is applied over another surface from which it is later stripped. The coating surface accepts faithfully the appearances of the casting surface. Therefore, this method is suitable to produce high-gloss finishes by casting against a highly polished surface, embossed materials by casting on pre-embossed casting paper. It is also useful in handling substrates which are difficult to coat because of their heat sensitivity or poor mechanical strength. Several quite different methods are known under the name of cast coating.

Cast coating of papers against a chrome plated drum is used to manufacture high-gloss clay coated paper. The process was invented by Donald Bradner in 1927 and has been extensively used for manufacture of glossy printing paper, such as needed for labels, high-quality illustrations, etc.[7]

Figure 4-24 shows the application of this coating. A kiss coater applies the coating over paper which is then pressed against a hot chromium plated drum. The coating dries in contact with the drum surface and accepts the

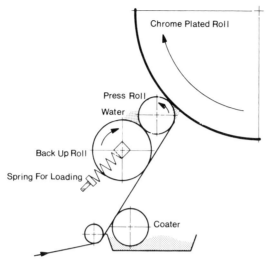

Fig. 4-24. Cast coating.

high finish of the chromed surface. The dried product is stripped away from the drum and wound up.

If the coating is not completely in contact with the drum surface, it will show non-glossy spots. Therefore, it is very important that no air is entrapped between the coating and the drum surface. Various techniques are employed to assure a complete contact. Maintaining a small pool of liquid at the point of contact, as shown in Fig. 4-24, helps to force the air out from the interface. The drum surface might be also pre-wetted or steamed just prior to contacting the coating.[8]

The surface of the chrome plated drum might be treated with a dilute solution of an oxidizing acid in order to impart the release properties. Several other methods are used to obtain an adequate release of the coating from the casting surface.

Several alternate methods are used to produce high-gloss paper. A process called cast calendering consists of re-wetting the surface of coated supercalendered paper and then contacting it with a highly polished surface of a roll. Another process is pre-cast coating. The coating is applied directly to the drying drum and transferred to the adhesive coated substrate after drying.

Similar cast coating methods are used to produce other products besides glossy paper. Polymeric films may be produced by casting over a stainless steel belt.

Casting on embossed release paper is a regular process to manufacture leather-like coated fabrics.[9] Polyurethane or polyvinylchloride coating is cast onto embossed release paper, dried and combined to a substrate using an adhesive. The product is stripped away from the release liner at the end of the operation and the release liner can be reused many times before discarding.

Pressure-sensitive labels are routinely manufactured by coating the adhesive over a release liner and then laminating the facing to the dried adhesive at the end of the line. Heavier release paper is easier to handle in the coating and drying operations than various papers, films and foils used for labels.

MENISCUS COATERS

The liquid coating is applied by lowering the roll, which carries the web, into the liquid and then raising it just above the liquid surface to form and maintain a meniscus. The amount of liquid deposited depends on the surface tension of the coating, coating viscosity, web speed. Very lightweight coatings can be obtained by this method. The process is slow at 20 m/min and the viscosity of the coating is low. The main application of meniscus coaters is in the photographic industry for coating of smooth plastic films.

Figure 4-25 shows a single-roll meniscus coater. The web tension must be maintained constant to insure that the web is held tightly against the roll. Also the pan must have an accurate liquid level control to maintain the meniscus between the web surface and the liquid.

Figure 4-26 shows a two-roll version of a meniscus coater. This unit simplifies the maintenance of the meniscus and is capable of faster speeds. Another version of the same coater has a slot die instead of the roll. Continuously flowing liquid maintains a constant level at the die exit. This type of a machine can be used to apply several liquid coatings one after the other.

BRUSH COATERS

Brush coaters were the first coating machines designed to imitate the coating application by hand brushes. First brush coaters were constructed in 1850's for coating of wallpaper with a clay filled coating. These machines are no longer used, but a brief description might have some historical interest. The brushes have been used for both application and finishing. Figure 4-27 shows the essential parts of a brush coater. The coating is picked up by a roll and transferred to the web by a brush. The reciprocating brushes are used to smooth out the coating. Brushing might help to work the coating into the paper and to produce a better bond.

Figure 4-28 shows a machine with reciprocating brushes built in 1949. Such coaters were still built in the late 1950's. The reciprocating brushes are

OTHER KNIFE AND ROLL COATERS 79

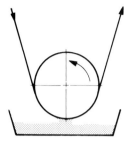

Fig. 4-25. Single-roll meniscus coater.

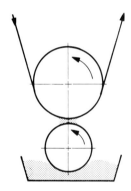

Fig. 4-26. Two-roll meniscus coater.

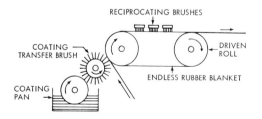

Fig. 4-27. Schematic diagram of a brush coater.

Fig. 4-28. Reciprocating brush machine. (*Courtesy Midland-Ross Corporation.*)

driven separately to accommodate the product requirements. Various natural and synthetic fibers were used for the bristles.

REFERENCES

1. Satas, D. Coating. *Handbook of Pressure-Sensitive Adhesive Technology* (D. Satas Ed.). New York: Van Nostrand Reinhold, 1982.
2. Middleman, S. *Fundamentals of Polymer Processing*. New York: McGraw-Hill, 1977.
3. Gartaganis, P. A., Cleland, A. J. and Wairegi, T. *Tappi* **61**, 4: 77-81 (1978).
4. Freeston, W. D., Jr. *Coated Fabrics Technology*. Westport, Conn.: Technomic, 1973, pp. 25-41.
5. Hwang, S. S. *Chem. Eng. Sci.* **34**:181-189 (1979).
6. Booth G. L. *Coating Equipment and Processes*. New York: Lockwood, 1970.
7. Bradner, D. U.S. Patent 1,719,166 (1929) (assigned to Champion Coated Paper Co.).
8. Casey, J. P. Pigmented coated processes for paper and board. *TAPPI Monograph Series No. 28* (1964), pp. 74-84.
9. Damewood, J. R. *J. Coated Fabrics* **5**:103-113 (1975).
10. Mushel, L. A. *Package Printing* (June 1979).

5
Rotary Screen Coating

John Eric Bell
J. Josephson, Inc.
So. Hackensack, New Jersey

The first rotary screen coater, the Aljaba machine, was constructed by A. J. C. de O. Barros in 1954. The key problems to the design of such a machine were finding a material for the screen and developing a method of applying the coating. Post-treated metal gauze, in a tubular form, was used for the screen and a special roller squeeze device was constructed.[1]

One major problem remained, the construction of a seamless mesh-like cylinder. The solution came in the late fifties from the combined efforts of a Dutch engraver H. de Vries and Veco Beheer Electro Forming/Photoetching B.V., Eerbeek, Holland.[2] The first public demonstration of this development occurred in 1963 when Stork Brabant of Boxmeer, Holland, printed textile materials using a truly seamless mesh cylinder and a Galvano type cylinder. Stork saw the potential of the seamless mesh (lacquer) type print cylinders and became their largest producer. This development permitted the use of metal or rubber type squeeze blades, and thereby produced some new solutions[3] to the basic points set out by Barros.

Peter Zimmer, Kufstein, Austria, an early pioneer, chose to follow and to develop the Galvano type process of rotary screen making often referred to as the Galvano Direct Design (GDD) process.

Lacquer screens, GDD type screens, and a variety of woven wire cylinders post-processed by a variety of plating techniques are used today.

SCREENS

A mother cylinder is made for each screen circumference and for each screen mesh number. The mesh pattern is precisely machined into the cylinder surface, producing a perfect seamless image around the cylinder.

82 WEB PROCESSING AND CONVERTING TECHNOLOGY AND EQUIPMENT

The lacquer screen is mass produced by a nickel plating process creating holes simultaneously with the formation of the nickel tube from the mother cylinder. This method of manufacture produces screens that are seamless, have a very accurate cross-sectional thickness and have equal hole size or mesh geometry, where the holes are essentially round. The mesh tubes are subsequently post-processed with a lacquer to seal off holes in the areas where coating is not required, hence the name lacquer screen.

Screen mesh is defined as the number of holes a screen has per linear inch; e.g., 25 mesh equals 25 holes equally spaced per linear inch. Up until 1978, lacquer screens were available from 9–120 mesh only. Due to the plating process, the hole size and screen wall thickness are interrelated. If a screen were required to have a thick wall, then at a certain point in the process the holes would seal, forming a tube with a pattern of dimples over the tube's inner surface (Fig. 5-1).

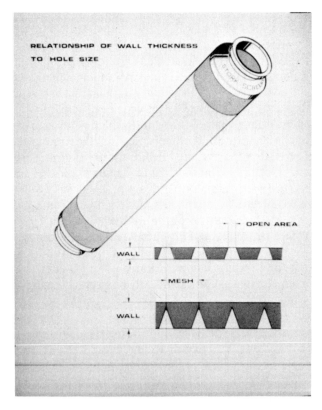

Fig. 5-1. Relationship of wall thickness to hole size in lacquer screens. (*Courtesy Stork Screens America, Inc.*)

Raster is the metric equivalent of mesh; i.e., the number of holes per centimeter. The Galvano screen maker may use the raster definition when referring to a screen mesh. Table 5-1 shows a range of lacquer mesh numbers available in certain screen sizes.

In 1978, the Penta screen was invented by Stork Screens, Holland;[4] screen sizes available are also shown in Table 5-1. These screens are made of nickel in much the same way as standard lacquer screens, but have hexagonal hole form. In standard screens, the hole size and free area decreases with increase in wall thickness, but the Penta screen has an almost constant hole size regardless of wall thickness, because the holes are not tapered (Fig. 5-2).

The hexagonal hole form combined with the special manufacturing process makes for a substantial increase in free area regardless of mesh. The hole size and free area are very important in rotary screen coating. By increasing the screen wall thickness, the screen life increases infinitely, but does not limit the application to thin coatings.

Table 5-2 gives an idea of the relationship between standard screens,

Table 5-1. A Representative Range of Typical Mesh Numbers Used in Screens of Various Sizes

STANDARD LACQUER SCREENS

COATING WIDTH	CIRCUMFERENCE OF SCREEN			
	640 mm	66.8 mm	820 mm	914 mm
610–2185 mm	9–11.2–17–20 25–40 60–80 100	40–60–70 80–100	25–40–60–70 80–100	40–60–70

PENTA SCREENS

COATING WIDTH	CIRCUMFERENCE OF SCREEN			
	640 mm	668 mm	820	914 mm
610–1930 mm	14 H 25 H 40 H 125 H 155 H 185 H	125 H	125 H	
3658 mm				14 H 25 H 40 H

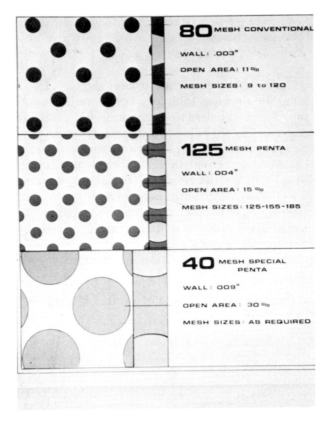

Fig. 5-2. Conventional, Penta and Penta Special Screens. (*Courtesy Stork Screens America, Inc.*)

Penta screens, and special Penta screens. This chart provides the basis for coating screen selection. As the mesh number increases in standard screens, the coating thickness decreases, and the resistance to flow increases due to decreasing free area. For heavily filled coatings, use of special Penta screens should be considered, because they have very high free areas with very thick walls, and large hole diameter offering less resistance to flow and to particulate matter.

Lacquer screens require a special lacquer post-treatment to make them ready for use. Engravers seal off the holes in the nickel tube, as dictated by the process, using a specially developed technique. Consider using a 1968-mm-long screen to apply a 1524-mm-wide coating; 222 mm at each end must be sealed off. This is done by coating the whole screen surface with a light-sensitive lacquer. After drying and masking off a 1524-mm band cen-

Table 5-2. Characteristics of Various Screens

STANDARD LACQUER SCREENS

MESH	NUMBER OF HOLES PER SQUARE INCH	HOLE DIAMETER, MICRONS (μm)	% FREE AREA	WALL THICKNESS, MICRONS (μm)
11.2	125	1000	17	220
17	290	525	12	220
23	400	425	14	220
25	625	375	12	220
30	900	300	11	220
40	1600	320	23	120
50	2500	200	15	100
60	3600	160	14	100
70	4900	125	13	100
80	6400	108	11	78
100	10,000	78	8	85

PENTA MESH

MESH	NUMBER OF HOLES PER SQUARE INCH	HOLE DIAMETER, MICRONS (μm)	% FREE AREA	WALL THICKNESS, MICRONS (μm)
125 H	15.625	78	15	100
155 H	24.025	65	13	100
155 DLH	24.025	50	7	110
185 H	34.225	50	11	87
215 H	46.225	40	10	87

SPECIAL PENTA MESH

MESH	NUMBER OF HOLES PER SQUARE INCH	HOLE DIAMETER, MICRONS (μm)	% FREE AREA	WALL THICKNESS, MICRONS (μm)
14	196	1000	30	400
14	196	1100	40	300
25	625	550	30	300
40	1600	350	30	300

tral along the screen, the uncovered lacquer is exposed to high-intensity light to initiate cure in the coating. This is followed by a wash cycle to remove the 1524-mm band of uncured lacquer before curing the remaining coating at high temperature to provide chemical and mechanical strength. At this point, the screen is fitted with special aluminum end rings to provide tubular rigidity and facilitate mounting in the coating machine.[5]

Galvano screens follow the same manufacturing cycle as lacquer screens.

86 WEB PROCESSING AND CONVERTING TECHNOLOGY AND EQUIPMENT

The nickel plating process of Galvano screens is essentially identical to that of the lacquer screen. The Galvano mother cylinder does not have the mesh pattern machined into its surface. The mesh pattern must first be made as a film, and the cylinder coated with a thin uniform coating of light-sensitive lacquer. Then the film is carefully and precisely wrapped around the lacquered cylinder, taking particular care in aligning the hole or mesh pattern where the film joins around the cylinder. The cylinder rotates while high-intensity light scans the cylinder surface to cure the lacquer visible through the film. Removal of the film and the uncured lacquer exposes the pattern of dots or the mesh pattern required. Plating the cylinder with nickel forms the screen where the ends of the tube are solid nickel, to which the end rings are later affixed,[6] following removal of the thin wall screen tube from the mother cylinder.

Coating screens are usually in the band of 9-40 mesh where either type of screen is suitable. However, the Galvano technique provides for an infinite array of meshes, hole sizes and free areas, including a mixture of hole sizes within one screen, opening up interesting possibilities for specialty processes.[7] Figure 5-3 shows a Galvano screen being installed.

Fig. 5-3. Placing a screen into a machine. (*Courtesy Stork X-Cel, Boxmeer, Holland.*)

Rotary screen development is an active area. The STK jet system[8] was introduced recently. Various improvements of lacquer and engraving technology have come out of the engraving industry.

Woven wire screens have been largely replaced by lacquer and Galvano screens. Nickel treatment of woven wire is interesting because it permits the creation of extremely fine mesh structures; i.e., 325 mesh screens with 18% open area used in the food and microprocessor industries.[9]

ROTARY SCREEN SYSTEM

While the screen is the most important part of the system, the doctor blade or squeegee blade mounted in the interior of the screen is next in importance. Figure 5-4 is a schematic view of the rotary screen system, where the squeegee blade presses against the inner wall of the rotating screen, and in turn against the web and counter roller. The compound is wedged between two relatively thin metal walls and is squeezed through the holes onto the material.[10] This is the most significant feature of this system, because the

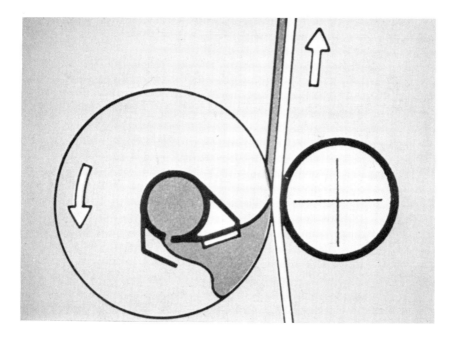

Fig. 5-4. Rotary screen system.

screen applies a uniform coating onto the surface of the web with almost zero coefficient of friction at the extrusion point. In this way, almost any web surface can be coated, regardless of its surface texture or cross-sectional thickness.

All rotary screen coating machines use screens of the same basic type; however, the squeezing devices differ. Blades, rollers and rods with or without magnets may be used. All of these devices are referred to as squeegees. The simplest is a stainless steel blade, supported by a holder in levers, designed to press the blade tip against the inner surface of the screen.

Figure 5-5 is a good example of a Stork PD rotary screen coater. The screen is located by its end rings in special bearing assemblies. It has an open coating width of 1524 mm and the squeegee levers, adjustments and auto level control are clearly visible, along with the paste feed pipe. Figure 5-6 shows the adjustment of the squeegee to increase pressure on the coating. This system is common to all PD series machines, where the holder acts as a supply pipe to distribute the coating to the screen interior. Variations on this basic concept come in the form of air expandable tubes to control blade deflections, double leaf construction blades, and rollers.

Fig. 5-5. Screen coater (PD III series machine, Stork X-Cel).

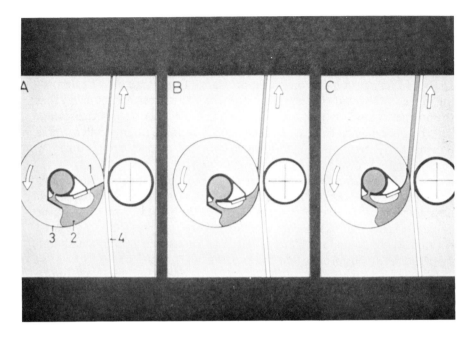

Fig. 5-6. Squeegee adjustment:
A—Short blade, low pressure, thin coating
B—Longer blade, higher pressure, heavier coating
C—Long blade, high pressure, heavy coating.

MACHINE MECHANICS

All the commercial rotary screen coaters adhere to the same basic design concepts. The screen is supported and driven so as to match the peripheral speed of the screen with the web, avoiding friction at the coating point.

Squeegee blades may range in thickness from 0.1-0.25 mm if stainless steel, but rubber, nylon, high-density polyethylene and other materials are also used successfully. The rods range in diameter from 6-15 mm.

The counter roller or support roll is a precision ground, chrome plated roll, driven or free. If the web is paper or similar material, no drive is required. If the web is a textile fabric, a reversible driven counter roller can be beneficial. Rubber covered counter rolls have been used in special circumstances.

Adjustment of the gap between the screen and counter roll is required to permit a range of web thickness. This can be achieved in a variety of ways.

This gap setting should never be more than the thickness of web being processed in order to avoid distorting the screen and losing control of the coating. One manufacturer recommends a gap setting of not less than 80% of the web thickness up to the web thickness.

The rod/magnet type arrangement provides a fixed range of shear according to the rod selected. The rod is located within the screen in such a way that the magnet is the only force to hold it against the inner wall of the screen. The rod is mechanically located on the center line of the counter roller, where the magnet is used to pull the rod against the screen to overcome the separating force of the compound.

The squeegee blade is usually moved or positioned mechanically. This construction provides continuous variation of shear force, with the added advantage of being able to position the blade tip past the center line of the counter roller to release heavier coatings. Short blade projections are used for thin coatings and long blade projections for heavy coatings. Small- and large-diameter rods produce the same effect.

In addition to these mechanical adjustments, it is important to accurately control the mass of compound within the screen. All machines are fitted with automatic level controllers to avoid overfill and to reduce operator attendance. If the compound level is not carefully controlled and allowed to fill the screen, then the radial force might cause the compound to pass through the screen mesh before it reaches the extrusion point. Similarly, if too little compound is in the compression zone, then insufficient shear force is developed and the coating might be irregular. Reference to the various figures gives a good guide as to the optimal compound level when the web is traveling in a vertical path past the screen.

SCREEN CHOICE

The screen, squeegee, machine settings and formulation for a given process must be carefully chosen in order to have a successful operation.

Assume the process is to ground coat wallpaper base stock with a polyvinylchloride plastisol coating. The vinyl coating must be 0.050 mm thick and 1524 mm wide, and must be processed at 55 meters per minute.

The formulation will contain filler and is of pseudo-plastic type having good shear thinning flow properties. Use of a 23 mesh screen produces the required coating. What can we expect if the screen is changed to a standard 25, 40, 60 and 80 mesh, or to a 125 mesh Penta, and finally to a 40 mesh special Penta type?

Referring to Table 5-2 and Fig. 5-2, we can relate the hole diameter, the wall thickness and the free area, expressed as a percentage of the total screen surface area. The percentage free area is a very important factor, since with free area increase, resistance to flow reduces, permitting easier passage of compound. The 23 mesh screen has 425 μm holes and 14% free area and is 220 μm thick. The 25 mesh screen has 375 μm holes and 12% free area and is 220 μm thick. The 40 mesh screen has 320 μm holes and 23% free area and is 120 μm thick. These three screens will produce the required coating. The thicker screens are a better choice from a durability viewpoint; their hole size is excellent for free passage of large particles or agglomerates. The 60 mesh screen has 160 μm holes and 14% free area and is 100 μm thick, and can produce a 0.050-mm coating. However, with such a small hole size, large particles and agglomerates will cause streaking problems. Generally, standard screens up to 40 mesh are suitable for coating compounds containing large particles. Screens above 40 mesh are more suitable for coating micro particle compounds, low solids and 100% reactive systems. Therefore, the standard 80 mesh screen and the 125 mesh Penta screen would not be suitable to form a coating in the above process. The 40 mesh Penta screen has 350 μm holes and 30% free area and is 300 μm thick, capable of coating the above system, but would apply a minimum thickness of 0.10 mm up to 0.25 mm. The special Penta screens would be chosen for their ability to lay down thick coatings from 0.19–0.50 mm with this type of compound, using the 40 H through 25 H to 14 H mesh screens. The upper coating limit with the very wide machine (3658 mm) is near to 0.30 mm.

A short, stiff squeegee projection, or small radius rod, is used with a comparatively small angle of compression. Squeegee selection will also influence the coating thickness as the following example explains.

Taking the same process as above with no filler present, a standard 60 mesh screen and a squeegee blade projection of 25 mm with the tip on the center line would not produce a coating. The blade is changed to a 28-mm projection and the tip positioned 2 mm above the center line, resulting in an acceptable coating between 0.038 mm and 0.050 mm. The ideal selection would therefore be a 23 or 25 mesh standard screen with a 26 mm squeegee projection on the counter roller center line. This would offer more latitude for formulation adjustment with filler and would ensure no screen distortion with assurance of excellent gauge control.

KNIFE/SCREEN MECHANICS

All knife coaters are scrapers of one type or another, where the web is pulled past a blade, either in contact or set at a fixed gap to the web. The resultant coating is directly related to the rheology of the coating, the blade shape, width and web speed.

92 WEB PROCESSING AND CONVERTING TECHNOLOGY AND EQUIPMENT

The compound is sheared directly between the web surface and the blade, where a broad foot knife profile creates a wedge effect or a compression zone. The hydraulic force generating from the compound's resistance to compression produces a separating force pushing the blade away from the web, and can often reach 10–16 kg/cm of web width or more. This explains the penetration of open or porous substrates. The blade profile and foot area are the major contributing factors affecting penetration and the web speed has little effect.[11]

Screens in thickness range between 0.0762 mm and 0.381 mm, and squeegee blades between 0.10 mm and 0.254 mm provide very little resistance to high rates of shear.

In a rotary screen machine, the resistance to the separating force can be as low as 0.32 kg/cm of web width. This is the force between the inner side of the screen wall and squeeze blade. This is illustrated in Fig. 5-7. A low force is particularly important for coating porous structures.

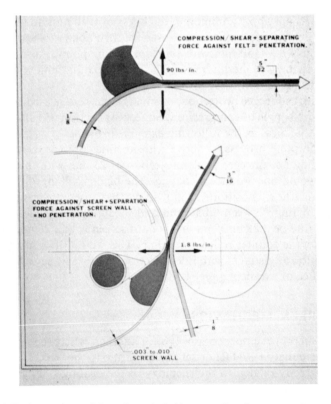

Fig. 5-7. Comparison of shear forces in knife-over-roll and screen coating processes.

The rotary screen has a fixed coating width dictated when made, or more effectively, a specially developed cold cure lacquer can be used to reduce the coating width. This lacquer can be removed with solvents. This feature of the screen eliminates edge dams. Splices pass through the screen machine with ease, due to the elasticity of the screen, eliminating the need for splice jumpers or operator attendance.

Low friction on the surface of casting paper and films improves the product yield. Rotary screens are used to apply the skin coat to release paper and for many other applications.

RHEOLOGY

Screen coaters are often used for vinyl plastisols and these coatings often exhibit non-Newtonian flow behavior. Dilatency must be avoided, no matter how small; Newtonian and pseudo-plastic behavior provides the most rewarding results. Typical viscosity-shear rate dependency of a vinyl plastisol suitable for rotary screen coating is shown in Fig. 5-8.

Screen release properties of compounds cannot be overstressed. Experience shows that a formulation can be easily developed to meet the viscosity-shear characteristics, but often the system will not leave the screen cleanly. The use of small quantities of lubricant and viscosity depressants is helpful.

Apart from polyvinyl chloride plastisols, the rotary screen coater is used to apply coatings of polyurethane latices, and various types of dispersions,

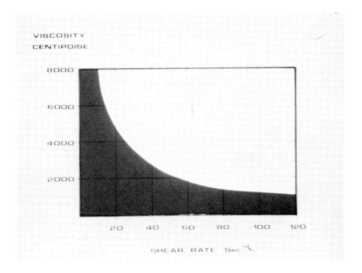

Fig. 5-8. Viscosity versus shear rate for a vinyl plastisol coating for a screen process.

including water-based pressure-sensitive adhesives, acrylics, solvent based inks, clay slurries, metallized coatings, 100% reactive ultraviolet curable systems, metal protective coatings, chemical and mechanical foam systems and silicone coatings. Figure 5-9 shows the flexibility of the rotary screen coater. By changing the screen and compound, a much wider product range can be made. Versatility of the rotary screen technique also appears in combination machine designs. Machines which combine a rotary screen with a knife-over-roll system are available.

Combining technologies can be rewarding to converters, but it is often difficult to introduce an additional operation due to web path, space and operation difficulties. However, due to the rotary screen containing the compound, it becomes easier to insert it into almost any location of a web process range as shown in Fig. 5-10.

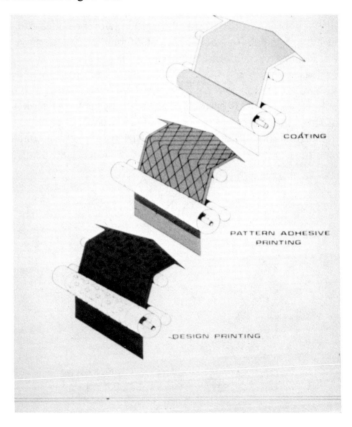

Fig. 5-9. Various applications of screen coating.

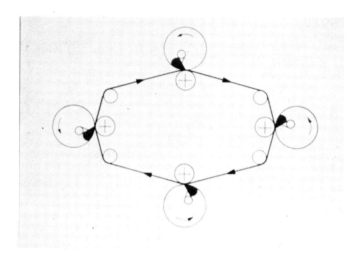

Fig. 5-10. Mounting of a rotary screen coating head in various locations in a web processing line.

Rotary screens are cleaned quite easily. Screen and squeegee washing machines are available for cleaning up these parts for the next color or product change. These machines are suitable for cleaning with water or solvents and it takes approximately 3 minutes to complete the job. A color change can be made within 5 minutes using spare screens or squeegee, or within 10 minutes, if there are no spares.

Polyvinyl chloride plastisols can be applied in thickness of 0.04–0.4 mm (under certain circumstances up to 0.6 mm); acrylics 8–140 g/m^2 out of 25–50% solids dispersion; polyurethanes 8–90 g/m^2 out of 20–35% solids dispersion; pressure-sensitive adhesives 4–160 g/m^2 out of 20–50% solids solutions or dispersions; clay slurries 10–25 g/m^2 out of 50% solids dispersion.

Cross-section coating accuracy can be expected within ± 5% on webs up to 2.6 m wide. Wider webs are processed (e.g., 4-m wide resilient floor coverings, where the construction of the screen support heads, squeegee and adjustment are different, to minimize deflection and to maintain the same coating accuracy. Wide width machines usually have a double side compound feed and control system using either one or two pumps.

Rotary screen coaters operate effectively over a speed range of 3–150 m/min.

Energy ratings are comparatively low, being 3 HP for machines up to 1200 mm in web width and 5 HP from 1200–2600 mm.

REFERENCES

1. Storey, Joyce. *The Thames and Hudson Manual of Textile Printing.* London: Thames and Hudson, 1974, pp. 133-135.
2. Caporale, L., Private conversation, Caporale Engraving, Inc.
3. Storey, Joyce. p. 140.
4. Sucheki, Stanley M. Penta® screen. *Textile Industries,* p. 71 (November 1978).
5. Storey, Joyce. pp. 145-148.
6. Storey, Joyce. pp. 142-145.
7. Stork Brabant, B. V. Holland. Unpublished data.
8. Zimmer, Peter. Private conversation, Peter Zimmer, Inc., Spartanburg, S.C., and Derek Turner, Zenith Engraving, Chester, S.C., June 1982.
9. Printing Systems, Inc. Company sales literature, Sparks, Nev.
10. Bell, Eric. *Rotary Screen Coater and Protective Surface Coating—Decoration—Creation,* AATCC, September 20-21, 1982, Marriott Hotel, Newton, Mass.
11. J. Bell, Eric. unpublished data.

6
Spray Coating

Donatas Satas
Satas & Associates
Warwick, Rhode Island

Spray coating is extensively used to paint various articles, but only rarely to coat continuous webs. It is mainly suitable for applying coatings over irregular surfaces, such as corrugated materials, and webs which are fragile and cannot be easily carried through a regular coating head. Binders for nonwoven fabrics are sometimes applied by spraying over loose unbonded fibrous mats, which are continuously fed from a web former. Thick mats can be bonded by spraying, avoiding squeezing that might compress the loose web.

ATOMIZATION

A jet of liquid breaks up into small particles if a certain speed is exceeded. This critical speed is related to surface tension and viscosity of the liquid.[1] The instability can also be induced by introducing the jet into a high-velocity stream of compressible fluid such as air. This method is the most suitable for obtaining small spray particle size. The mechanism of atomization is a two-step process: formation of ligaments by air friction followed by the collapse of ligaments forming drops. The atomized particle size decreases as the air velocity increases. The lower particle size limit is about 5 μm, although some smaller particles (1 μm) can be also present. If a larger particle size is required (50–300 μm), other atomization methods should be used, such as airless spraying, or spinning disk. Pneumatic atomization, however, is the only known method to produce a fine spray in which diameters of all droplets are less than 15 μm.[2]

Polymer solutions may have viscoelastic properties and their atomization

98 WEB PROCESSING AND CONVERTING TECHNOLOGY AND EQUIPMENT

might proceed differently. Besides drops, elongated ligaments and various other irregularly shaped bodies might be formed. The ligaments, which are formed during spraying, disintegrate into drops when a Newtonian fluid is sprayed, but remain in case of spraying of viscoelastic fluids. Fibrous, porous polymeric coating can be produced utilizing this property.[3,4]

In addition to the pneumatic and pressure nozzle atomization, liquids can also be atomized by centrifugal swirl nozzles, spinning disks, impingement and electrical charge.

AIR SPRAYING

Air guns are most often used for spraying of liquid coatings. A high degree of atomization and good control of spray uniformity are the reasons for the popularity of this method. The disadvantage is considerable overspray, unless electrostatic equipment is used in conjunction with air atomization, and relatively high demand for energy in the form of compressed air.

Figure 6-1 shows the cross-section of a spray gun. Liquid is introduced into a high-velocity air stream discharged through an annular orifice around the opening for the liquid. The liquid is atomized and carried to the target by the air. The spray pattern is round and the spray density is somewhat higher in the center of the spray cone. A needle is placed in the fluid outlet. It closes and opens the fluid outlet and is air operated for automatic guns. The needle allows the interruption of the fluid flow at the end of the stroke and it also keeps the fluid nozzle clean. The fluid tip is made from a hardened alloy. If abrasive materials are sprayed, such as filler containing coating, Carbaloy fluid tip and needle are used.

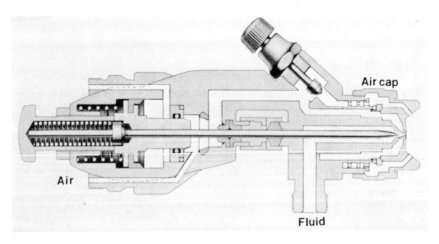

Fig. 6-1. Cross-sectional view of an air atomizing automatic gun. (*Courtesy DeVilbiss Co.*)

The air gun is usually equipped with a cap which has drilled air passages. The air delivered by the cap shapes the spray cone into a fan. Fan shaped spray is easier to overlap and a more uniform coverage is obtained. About 60% of the air energy is used for the atomization and 40% for the pattern formation. Air pressure as low as 100–170 kPa is sufficient for atomization; somewhat higher pressure (about 200 kPa) is usually used with spray guns.

Decrease of viscosity helps to decrease the atomization air pressure and to obtain a finer spray. Therefore, solvent-based coatings sometimes are heated in a heat exchanger placed before the spray guns. The lines leading from the heater to the guns might be traced with hot water to maintain the temperature. Increased temperature also helps the drying of the coating.

For moving web coating, automatic guns are mounted on either a reciprocal or a rotary carriage. Several guns may be mounted on one carriage. Spraying from several guns and continuous movement of the carriage across the web, increases the uniformity of the coating. The automatic guns are shut off by air operated needle at the end of each stroke. This decreases the overspray and keeps the spray nozzle clean. Traverse carriage, such as shown in Fig. 6-2, gives a more uniform spray distribution across the web, than the rotary carriage. Because of the curved path the latter gives a higher spray density at the edges.

Fig. 6-2. Reciprocating spray coater. (*Courtesy DeVilbiss Co.*)

Instead of mounting the spray guns on a moving carriage, they are sometimes mounted on a rod which rotates back and fourth around a fixed point. This is much simpler set up than a moving carriage, but the spray uniformity is not equal to that achieved by traversing guns. The best coating uniformity is achieved by reciprocating guns with a fan spray pattern where the long axis of the fan is in the direction of the web movement. The rate of cross-machine reciprocating movement should be fast relative to the web speed in order to have a good overlap of the spray pattern.

HYDRAULIC SPRAYING

Forcing a liquid through a nozzle may break it up into drops, if a certain velocity is exceeded. The air-less spray guns are based on this principle. Such guns are not useful for slurries, because solid particles clog up the orifice, they cannot be used for atomization of higher viscosity fluids, because of the excessively high pressure that is required to cause atomization. While the droplet size is larger than that obtained with pneumatic spraying, air-less spray is very useful and economical method of atomization for low-viscosity liquids, such as water, or aqueous emulsions of binders.

Many different nozzle designs are available for air-less spraying. Swirl-type nozzle imparts a spinning motion to the fluid before it is discharged. The rotating motion is imparted by providing a spin chamber with fluid inlets set at an angle. This type of a nozzle is shown in Fig. 6-3. Another design forces the liquid through tangential slots or helical grooves before it leaves the nozzle through the outlet. It is also possible to atomize a liquid by forcing it through a non-circular orifice. Non-uniform stresses are created in the liquid stream which cause the break-up of the liquid column.

Impact atomizers cause the break-up of the fluid stream by impinging it against a barrier. The spray can also be shaped into a fan by such spraying heads (Fig. 6-4).

ELECTROSTATIC SPRAY GUNS

Liquids can be broken up into small particles by electrostatic forces alone, but such method of atomization is not efficient and it is rarely used. Already atomized liquid particles can be electrically charged in order to direct them to a target minimizing the overspray. Electrostatic spraying is very useful and it is widely employed for coating of irregularly shaped objects, but it is less often used for continously moving webs.

Liquid spray atomized by air, or other means, is introduced into an electrostatic field where each particle receives a static charge. The electrostatic field is created by a cage type electrode consisting of fine wires supported on

SPRAY COATING 101

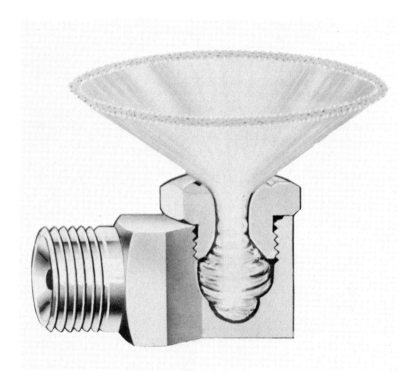

Fig. 6-3. Wide spray swirl nozzle. (*Courtesy Spraying Systems Co.*)

a metal frame. High-voltage (100,000 volts), low-amperage (5 milliamperes) negative potential is applied to these wires.

Another way to produce charged spray particles is to feed the coating material at a controlled rate to a point at which a highly concentrated electro-

Fig. 6-4. Impact spray nozzle. (*Courtesy Spraying Systems Co.*)

102 WEB PROCESSING AND CONVERTING TECHNOLOGY AND EQUIPMENT

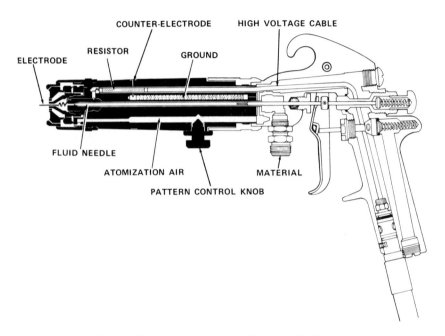

Fig. 6-5. Electrostatic spray gun. (*Courtesy DeVilbiss Co.*)

static field is formed. This point is located on the end of the spray gun (Fig. 6-5).

The work to be coated is grounded and must be conductive for effective application of electrostatic spray coating. If the material is nonconductive it must be treated to impart some conductivity.[5] There are several methods that can be used for this purpose:

a. Wetting the surface with water
b. Heating the surface
c. Backing with conductive material
d. Depositing a conductive coating; i.e., quarternary ammonium salt solution or gel
e. Sulfonating the surface (for polymers that have aromatic groups on the surface)
f. Use of conductive pigment, such as carbon black.

ROTARY DISKS

Spinning disk atomizers have been widely used for spraying operations, especially in the chemical processing industry. The mechanism of drop formation from a spinning disk has been discussed by Frazer *et al.*[6,7] The method is suitable to atomize slurries and viscous liquids.

Bell type rotary atomizers are used in combination with electrostatic field. High-speed atomizer rotating at 10,000-40,000 rpm discharges finely atomized particles from the edge of the bell. These particles are also electrically charged.

ROLL FLINGER TYPE COATERS

These coaters have been developed at the same time as air-knife coaters and were intended for the same application: clay coating of paper. While the air-knife coating became important, the flinger type coaters were largely forgotten, although it was demonstrated that they are suitable for quality coating.[8]

The coating principle is shown in the Fig. 6-6. The web is pre-wetted in the kiss roll coating station and then enters the flinger coater. The solution is applied by a nozzle to a fast (1000 rpm) rotating grooved roll A. The coating from roll A is splashed on the surface of roll B, which could be running as fast as 2000 rpm, and then onto the web surface supported by a smooth roll C.

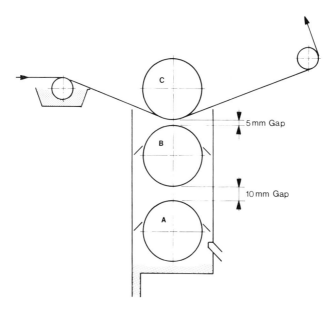

Fig. 6-6. Roll flinger coater.

AUXILIARY EQUIPMENT

Spraying requires various auxiliary equipment consisting of coating reservoir from which the coating is delivered to the spray gun. Pressure feed tanks may be used, or the coating may be pumped by positive displacement pumps with accurately controllable pumping rate. Air-less spraying requires high-pressure pumps.

The overspray must be collected from the exhaust air stream and scrubbing by a water curtain might be required to remove the particulate matter.

Spraying by air atomization requires compressed air and the air compressor is an important part of the investment.

REFERENCES

1. Muirhead, J. *Science and Technology of Surface Coating*. (B. N. Chapman and J. C. Anderson, (eds.). London: Academic Press, 1974, pp. 248-256.
2. Marshall, W. R., Jr. Atomization and spray drying. *Chemical Engineering Progress Monograph Series No. 2,* Vol. 50 (1954).
3. Satas, D. *Ind. Eng. Chem.* **57**: 38-42 (1965).
4. Satas, D. U.S. Patent 3,232,819 (1966) (assigned to Kendall Co.).
5. Spiller, L. L. *Science and Technology of Surface Coating*. (B. N. Chapman and J. C. Anderson, (Eds.). London: Academic Press, 1974, pp. 28-42.
6. Fraser, R. P., Dombrowski, N. and Routley, J. H. *Chem. Eng. Sci.* **18**: 315-321 (1963).
7. Fraser, R. P., Dombrowski, N. and Routley, J. H. *Chem. Eng. Sci.* **18**: 323-337 (1963).
8. Booth, G. L. *Coating Equipment and Processes*. New York: Lockwood, 1970, pp. 73-74.

7
Calendering

Donatas Satas
Satas & Associates
Warwick, Rhode Island

Calendering is a very old technique developed for processing of rubber. It has been in use for some 150 years. Its major application has become processing of polyvinyl chloride, although it is still used for rubber as well as for various thermoplastic polymers. It is particularly useful in processing high-viscosity polymers, especially the ones susceptible to thermal degradation, or filled with a large amount of fillers, which increase the viscosity beyond the capability of other processes.

Manufacturing of films and sheets is the main use of calendering. Coating and laminating are less important applications. Calender stack is also a standard part of every paper making line: it is used for surface coating and finishing of paper. Our interest is limited to coating and laminating applications outside of paper manufacturing.

In calendering the thermoplastic polymer softened by heat is forced through a narrow gap between two corotating rolls (Fig. 7-1). The pressure exerted on the polymer shapes it into a flat sheet of uniform thickness which is carried by a faster rotating roll usually at a lower temperature. Calender can provide a variety of speed differentials between rolls. These differentials control the amount of work done on the polymer and help to regulate the heat build-up in the bank.

While one nip calendering is possible, usually the heat-softened polymer mass passes through one or two additional nips. Three-roll (two-nip) calenders are usually used for rubber, four-roll calenders (three-nip) are employed for vinyl.

The nip pressures might reach 360–630 kg/linear cm for soft polymers to 1000 kg/linear cm for higher compressive modulus materials.

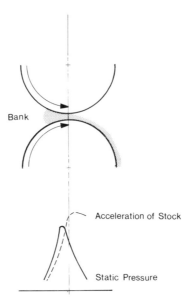

Fig. 7-1. Calender nip and the distribution of static pressure and stock acceleration.

Large amount of polymer can be processed by a calender, as compared to extrusion, because calendering requires lower input of mechanical energy. A calender can process 400–4000 kg/hr of polymer. The amount of polymer to be processed determines the production rate, not the calender speed. Typical speeds attainable on a calender are 50 m/min for a heavy (0.25 mm thick) sheeting, 100 m/min for a thin film; glossy rigid material is produced at a slower speed of 10–30 m/min. Extra wide machines up to 2.6 m have been constructed for rigid and up to 3.4 m for flexible vinyl film calendering. Mass preheated on a roll mill, or on an extruder, is fed into the first nip. The banks should be kept as small as possible. For best uniformity the feed bank should be 10–13 cm in diameter and it should be continuously rotating to prevent stagnation and cooling. Second and third banks should be no larger than 1.5 cm in diameter.[1]

Mathematical models for calendering have been reviewed by Middleman,[2] Tadmor and Gogos[3] and several other authors. It is difficult to construct a sufficiently accurate and meaningful model, and a rigorous theoretical analysis incorporating all practically significant effects becomes quite complicated. The most commonly used model has been proposed by Gaskel.[4] The Finite Element Method (FEM) is a useful tool in analyzing the flow of non-Newtonian fluids in complex geometries, such as the case of calendering. The calendering process was analyzed with FEM by Kiparissides and Vlachopoulos.[5]

In addition to the formation of the film, a process which involves the flow of polymer, the polymer film or coating also acquires a finish in the calender as a result of interaction between the roll and coating surfaces.

CALENDER TYPES

Coating and laminating of fabrics and other substrates is carried out by several processes.[6] Comparison of the main characteristics of vinyl calender coating with other coating techniques is given in Table 7-1. Amongst plastic materials vinyl is most important for calendering. Smaller quantities of polyethylene, ABS and few other plastics are also processed by calendering. PVC calendering requires higher temperatures (150–190°C) than rubber (100–130°C).

Calendering requires a high capital investment, as much as five times the investment required for an extrusion or plastisol coating line. Less expensive vinyl resins are used for calendering and this can be an important advantage offsetting the higher equipment costs. Rigid vinyl coatings can be easily handled on a calender, but are not possible to produce by plastisol coating, unless solvent is added to decrease the viscosity. Plastisol when fused releases fumes and organosol releases solvent vapor. Highly plasticized coatings are more difficult to process on a calender.

Extrusion coating is best suited for heavy vinyl. In lighter gauges the uniformity of an extruded coating is poorer than that of calendered one. Gauge adjustments are more difficult and take a longer time on an extruder than on a calender, or a plastisol coater.

Calenders are sturdy, and long lasting pieces of equipment and calender lines are rarely scrapped. They can be renovated and upgraded by addition of new accessory equipment.[7]

Many different arrangements of calender rolls are used. The most often employed arrangements are inverted "L" (Fig. 7-2a), generally used for vinyl calendering, and inclined "Z" (or "S," see Fig. 7-3) used for rubber calendering. "Z" arrangement has been used since 1893, inverted "L" is

Table 7-1. Comparison of Various PVC Coating Methods

	CALENDERING	EXTRUSION	PLASTISOL COATING
Coating thickness, mm	0.05–1	0.1–3	0.0025–0.3
Average coating accuracy, %	3	10	7
Machine output, kg/hr	400–4000	350	350
Resin cost	Lowest	Higher	Highest

108 WEB PROCESSING AND CONVERTING TECHNOLOGY AND EQUIPMENT

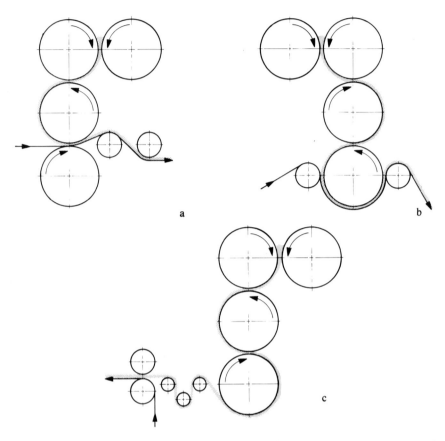

Fig. 7-2. Calender lamination: a—Nip lamination in an inverted "L" calender b—Inverted "L" calender with a squeeze laminating roll c—Remote squeeze roll lamination.

somewhat newer, it supplanted vertical stack arrangement used in early 1920's. The inverted "L" arrangement simplified the adjustment of two top rolls in a vertical calender stack.

LAMINATING

There are several processes used for calender coating and laminating.

Nip lamination is shown in Fig. 7-2a. This method has been adopted from rubber calendering. The nip between rolls 3 and 4 is used for lamination, sacrificing one working calender nip. This results in a decrease of calender capacity. The control of the coating strike-through into the fabric is difficult in this method, but the substrate helps to pull the material through the

CALENDERING 109

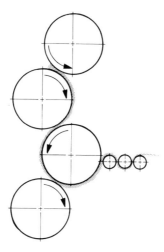

Fig. 7-3. Inclined "Z" calender.

calender making it easier to handle soft and extensible materials such as shown in a silicone rubber coating line (Fig. 7-4).

The disadvantage of nip lamination is overcome by installing a squeeze roll (Fig. 7-2b). The lamination takes place between the squeeze roll and the #4 calender roll, rather than between rolls 3 and 4. This way an additional working nip is added to the calender. The substrate helps to pull the laminate off the last roll. The first nip is the feed pass, second-metering pass, the third nip forms, gauges and finishes the vinyl film.

Lamination in a calender nip is accomplished at a higher pressure that the lamination by a squeeze roll. Calenders for laminating purpose typically operate at a pressure of 25–250 kg/linear cm on a fixed, electromechanically

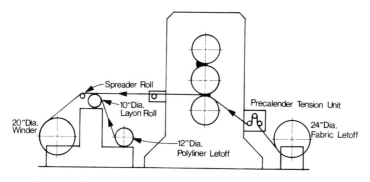

Fig. 7-4. Silicone rubber coating line, 12″ x 52″—three-roll, 10 m/min calender.

adjustable gap between the calender rolls. If necessary higher pressures (up to 400 kg/linear cm) can be employed. Hydraulically or pneumatically loaded rubber covered squeeze roll operates in the range of 3.5-100 kg/linear cm. Higher pressures may cause a failure of the rubber bond to the metal core.

Another version of squeeze roll lamination is shown in Fig. 7-2c. The laminating nip is located away from the calender and the substrate is laminated after the calendered film has been removed from the last calender roll. The squeeze roll is not in contact with hot cylinder roll and the rubber cover lasts longer. The substrate does not help to pull the calendered film and handling of the web is more difficult. If the film cools off, it might be more difficult to obtain adequate penetration of the polymer into the substrate. The substrate, however, is not exposed to the hot calender roll and less temperature resistant substrates can be used by this method.

Calender coating and laminating in line with vinyl film formation has been decreasing in favor of a separate laminating step. Vinyl film can be produced on a calender at speeds as high as 70-100 m/min, while calender coating or calender laminating is a much slower operation carried out at 15-45 m/min. Free vinyl film is recoverable, while the laminate is not. Because of the higher production speeds and lower waste, making the film separately and then laminating on different equipment can be more economical than in-line laminating. Vinyl film is laminated to various substrates by adhesives, or by using heat and pressure.

PRESSURE-SENSITIVE TAPES

Pressure-sensitive tape manufacturing by calendering has been reviewed by Eighmy[8] and by Satas.[9] Calendering is used to manufacture pressure-sensitive adhesive tapes by coating the adhesive mass onto a cloth backing. Three roll calenders are mainly used for this application. The typical calendering conditions for an adhesive coating operation, as shown in Fig. 7-5a, are: top roll at 130°C rotating slowly, center roll at 38°C, bottom roll at 95°C. The adhesive is coated at 10-100 m/min. A photograph of such calender is shown in Fig. 7-6.

Rubber based pressure-sensitive adhesives tend to follow the faster of the two rolls. Most of the polymers follow the hotter of the two rolls, except styrene-butadiene rubber (SBR) and polyolefin thermoplastic elastomers (EPDM).

The most popular calender for pressure-sensitive tape manufacturing is a 24" diameter by 66" width, three-roll vertical unit. Newer machines might have an angular roll arrangement (Fig. 7-5b). Rolls are loaded at 90-600 kg/linear cm, which is slightly lower than required for natural rubber and

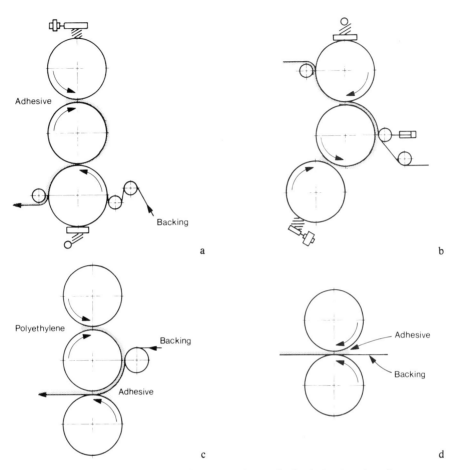

Fig. 7-5. Pressure-sensitive adhesive calenders: a—three-roll calender b—Angular roll arrangement c—Simultaneous polyethylene and adhesive coating d—Two-roll calender.

other elastomers, or polyvinyl chloride calendering to the same thickness. The higher loads are required for lower mass thicknesses (0.075–0.150 mm).

Fabric can be coated simultaneously with polyethylene on one side and pressure-sensitive adhesive on the other side, as shown in Fig. 7-5c. Polyethylene tapes without a web support can be also made by this method. The largest volume pressure-sensitive tape produced by calendering is pipe wrap tape used to protect steel pipe from corrosion.[10] This tape is made by simultaneously calendering polyethylene backing and butyl rubber based pressure-sensitive adhesive mass.[11]

Calendering of adhesive to a thickness below 0.15 mm can result in a coating containing many pinholes. To eliminate the pinholes, greater tempera-

Fig. 7-6. 24" x 64" three-roll calender for making pressure-sensitive adhesive tape. (*Courtesy Farrel Machinery Group.*)

ture differentials and higher friction ratios are used. Instead of a ratio 5/4, a ratio 5/1 or even 20/1 might be employed.

Adhesive can also be calendered on a two-roll calender as shown in Fig. 7-5d. The top roll might be at 50°C and the bottom at 170°C. An adhesive coating line is shown in Fig. 7-7.

FRICTION CALENDERING

Electrical friction tape and similar products are made by frictional calendering. Such tapes have been an important product in the past, but have been supplanted by other tapes. A friction calender is shown in Fig. 7-8. The adhesive mass in the friction calendering follows the middle roll which is running at a higher speed and forms a small spinning bank at the bottom nip. The bottom roll is running at a 70–80% slower speed than the middle roll and the fabric follows the slower bottom roll. The frictioning which drives the mass into the fabric occurs at this second nip. In the case of normal vinyl lamination to a substrate the bottom roll is run at the same speed as the middle one.

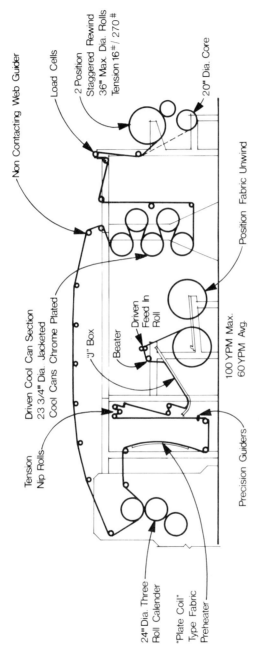

Fig. 7-7. Tape and coated fabric calender.

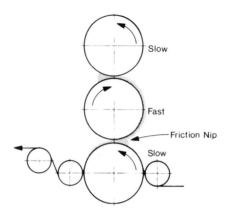

Fig. 7-8. Friction calender.

ROLLS

Chilled iron rolls are used for calendering. Cast iron can yield either a gray or white appearing structure, depending upon the rate of cooling. In the process of casting of chilled iron rolls, the surface of the roll is cooled at a fast rate and this gives a white structure, while the center of the roll cools at a slower rate and gives a gray structure. The principal constituents of the white structure are cementite and pearlite and it has no free graphite. This structure is very hard and it has good wear-resistant properties, but it is brittle. The gray cast iron has some free graphite and is less brittle. The depth of the white chilled surface is controlled by the rate of cooling. It should be sufficiently deep to allow the roll to be ground, but it should not be so deep as to decrease the roll strength. Smaller diameter rolls might have the white structure extend 10 mm from the surface, while larger rolls might have the depth of chilled surface extend to 50 mm.

The new rolls are of 68/72 Shore scleroscope hardness. During use the roll surface is further work-hardened. The rolls are ground within 50 μm and then further finished within 12 μm.[12]

The calendering rolls are drilled for heating. Drilled, rather than chambered, rolls are preferred. The drilled roll, as shown in Fig. 7-9, has holes located near the surface and this, along with the high circulation rate of the heating fluid (500–1000 1/min), or of hot water (300–500 1/min), helps to maintain the roll temperature uniform across the width. The temperature variation across the calender roll is kept within 1°C on an empty machine. Drilled holes are also more responsive to the changes of fluid temperature. These drilled passages are connected to the central bore. The rolls are fitted at the end with a rotary union and a syphon pipe.

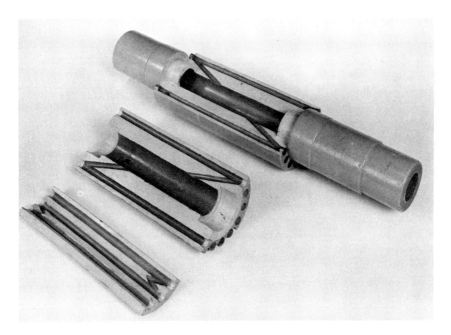

Fig. 7-9. Cutaway of a drilled roll. (*Courtesy Farrel Machinery Group.*)

At low calendering speeds heat is transferred from the roll to the polymer. At high speeds the reverse might be true. A high degree of mechanical work put into the polymer might raise its temperature above that of the roll surface.

Rolls can be set out of round permanently by nonuniform heating. Therefore, the passages for the heating fluid must be kept free of obstructions, and prolonged stops in excess of a few minutes at temperatures above 135°C should be avoided. The temperature change should be slow. It should not exceed 3°C/min up to 180°C and 1°C/min above this temperature.[1]

Stock should not be allowed to run out: it separates the rolls and prevents surface damage. Metal detectors are used to prevent a piece of metal in the stock from entering the calender. Nonuniform stick-slip rewind force should be avoided. Rolls require regrinding every three to five years. More frequent regrinding might be needed, if precision work is required.

Roll or sleeve bearings are used for calender rolls. High pressure on a sleeve bearing might interrupt the lubricant film, causing chattering and damage to the bearing. At high loading and slow speeds, roll bearings are preferred. Double-row, tapered type bearings are generally used. They may be stabilized for high-temperature calendering conditions. Calenders operating above 120°C should be equipped with temperature controlled circulating fluid lubrication.

Motorized roll gap adjustments may be used. Rolls 1, 2 and 4 are provided with single or two-speed motorized screwdowns for adjusting roll speeds. Motorized roll adjustment is used to control the nip between #1 and 2 rolls. The gap is adjusted by electrical motors at the rate of 0.5–2.5 mm/min for each end of the roll driven individually.[8]

High separating forces are developed between the calender rolls. These forces may cause roll bending and consequently nonuniformity in the thickness of calendered coating. They also can reduce the oil film thickness in sleeve bearings and change the gap. The magnitude of separating forces depends on the modulus of the polymer stock. Rubber calendering develops lower separating forces than vinyl calendering. Increase of the temperature decreases the compressive modulus of the polymer and decreases the separating forces. A difference of 15°C in stock temperature in some polymers results in a change of 50% in roll separating force.[1] Separating forces increase with decreasing coating thickness. A force of 140 kg/linear cm is considered to be low and is met in case of thick rubber stock. In pressure-sensitive adhesive calendering, natural rubber stock generally causes higher separation forces than other pressure-sensitive stocks. The separation forces can be as high as 2000 kg/linear cm. This approaches the pressures developed in the nip of a light duty steel rolling mill. Generally the increase of the stock viscosity, size of the bank, or decrease of coating thickness below 0.18 mm increases the separating force exponentially.[8]

The uniformity of nip pressure can be checked by obtaining an impression on aluminum foil, carbon paper, rubber sheet in combination with NCR paper, or by the rubber pad method. The rubber pad method involves inserting rubber pads of uniform thickness along the entire nip, leaving small separations between the rubber pads sufficiently wide for a filler gauge to fit. The gap between the rolls is checked at these locations at various levels of compression. Other methods to measure the nip pressure have been developed[13] and various devices to control it have been devised.[14]

High separation forces cause roll deflection and the resulting non-uniform gap causes uneven caliper of the calendered coating. Several methods are available to compensate for the roll deflection:

- Roll crowning
- Roll bending
- Roll crossing.

Roll crowning is an easy and reliable method to compensate for the roll deflection, except that the crown is designed for one load level only. If the calender is run at various loads, crowning is insufficient and continuous methods for deflection compensation are required, such as roll bending or roll crossing. Earlier calenders, and the less elaborate machines of today, require crown adjustment in the last nip. More elaborate machines have crown adjustment at each of the nips.

The calculations of the crown have been reviewed.[14] Stone and Liebert[15] give the following formula for calculating of maximum deflection.

$$y_{max} = \frac{5ql^4}{384\,EI} \left[1 + \frac{24}{5} \left(\frac{h}{l} \right) + \left(2\,\frac{d}{l} \right)^2 \right] \qquad (1)$$

where:

- y_{max} = maximum deflection
- q = total nip pressure
- l = web width
- E = modulus of elasticity of iron roll
- I = moment of inertia of roll body
- d = diameter of roll body
- h = distance from centerline of neck bearing to edge of web (roll neck overhang).

The geometry of the roll bending is shown in Fig. 7-10. The first term in the bracket of Eq.1 represents bending due to the uniformly distributed load, the second term is due to the overhang of the neck bearing, and the third term is the deflection due to shear. The true deflection curve is the fourth order polynomial; 150° sine curve is a close approximation of the true deflection curve.

Roll distortion can be corrected by roll crossing and by roll bending. These methods allow for a continuous correction to suit varying calendering conditions.

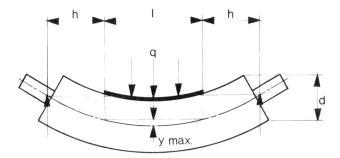

Fig. 7-10. Deflected bottom roll.

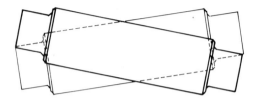

Fig. 7-11. Roll crossing.

Roll crossing is accomplished by moving the bearing boxes so that the axes of the two rolls cross as shown in Fig. 7-11. This increases the gap at the ends of the rolls, leaving the gap in the center the same. On a typical four-roll calender, movement of roll 3 provides a diametral crown effect up to 0.5 mm in two nips.[16] This is sufficient for most calendering jobs. The shape of the roll crossing curve is different from the deflection curve and this gives the "oxbow" effect. The crown correction is not linearly related to the bearing box movement and this complicates the application of automatic control.

Roll bending is accomplished by applying a bending load on the auxiliary bearing housings mounted on the roll extension outside the main bearings. Hydraulic cylinders fastened to the calender housing are used for this purpose. The bottom roll as shown in Fig. 7-12 bends up correcting the gap non-uniformity. Roll bending is sometimes used to fine-tune the effect of roll crossing. The amount of bending correction is limited to 0.05 mm on the gap change, because of the excessive stresses at the roll necks.[17]

A target on the machining accuracy for a good calender is for a maximum TIR of 0.0025 mm on the calender rolls, journals and bearings as a system measured unloaded. When running at an elevated temperature this increases to about 0.008 mm.[1] Sometimes rolls are ground hot to increase the accuracy. A good calender can hold a flat sheet to 0.005 mm variation in thickness for a 0.05–0.5 mm thickness range, or even 0.0025 mm for a 0.05–0.18 mm thickness range.

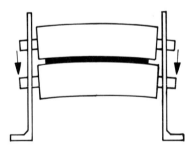

Fig. 7-12. Roll bending.

Induction Heated Roll

This type of roll is a more recent development. Such rolls are suitable for various polymer processing applications, including calendering.

The roll consists of an induction coil incorporated inside the roll, which generates alternating flux. The rotating roller located concentrically to the induction coil is heated by the flux. The jacket within the roller contains vapor and liquid of a heat transfer fluid. Continuous condensation and evaporation, taking place within the jacket, helps to maintain a uniform temperature across the face of the roll. The temperature is maintained within ±1%. Figure 7-13 shows such an induction heated jacketed roll.

DRIVES

The simplest drive arrangement is a single drive motor with power transferred to the individual rolls by gears, giving a fixed speed differential between the rolls.

Multimotor drives are often selected for modern machines. Continuous adjustment of speeds is then possible. Inverted "L" calenders might often have two drives: one for rolls 1, 2 and 3 and another for roll 4. Top and center rolls (2 and 3) absorb more of the total power, since they are servicing two nips. Offset and bottom rolls take less power. A typical 32" x 96" four-roll PVC calender, operating at 100 m/min with flexible and semi-rigid vinyl compounds, requires 550–750 HP.[17] The drive horsepower depends on roll diameter, roll face width, roll surface speed and rheological properties of the coating at the operating temperature. A typical 24" x 66" three-roll pressure-sensitive tape calender would have the top roll driven separately by a 50-HP motor and the center and bottom rolls would be driven by a 150-HP motor.

Fig. 7-13. Induction heated jacketed roll. (*Courtesy Tokuden Co., Ltd.*)

The normal rating for this size calender, which forms the mass in the top nip and laminates in the bottom nip, is 1.3-2 HP/(m) (min).

ACCESSORIES

Stripper rolls are used to remove the web from the last calender roll. Coated webs are much easier to remove than hot vinyl film. The original stripper roll was a heated large-diameter roll next to the calender. A roll rotating in the opposite direction to the web travel was used. This roll would be followed by another, in order to maintain proper tension.

Currently most stripper roll assemblies consist of three 15-25-cm-diameter hard chrome plated, or Teflon covered, rolls located close to the last calender roll. The closer these rolls are to the calender, the lower is the draw.

Vinyl embossing is done in line with calendering in a nip between an engraved steel roll and a rubber roll.

Cooling cans are driven to minimize the tension on the material. The cans closest to the calender are at a higher temperature. The temperature is controlled by running water in the reverse direction, or by individual temperature controls. The cooling cans are of single or double shell construction.

Dams are used to constrain the mass on the calender. They are equipped with Teflon scrapers and may be heated or cooled to prevent sticking, or to decrease the polymer degradation.

A hydraulic splice relief cylinder is added when the bottom roll is used for lamination. It allows opening the gap and letting the splice pass without breaking the web.

REFERENCES

1. Rosato, D. V. *Plastics World* **37** (7) 64 (1979).
2. Middleman, S. *Fundamentals of Polymer Processing.* New York: McGraw-Hill, 1977.
3. Tadmor, Z. and Gogos, C. G. *Principles of Polymer Processing.* New York: John Wiley & Sons, 1979.
4. Gaskell, R. E. *J. Appl. Mech.* **17**: 334-336 (1950).
5. Kiparissides, C. and Vlachopoulos. *J. Polym. Eng. Sci.* **16**: 712 (1976).
6. Perlberg, S. E. and Burnett, P. P. A. *Encyclopedia of PVC,* Vol. 3. L. I. Nass (Ed.). New York and Basel: Marcel Dekker, 1977, pp. 1361-1414.
7. *Calender Upgrading and Repairs.* Farrel Machinery Group, Ansonia, Conn.
8. Eighmy, G. W., Jr. *Proceedings Adhesive and Coating Technology, PSTC Technical Seminar,* June 18-19, 1980, Rosemont, Ill., pp. 144-171.
9. Satas, D. *Handbook of Pressure-Sensitive Adhesive Technology.* D. Satas (Ed.) New York: Van Nostrand Reinhold, 1982, pp. 527-532.
10. Harris, G. *Handbook of Pressure-Sensitive Adhesive Technology.* D. Satas (Ed.) New York: Van Nostrand Reinhold, 1982, pp. 450-462.

11. Morris, J. F. U. S. Patent 2,879,547 (March 31, 1959) (assigned to The Kendall Co.).
12. Tait, G. E. *Metallurgical Factors Affecting Calender Roll Performance*. Publication of the Dominion Engineering Co.
13. Kuehn, H. E. *Paper Trade Journal* **144**(29): 36 (July 18, 1960).
14. *Calendering and Supercalendering*. Lockwood Trade Journal, New York, 1963.
15. Stone, M. D. and Liebert, A. T. *TAPPI* **44**(5): 308 (May 1961).
16. *Plastic Calender Lines, Bulletin No. 233*. Farrel Machinery Group, Emhart, Ansonia, Conn.
17. Meienberg, J. T. *Modern Plastics Encyclopedia, 1981-1982,* Vol. 58, No. 10A:243-244 (October 1981).

8
Extrusion

Kenneth A. Mainstone
The Black Clawson Company
Fulton, New York

Extrusion coating describes the continuous combining of a melted thermoplastic film with a moving web substrate in the nip formed between a heat removing chill roll and a pressure loaded, resiliently covered nip roll. In wrapping the chill roll, the heat is transferred from the melted thermoplastic allowing the web to be stripped off the chill roll together with the now solid plastic film combined as a surface coating. Extrusion laminating is a similar process in which two separate webs are combined together with the thermoplastic melt in-between to form a sandwich lamination of the three layers after cooling on the chill roll (Fig. 8-1). In both cases, the thermoplastic melt is delivered from an extruder to a longitudinal slot orifice die across the width of the web at a rate appropriate to the required coating thickness and

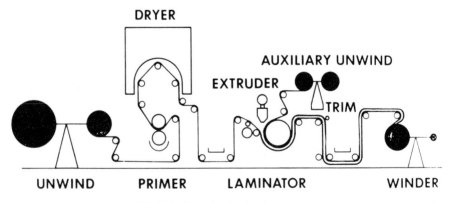

Fig. 8-1. Extrusion laminating process.

from which it is drawn down by the substrate into the nip with the chill roll.

Extrusion coatings are functional and provide such properties as moisture vapor barriers, liquid barriers, gas transmission barriers, grease barriers, heat sealing surfaces, surface friction modification, variable light reflection surfaces, transparency, opacity, scuff resistance—all dependent upon the types of coating materials used and their processing conditions. The vast majority of extrusion coatings and laminations are used in food packaging applications. Almost anything one picks up in the supermarket in a flexible package, from dry soup to nuts, is contained in an extrusion coated or laminated construction. The constructions are engineered to contain the foodstuff, retain the flavor, protect liquids from getting out or moisture from getting in, provide a convenient sealing system for closing the formed package without leakage, provide or protect a printed surface and frequently provide the sales appeal for the contained product.

The commercial application of the extrusion coating process started in the late 1940's when polyethylene started to become available to the private sector. Previously, throughout the 1940's, it had been assigned by the U.S. Government almost exclusively for the military applications of insulation for electrical, radar and communications wire and cable. Domestic manufacture of the polymer started around 1943 after its invention and initial development in England in the 1930's, just in time for World War II.

Polyethylene was soon recognized to possess many useful properties applicable to the needs of the paper industry for a variety of flexible packaging functions. A barrier to moisture, oil and grease and inert to most chemicals, it had toughness and flexibility, was not toxic and had essentially no taste or odor. Its flexibility extended to temperatures well below zero and its softening temperature was conveniently higher than any hot weather, about 80°C. For closing a package, the heat sealability of polyethylene in the 150°C temperature range was a very useful property as was the fact that polyethylene tended not to stick to the contents of a package. What has been called the marriage between the paper industry and the plastics industry was soon consummated in the production of extrusion coated and laminated flexible packaging constructions.

If there were a loser in those early days in the development of polyethylene coating, it would have been paraffin wax, particularly during the late 1960's and throughout the 1970's when the huge tonnage of polyethylene coated paperboard came on-stream to replace wax coated board for milk cartons.

MATERIALS

Usually the substrate web provides the strength or rigidity for the extrusion coated or laminated construction. For example, the paperboard provides the structural integrity of the milk carton and the polyethylene coating on each side provides the functions of liquid retention and heat sealing the container. Substrates used in extrusion coating include all grades of paper and paperboard, cellophane, plastic films such as biaxially oriented polyester and polypropylene, nylon film, woven and nonwoven textile webs and metal foils. Any of the same substrates can also be used as the secondary web in a sandwich laminated construction although by far the most common is aluminum foil in a thickness of about 7.5 μm. Thermoplastic polymers commonly employed in extrusion coating are polyethylene and various copolymers, polypropylene, polyester and polyamide and the choice of which to use depends upon the physical properties required of the coating.

ADHESION

One of the most critical factors in an extrusion coating or lamination is the adhesion of the polymer to the substrate. Different end uses will have different demands for the degree of adhesion. For example, some relatively low adhesion requirements will allow the coating process to run at speeds in excess of 600 m/min, while some relatively high adhesion requirements will limit processing speeds to much less than 300 m/min unless some adhesion promoting system is employed.

There are a variety of adhesion promoting systems and their selection usually depends upon the coated or laminated construction required and the substrates and polymers to be employed. Oxidation of the polymer melt in the drawdown distance from the die to the substrate is one such system and has some range of adjustability in that melt temperature can be increased to increase the level of oxidation and the drawdown distance can be increased for longer exposure to air. This method, however, soon runs into limitations since, although adhesion to the substrate is improved, the excessive oxidation of the melt leads to the problem of odor in the polyethylene, which can be unacceptable in food packaging, and the problem of loss of the heat sealing characteristics of the polymer inhibiting the forming and sealing of the package.

A more recent development of this oxidation method, now finding commercial acceptance, involves the use of an ozone generator from which a high concentration of ozone is delivered through a nozzle to impinge upon only the side of the melt to be joined to the substrate. This system permits lower melt temperatures to be used, increases the adhesion of the polymer to the substrate and reduces the oxidation on the opposite surface of the polymer so that heat sealing characteristics are also improved.

Other systems of adhesion promotion involve surface treatment of the substrate web. For many paperboard coating applications the board surface is exposed to the direct impingement of a gas flame burner, which improves the adhesion with the polymer. For many paper coating operations, the paper surface is exposed to the bombardment of a high-frequency electrostatic corona having a similar effect to the flame burner. For many film coating applications, the surface of the film substrate is precoated with a solvent solution of a chemical primer. The solvent is evaporated and exhausted in a drying oven and leaves the adhesion promoting primer on the surface to be combined with the polymer at the extrusion coating station. There are many proprietary formulations for such primer coatings, the most commonly used being based on polyurethane or polyester type adhesives in a hydrocarbon solvent solution. The solvent also helps overcome the wettability problem of plastic films. Water or water-alcohol based primer coatings are similarly employed for suitable substrates, such as paper and paperboard, the most commonly employed formulation being a solution of polyethylene-imine at 1 or 2% solids. Primer coating formulations are engineered by the suppliers to suit the specific applications of substrates and polymers to be combined and the ambient conditions and life expectancy of the end product.

PRIMER COATERS

The most commonly employed machine for the application of primer coatings onto a substrate is the direct gravure coater. The quantity of primer deposited on the web is controlled by the depth of the pattern of the engraving on the gravure roll and by the percentage of solids contained by the liquid vehicle of the solution. The liquid vehicle is evaporated in a drying oven immediately following deposition on the web. Typical vehicles are water, alcohol and hydrocarbon solvents.

Another type of primer coater found in extrusion coating machine lines is the three-roll squeeze coater for aqueous polyethylene-imine (Fig. 8-2) coating. This machine does not employ an engraved roll or a doctor blade. The three rolls arranged vertically are steel, rubber covered and steel, and all are smooth surfaced. The bottom roll rotates at about 10% of line speed, picking up the coating from a pan, transferring it to the rubber roll, which is nipped with the top roll with the substrate web in between, so that as the rolls rotate the coating is deposited upon the surface of the web. The intimacy of contact between the rolls is adjusted by hand-wheel operated wedge spacers at each side. The purpose of the design is to deposit a minute quantity of the primer on the web, no more than is needed, about 0.1 g/m^2 dry coat weight. In cases where the basis weight of the web is high enough, as in paperboard, the addition of the moisture to the web by priming is insignificant and the evaporation of the liquid vehicle is unnecessary. This saves the investment, space and operating cost of a drying system.

Fig. 8-2. Three-roll primer coater.

EXTRUDERS

The function of the extruder is to receive the solid pellets of the polymer, usually at ambient temperature and convey, melt and mix the material to produce a homogeneous melt of uniform temperature and uniform pressure at the discharge end, where the melt is fed through a right-angled adapter down into the die. Within the die the polymer spreads out to the full width of the machine to emerge from the exit slot to be drawn down into the coating nip with the substrate web. The die slot is adjustable to regulate the cross-machine thickness uniformity of the polymer and to provide an effective slot width to be compatible with the width of substrate web to be coated. When off-coating, the extruder is retracted from the machine line on a motorized carriage.

An extruder is driven by a variable speed drive, usually a dc electric drive, providing output adjustment by speed adjustment (Fig. 8-3). The speed of the extruder drive can be referenced to the speed of the web traveling through the machine as a method of controlling total coat weight. A ratio potentiometer provides the machine operator with an adjustable range of extruder output corresponding to the web speed. A trimming adjustment is also necessary since extruder output is not exactly linear with speed change. The

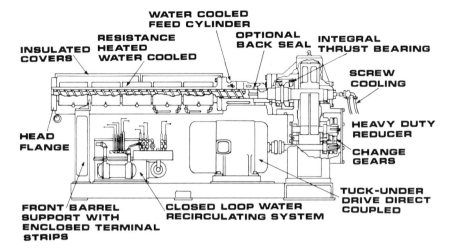

Fig. 8-3. Schematic diagram of an extruder.

drive motor is coupled to the high-speed input shaft of a gear reducer having either a two- or three-stage reduction with ground helical gearing being preferred as a practical combination of efficiency and quiet running. The gear reducer provides the required ratio between the typical motor speed of 1750 rpm and the typical extruder screw speed of about 240 rpm. The sizing of the drive train speed, torque and horsepower is calculated for specific applications. Some gear reducers feature the facility to exchange a gear set for a different ratio so that the machine can be regeared, for example, to produce a higher torque at a lower speed while still using the connected horsepower. While the electric drive can be adjusted from full speed to zero for stopping the machine, it usually accommodates an operating speed range of about five to one relative to top speed in process conditions. The drive is usually interlocked for protection against alarm conditions such as overload, failure of lubrication circulation or failure of the heat exchange system for cooling the lubricant in the gear reducer.

The output shaft assembly of the gear reducer includes a thrust bearing which restrains the rearward thrust resultant due to the work performed by the extruder feed screw. The thrust bearing is usually housed integrally with the gear reducer although some of the bigger extruders are equipped with a separate independent thrust housing to facilitate maintenance and replacement upon failure. The termination of the output shaft of the gear reducer is a hollow bore into which the drive shank of the feed screw is socketed and keyed.

The polymer is introduced into the extruder by way of a feed hopper, which is typically sized to contain about a half-hour's supply. The polymer form is usually a cylindrical pellet measuring about 3 mm diameter by about 3 mm long. The pellets have a bulk density of about 560 kg/m^3 and, being free flowing, enter into the circular feed opening of the extruder by gravity. To prevent extruder heat backing up from the downstream barrel to the feed hopper, the feed cylinder is jacketed for cooling water circulation. If the heat were allowed to back up, it could cause the pellets in the hopper to soften and stick together to form a dome-shaped bridge which would then prevent further flow.

Rotation of the feed screw conveys the pellets forward into the barrel of the extruder. The size of an extruder is described by the nominal dimension of the bore of the barrel and the ratio of its length to diameter. Currently popular L/D ratios are about 30:1, giving a practical combination of dwell time and melt homogeneity. The size of the extruder is selected by the output rate required. Different polymers at different temperatures will greatly vary the output rate capacity of any extruder, but some typical examples are given in Fig. 8-4.

The cylindrical barrel through which the polymer is conveyed is heated throughout its length in about six separate zones of temperature control, increasing in temperature from the rear to the exit end. The most commonly employed heating method is electric resistance heating. Most machines are also equipped with water cooling on most zones. The temperature controller for each zone will then call for additional heating or cooling, depending upon the relationship of the actual temperature to the desired set-point temperature. The early galvanometer type temperature control instruments gave way to digital type instruments, which now are giving way to microprocessor based control instrumentation. Modern machines are equipped with microcomputer systems in distributed control environments equipped with cathode ray tube display of process data, hard copy print-out of information and production reports and memory storage systems providing recall of process conditions for a variety of product requirements (Fig. 8-5).

While most extruders are relatively similar in construction, the heart of the process is in the technology of feed screw design to optimize the performance of the process relative to the rheological behavior of the wide variety of polymers available. The feed screw for each machine is specifically designed for the intended application and the specific polymers intended to be used.

EXTRUDER SIZE	2 1/2 IN. WITH 75 HP	3 1/2 IN. WITH 125 HP	4 1/2 IN. WITH 200 HP	6 IN. WITH 400 HP	8 IN. WITH 600 HP
OUTPUT	325 LB/HR	625 LB/HR	1200 LB/HR	2200 LB/HR	3800 LB/HR

Fig. 8-4. Extruder output for various extruder sizes and drives.

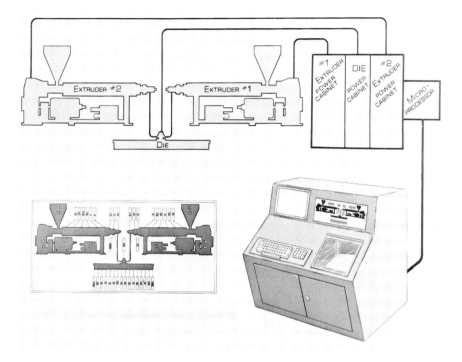

Fig. 8-5. Microcomputer control system.

In normal operation, the heat required to continuously melt the polymer is supplied by the extruder drive via direct conversion of the mechanical energy into thermal energy. This process occurs by virtue of the frictional forces between the feed screw, the polymer and the barrel wall. The magnitude and duration of these frictional forces differentiate one extruder design from another. While the outward appearance of most extruders may be similar (Fig. 8-6), the internal features often contain radical departures with respect to screw design and connected speed and torque. These differences are most important when a designer intends to optimize the extruder performance for a particular polymer or range of polymers.

DIES AND ADAPTERS

An adapter is connected at the exit end of the extruder. It makes the right angle transition of the melt flow vertically downward into the die. The adapter normally includes a filter screen which is removable for changing out of the side. The adapter is equipped with an externally adjustable valve which, by providing resistance to flow, creates additional back pressure in

Fig. 8-6. Extruder and die.

the machine. All the parts of the polymer conveying system are heated and controlled in zones. The downspout of the adapter may contain a static mixing device which repeatedly divides and twists the melt stream and by mixing breaks up the tendency for temperature stratification from the outer edge to the center of the melt flow. Thus a thermally homogenized melt is presented to the entry of the die, which is desirable for maintaining thickness uniformity of the applied coating. Instrumentation in the adapter includes at least one transducer providing melt pressure indication and a thermocouple providing melt temperature indication. Melt pressure upstream of the valve and filter screen usually is between 10 to 25 MPa. Melt temperature for extrusion coating would usually be between 230° and 330°C, depending upon the type of polymer being extruded and the type of product required.

The polymer melt enters the die at the center and spreads out to the full width of the internal manifold. The manifold is sized to accommodate this flow without inducing excessive pressure drop from the center to the ends. Some dies have a flow manifold which drapes downward like a coathanger to provide less resistance to flow at the ends than at the middle and so to accommodate the pressure drop from the center to the ends. These are known as coathanger dies (Fig. 8-7). Dies having a straight, cylindrical bore, flow manifold are known as keyhole dies since in a cross-sectional end view the flow passage suggests a keyhole shape. Dies installed in the extrusion coating industry range in width from about 60 cm for laboratory machines up to about 400 cm for production machines. They are mostly resistance heated in temperature control zones of about 20–30 cm increments. The flow passage

Fig. 8-7. Interior of a coathanger die with external deckles and internal bead control rods.

below the manifold diminishes to a preland slot of about 2.5 mm and then to a final land slot of about 0.75 mm. One of the two lips forming the exit slot of the die is adjustable across its width in 2.5- or 5-cm increments to regulate the cross-machine thickness uniformity of the emerging melt. The melt emerges at a thickness of about 0.75 mm and is then drawn down by the high speed of the substrate web entering the laminator nip, so that the eventual thickness is reduced to the coating desired. This could range from about 0.005 mm thick as for a single-serving sugar pouch up to about 0.1 mm thick as for a poly-coated table placemat. The rheological characteristics of the polymer need to be compatible with the rate of drawdown required.

As the melt is drawn down from the die, it also reduces slightly in width. This is known as neck-in. Neck-in at each end of the die is usually in a dimension of about 2.5–7.5 cm. When neck-in occurs, the edges of the melt increase in thickness to produce an edge bead three or four times the thickness of the film. Internal deckles are often employed to produce a starve-feeding effect at both ends of the die to reduce the thickness of the edge beads. External deckles are also provided for adjustment of the effective length of the exit slot of the die to be compatible with the width of the substrate web to be coated. The extrusion coating is laid down wider than the substrate web, so that the edge beads are beyond the edges of the substrate and they are subsequently trimmed off with slitter knives. This is called over-coating. To prevent the overcoating sticking to the rubber covered pressure roll of the laminator nip, a strip of Teflon tape is circumferentially wrapped around and stuck to the roll at each side, spaced to suit the web width. A more sophisticated execution of this technique employs a pair of endless Teflon tapes running on guiding and tensioning pulleys with either handwheel-operated, or motorized lead-screw adjustments for changing their spacing to suit changes in web width. The tapes then run continuously, wrapping the rubber covered nip roll and partially wrapping the chill roll since the cooling increases the effective life of the tapes. For at least ten years, manufacturers have tried to eliminate the necessity for Teflon tapes by covering the rubber covered nip roll in a fluorocarbon sleeve. This device was singularly unsuccessful, since the adhesive bond between the rubber and the sleeve would not last. Around 1980 a new supplier of such fluorocarbon sleeves introduced his product and showed life in continuous operation of many months.

Almost all dies are manually adjusted for cross-machine coating thickness uniformity in response to the operator's interpretation of the data provided by a scanning nuclear or infrared thickness measuring system. The operator turns the adjusting bolts with a wrench either to open or close the die gap. In 1962 an American machinery builder introduced a remotely controlled die where the operator, interpreting the coating thickness profile data displayed by a measuring gauge, used a panel of switches to command a remote mechanical actuator to adjust the required die bolt. This idea did not find commercial popularity in extrusion coating until the beginning of the 1980's by which time the development of microcomputer technology was able to displace the man's interpretation of data and his reaction to it, by closing the loop between the measuring gauge and the die adjustment with an algorithm. A concurrent development was the use of thermally expandable studs as the mechanism for adjusting the die gap as an alternative to adjusting screws.

COEXTRUDERS

Among the most interesting developments in the extrusion coating and laminating process during the decade of the 1970's, and fast growing in the 1980's, is coextrusion. Coextrusion is the process of simultaneously extruding two or more different polymer layers from two or more extruders through one die system to produce a multi-layered coating.

There are many valuable advantages to be obtained from coextrusion in terms of both functionality and economy. The process allows a very thin layer of an expensive functional copolymer to be combined with a relatively inexpensive polymer for the bulk of the coating so that, for example, a surface heat sealing advantage can be gained with a thickness adequate for the function but too thin to be drawn down from a die by itself unsupported. The coextrusion system combines the two layers together within the die so that the thick one supports the thin one in the drawdown to the chill roll.

A similar example could be the coextrusion of 0.04 mm low-density polyethylene with 0.004 mm of ethylene-acrylic acid copolymer (EAA) to gain improved adhesion to aluminum foil without excessive temperatures.

The less expensive bulk of the LDPE provides the support for the thin layer of EAA in the drawdown. Yet another example could be the coextrusion of a layer of ionomer between layers of LDPE and nylon. The ionomer then provides the adhesion between the polyethylene and the nylon, which otherwise would not adhere to each other. The coextrusion of such a construction could also be an alternative to the process of chemically priming a polyethylene film and extrusion coating it with nylon.

The coextrusion process offers an attractive alternative to the use of solvent primers. It does not require explosion proof apparatus, eliminates energy cost for evaporation of solvents, decreases floor space for machinery, web handling length within the machine and manpower for crewing and inventory of substrates.

There are three fundamental concepts of design of coextrusion systems for extrusion coating. They are the *combining adapter,* in which a small rectangular sandwich of the polymer layers is formed in a feed-block immediately before entering the die; the *dual slot die,* in which the full width polymer layers join together outside the die in the drawdown to the extrusion coater nip; and the *multi-manifold die,* in which the full width polymer layers combine together within the die prior to the exit slot.

COMBINING ADAPTER

By far the most commonly used coextrusion system for extrusion coating is the combining adapter, also known as the feed-block. The convenience of the combining adapter is that it is used in conjunction with a conventional, single manifold, single slot die having the same single set of die adjusting bolts so familiar to the operator. The only difference in the die itself is that the entry port in the top is machined in a rectangular shape instead of the usual round hole. The rectangular shape matches the dimensions of the rectangular sandwich formed in the combining adapter of the polymer layers fed from each extruder. Figure 8-8 illustrates the concept where the top half of the diagram represents the combining adapter and the lower half represents the die. Figure 8-9 shows the two assembled.

The single set of die adjusting bolts allows the operator to achieve overall gauge uniformity of the coextrusion exactly as he would for mono-extrusion. Interlayer uniformity of the discrete layers of the coextrusion is obtained inherently with the laminar flow behavior of polymers in such die systems, providing that the combinations of resins selected are not too dissimilar in flow behavior to produce such uniformity. There are several factors which can affect the interlayer gauge uniformity in the cross-machine direction in a multi-layered coating produced by the combining adapter system, even though the overall gauge uniformity of the coextrusion may be well within

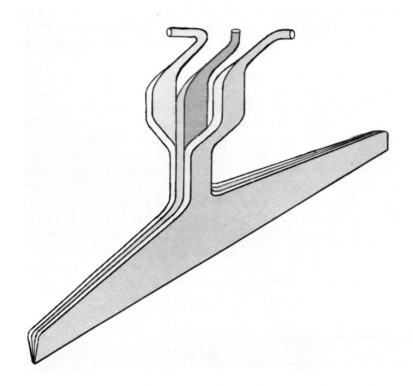

Fig. 8-8. Schematic diagram of combining adapter system of coextrusion.

acceptable tolerance. For the most part, these several factors all equate to the viscosity differential between adjacent layers, since the lower viscosity material will always tend to displace the higher viscosity material toward the two ends of the die.

The relative viscosities do not have to be similar. In many cases, viscosity ratios in a range of up to 3:1 or more will produce commercially acceptable interlayer gauge uniformity. The main thing to keep in mind, particularly in the initial selection of a coextrusion machinery system, is the interaction of all the variables and their accommodation of the process and product variables intended for the production operation.

One should first consider the polymer combinations intended for coextrusion and compare their relative viscosities at processing conditions of melt temperature and shear rate. As the size and shape of the flow channel diminishes to increase resistance to flow in the passage of the melt stream from the combining adapter and through the die, so will the shear rate increase. As the shear rate increases, the apparent viscosity of the melt will

Fig. 8-9. Three-layer coextrusion combining adapter with single manifold die.

reduce. But different polymers change viscosities at different rates when exposed to similar changes in shear rate. Prediction of viscosity differentials under processing conditions is a first step in the selection of a coextrusion system compatible with the intended product. Some latitude of adjustment is usually available. Some work is also being done on the original deformation of the layered sandwich in the combining adapter to compensate for subsequent flow variation in the die.

The use of melt temperature differentials is limited in a combining adapter system, largely due to the long dwell time between combining of the melts in the adapter and exiting from the die slot in full width. The long dwell time allows heat exchange between layers and the consequent change in temperature can cause a change in viscosity and a change in interlayer gauge uniformity. Die design influences this dwell time in contact. A die having a relatively small inventory manifold design and a coathanger drape, to compensate for pressure drop from the center to the ends, will minimize dwell time in contact by increasing the velocity of the melt stream through the system. A keyhole type die with a straight cylindrical manifold usually has a large bore containing excessively large inventory of melt to try to reduce the effect of pressure drop over the length of the die. This type of die increases the dwell time in contact and reduces the latitude available for temperature differentials as a process control function. The overall width of the die and the polymer throughput rate are other factors affecting dwell time in contact. These are not necessarily limitations, since combining adapter coextrusion systems are in commercial operation in widths up to 350 cm in the somewhat similar process of chill roll film casting.

Layer proportions in many combining adapter systems offer a broad range of adjustment, many within a range of 5–95% and some reporting skin layers of less than 5%. Very thin skin layers are also reported to extend significantly the viscosity ratio range between adjacent melts since the thin skin virtually has no place else to go but to stay in line. There is just not enough of it to be squeezed out and displaced by the higher viscosity bulk of the construction.

The internal design of many combining adapters includes mechanical adjustments such as choker bars or flow restrictor vanes to adjust the size of the aperture for each melt stream to control melt velocity relative to the layer proportions at the point of combining (Fig. 8-10).

Despite the apparent limitations described above, the combining adapter system continues to be by far the most popular system for production operations of coextrusion in extrusion coating.

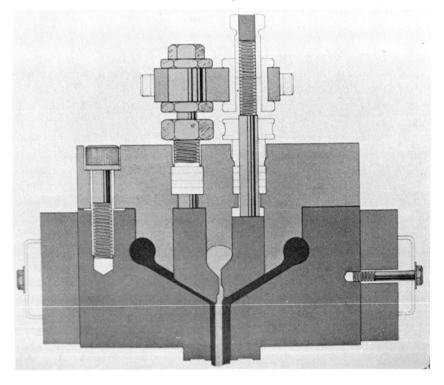

Fig. 8-10. Schematic diagram of the interior of three-layer coextrusion combining adapter.

Probably the most popular feature of the combining adapter system is its simplicity of operation compared with the other two systems. Since the interlayer gauge uniformity is achieved inherently and predictably and is not mechanically adjustable anyway, the operator's responsibility is confined to maintaining overall gauge uniformity with one set of die bolts just as he would in a mono-extrusion operation. Of course, he now has two or three or four extruders to monitor, but the current trend to microcomputer control of extrusion variables simplifies the task. Another popular feature of the combining adapter system is the flexibility available for various layer constructions. For example, by changing the feed pipe connections between a three-layer combining adapter and two extruders, one can produce A-B, B-A, A-B-A, and B-A-B layer constructions. With three extruders the range is progressively increased. An alternative to changing the feed pipe connections is available in combining adapters containing an exchangeable manifolding cartridge with which layer configurations can be changed in a few minutes. Yet another popular feature of the combining adapter system is the relative ease of disassembly for cleaning compared with the other two systems. Some color concentrates at elevated temperatures have an affinity for the walls of the flow cavity which necessitates occasional cleaning. A single manifold die is much simpler to disassemble and clean than the multi-manifold die employed by the other systems.

DUAL SLOT DIE

The dual slot die coextrusion system employs two independent entry ports, two independent full width flow manifolds and two independent exit slots from which two separate melt films emerge to combine together in the drawdown between the die and the nip of the extrusion laminator (Fig. 8-11).

Since there is essentially no dwell time in contact between the two melt layers, very different melt temperatures can be maintained within the practical limits of thermal insulation of the two halves of the die. Each die half is independently zoned for temperature control. The most useful commercial application for the dual slot die system is in the coating of paperboard for liquid packaging, such as the milk carton. The application requires a relatively high temperature, in excess of 300°C, to gain adhesion between the polyethylene and the paperboard. But such a temperature severely diminishes the heat-sealing characteristics of the polyethylene needed for forming and sealing the carton. So the dual slot die is used to apply a 325°C layer against the surface of the paperboard and a 305°C layer on the outer surface to gain both, the required adhesion and heat-sealability.

Since the two melt films are spread to their full widths independently, very different melt viscosities can be accommodated within the practical limits of

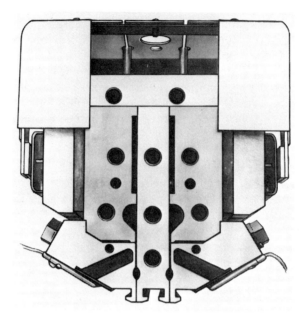

Fig. 8-11. Dual slot coextrusion die.

the drawdown characteristics of the melt. Differences in neck-in dimensions are adjustable with independent external deckles on each die slot and internal deckles provide adjustment of the thickness of the edge beads.

There are also some disadvantages to the dual slot die. Since the two melts emerge full width from two separate die slots, it is necessary to provide independent mechanical adjustment of gauge uniformity for each layer. So the operator is faced with two sets of die bolts, one on each side of the die, the decision of which side to adjust when a gauge deviation is indicated and the problem of access to the die bolts on the chill roll side.

A limitation of the dual slot die is that two slots produce two layers which are drawn down separately and so minimum layer thickness is subject to the drawdown characteristics of the unsupported single layers. The addition of a combining adapter to one of the entry ports can overcome this limitation by producing a supported thin skin and also provide additional latitude of layer combinations.

MULTI-MANIFOLD DIE

The multi-manifold die, as its name implies, contains two, three or four independent entry ports and full width flow manifolds which are arranged to combine the melt streams together at their full width prior to emerging as a coextrusion from a single exit slot (Fig. 8-12). The exit slot is equipped with one set of die bolts for adjustment of gauge uniformity of the overall coextruded thickness. Since the melt streams flow to full width independently before combining, it may be necessary to mechanically adjust the gauge uniformity of each layer independently. Consequently, such dies may be equipped with a full width adjustable chokerbar for each manifold. This now becomes a very complicated die in terms of the operator's adjustment and presents a major project in disassembly and reassembly for cleaning.

However, this die offers much more latitude than a combining adapter for greatly differing melt viscosities and, depending upon the number of channels and the facility for independent temperature control zoning each side, a little more latitude for melt temperature differentials since the dwell time in contact is considerably reduced in comparison with the combining adapter.

Coextrusion will remain the most technically growing aspect of extrusion coating for the next several years, as the market catches up with the presently available technology and technology advances to create new constructions, new copolymers and new techniques. The combining adapter will continue to be the most favored system of coextrusion due to its simplicity of operation, versatility of layer constructions, flexibility for changing layer constructions and ease of disassembly for cleaning. Polymer combinations will more likely be tailored for compatibility with this system rather than select-

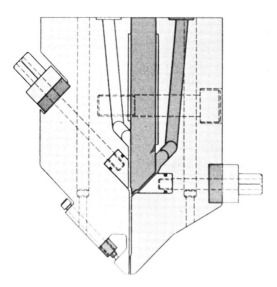

Fig. 8-12. Multi-manifold internal combining coextrusion die.

ing another system. The dual slot die will continue to find the application of board coating for liquid packaging, but is not likely to expand its range of application beyond such differential temperature usage. The multimanifold, internal combining die, will never be a popular system and will only be used as a last resort for applications involving viscosity differentials beyond the capacity of the combining adapter.

WEB HANDLING

The term web handling describes the complete machine line web conveyance system from new rolls entering the unwind through to finished rolls leaving the winder. The web conveyance system encompasses all the functions of splicing-in new rolls at the unwind, tension control, edge guiding, speed regulation, coating, drying, laminating, trim slitting and roll changing finished rolls at the winder.

Full speed, continuous unwinding-splicing mechanisms are commonly employed since most extrusion coating and laminating machines operate at speeds between 100 and 400 m/min and some as high as 600 m/min. Unwind tension control systems include pneumatic brakes for heavier gauge substrates and regenerative dc braking for light gauge substrates, where the required tension horsepower may be less than the friction-loss horsepower of the machine. Throughout the machine line, systems designed to handle heavier substrates would be likely to employ speed-regulated drive systems with adjustable torque at each section, while lines designed for light gauge applications would be likely to employ a tension regulated system, where each driven section is controlled by a tension sensing device, such as a dancer roll or a tension force transducer.

Tension isolating pull roll sections are used where the need for controlling tension differentials occurs in the machine line. For example, an infeed pull roll section after the unwind can generate the higher process tension, allowing the benefits of reduced torque at the unwind spindles; a pull roll section after the laminator can provide the additional tension in a short span to aid strip-off from the chill roll and a pull roll section preceding the winder will allow taper-tension winding without reflecting the tension reduction back upstream.

Automatic edge guiding systems, such as unwind sidelay bases or steering rolls, are used to maintain the uniform tracking of the web relative to the melt drawdown from the die.

Direct gravure, offset gravure or three-roll squeeze coaters are used for the application of chemical or adhesive primer coatings and the solvent or aqueous vehicles are evaporated in a hot-air dryer. Dryers come in several configurations such as a drum support, or an arch of idler rolls and the currently popular floatation type, where the web is supported by the air flow at the multiple nozzles without contacting any roll. In-line printing is performed on some extrusion coating and laminating machines with some more sophisticated lines offering up to three color registered rotogravure printing in-line.

The extrusion coater-laminator section is usually the lead section of the line establishing process speed with all other sections regulated relative to it. The coater-laminator section contains the chill roll, the diameter of which is sized according to the quantity of heat to be removed from the polymer in terms of kcal/hour/100 mm of width of the roll face. Polyethylene at an application temperature of 330°C contains about 245 kcal/kg and at a strip-off temperature from the chill roll of 52°C contains about 28 kcal/kg. The chill roll, therefore, has to remove 217 kcal/kg. An average size extruder puts out about 1000 kg/hr, giving a heat transfer load of 217,000 kcal/hr. An average width machine is 1500 mm giving a heat transfer rate requirement of 14,500 kcal/(hr) (100 mm of face). In this example, a typical design of chill roll would need to be about 75 cm in diameter with a water throughput rate of about 2500 l/min at 18°C to remove this heat while keeping the water-in/water-out temperature differential within about 1.5°C. Different manufacturers have proprietary designs for their chill roll constructions and materials for which they claim varying measures of capacity and efficiency. Most machines installed have chill rolls ranging from 50 cm in diameter with a water recirculation rate of about 1200 l/min up to 90 cm in diameter and 3200 l/min accommodating a range of extrusion rates of approximately 15–80 kg/(hr) (100 mm of face).

The typical construction of a chill roll includes an inner cylindrical shell having a relatively thick wall designed to reduce deflection across its length and a cylindrical outer shell having a relatively thin wall to enhance heat transfer (Fig. 8-13). The annular space between the two concentric shells is quite small, less than 2.5 cm and contains spirally wound flutes from one end to the other, creating a spiral passage through which the cooling water flows. Rotary unions provide entry and exit of the water at the roll journals. Water velocity through the roll would be about 4–6 m/sec, providing sufficient turbulence to promote efficient heat transfer without generating excessive pressure drop. Corrosion of the internal flow passage is a common problem with chill rolls. Some rolls are internally plated for corrosion resistance, some are designed for disassembly for cleaning, but in any case the water should be treated to minimize mineral content. The water is pumped at a constant rate through a loop including the chill roll. As the water temperature tends to rise above the set point, a modulating valve operated by a temperature sensor allows warmer water to exit the loop and cooler make-up water to enter the loop maintaining the desired temperature set point. The warmer water is

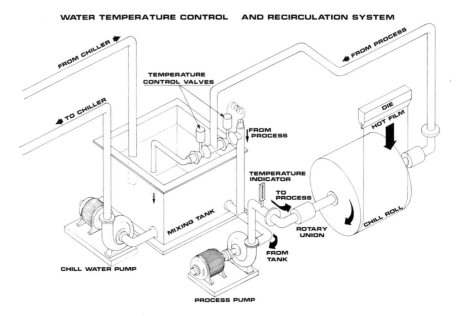

Fig. 8-13. Water temperature control and recirculation system.

then recirculated through a refrigeration system and returned to the cool, make-up water reservoir. The typical set point for the process water temperature is between 16° and 21°C. Colder temperatures are avoided, since they would promote condensation on the outer face of the roll, which could cause the edges of the web substrate to swell. Some machines in humid climates are equipped to blow dehumidified air onto the periphery of the chill roll to combat condensation. The outer surface of the roll face is chrome plated for corrosion resistance and surface textured to impart either a glossy or a matte finish to the plastic coating.

The rubber covered roll, which is air cylinder loaded to form the extrusion coating nip with the chill roll, is sized in diameter to avoid excessive deflection since common practice does not include crowning. A larger roll, however, tends to have a broader nip imprint, due to flattening of the rubber in the nip, and this can cause stretching of an extensible web and breaking of an inelastic web, such as an aluminum foil. Consequently, machines intended for such laminations, as films to foils, or lightweight papers to foils, are equipped with a small-diameter rubber covered nip roll of about 10–15 cm. Since such a small roll would easily deflect upon the imposition of loading to its journals to produce the typical design force of 25 kg/ linear cm, a larger back-up roll is provided instead, which is itself pressure loaded and in turn nips the rubber roll against the chill roll. This is called a flexible packaging nip configuration, and machines are built this way for laminations described and also with the larger rubber roll for paper or paperboard coating applications. A larger roll with a single nip will have a longer life than a smaller roll in a double nipped configuration.

Process variables require that both the vertical and horizontal dimensions from the extrusion coating nip to the exit slot of the die be adjustable. Varying the vertical distance changes the dwell time in which the melt is exposed to air and so varies the degree of surface oxidation of the melt, which, as previously described, affects adhesion, heat-sealability and odor. Varying the horizontal distance adjusts the dwell time in which the melt is in physical contact with the substrate prior to meeting the chill roll and will have an influence on the degree of mechanical adhesion between the two. If the substrate is a thermoplastic film, it will melt in extended contact with the hot polymer so the horizontal distance from die to nip will be adjusted to minimize the contact time. Some machines are supplied with these two adjustments in the laminator, which by hand-wheel operation or by motorized actuation will move the nip vertically and horizontally relative to a fixed die position. Some machines are supplied with these two adjustments in the ex-

Fig. 8-14. Extrusion coating operation.

144 WEB PROCESSING AND CONVERTING TECHNOLOGY AND EQUIPMENT

truder carriage, which by motorized actuation will move the die vertically and horizontally relative to a fixed laminator nip position. Preference for one method or the other may be claimed, but they both have the same effect.

The extrusion coating section is arranged in a machine line to coat either one side of a web or the other, usually not both. The facility to coat either side might be provided by an unwind machine able to unwind a roll in either direction. An extrusion coating machine can also be provided with a coating nip assembly on both sides of the chill roll and the facility to arrange the die and nip configuration to coat either side of the web. Another device facilitating coating either side of the web is a turning bar which, as its name implies, will turn the web over upside down within the machine line. The web traveling downstream makes a 90° turn around a bar which is angled at 45° to the machine direction, which turns the web upside down but going in the wrong direction, perpendicular to the required flow. The web then makes a 180° turn around a roll, now right side up and traveling in the other direction perpendicular to the required flow. The web then makes another 90° turn around a bar which is angled at 45° to the machine direction which again turns the web upside down but now traveling in the original downstream direction (Fig. 8-15). The turning bars are air-greased, meaning that they have small perforations through which air is blown so that the web rides around them on a cushion of supporting air and does not contact the stationary metal bars. Turning bars are useful devices. Some machine lines might have two or three of them providing much flexibility of which side of the web is to be coated in multiple coating stations.

Some machine lines employ two extrusion coating stations in tandem al-

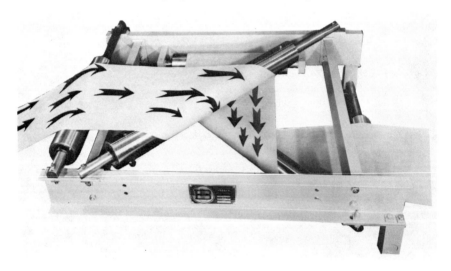

Fig. 8-15. Turning bar.

EXTRUSION 145

lowing either two-side coating in one pass or extrusion lamination plus one-side coating. Frequently the second station will be equipped for coextrusion. A typical product could be paper-poly-foil-poly-poly. The "polys," of course, can all be different polymers depending upon the required function. A third station can provide the ability for extrusion lamination plus two-side coating such as for a construction of poly-board-poly-foil-poly-poly. There are also machine line installations of four stations where the last poly layer is applied independently.

Machines designed essentially for paper and board applications will usually employ a single drum surface winder. These can be equipped with automatic cut-off and roll change mechanisms and run in a speed range of about 150–600 m/min with rewind roll diameters in a range of about 100–210 cm. Machines designed for flexible packaging substrates such as films, foils and lightweight papers will employ a turret type center winder with an automatic cut-off and roll change mechanism. Rewind roll diameters are usually in a range of about 60–125 cm at speeds of about 150–450 m/min.

SAFETY

Machinery builders try to design safety for the operator into their equipment. Safety systems include physical barriers, function limitations, warning notices and reaction devices. Inrunning nips between rolls have barrier type

Fig. 8-16. Extrusion coating line viewed from winder.

guards which prevent an operator inserting his person into them. A light beam, which if interrupted will put a machine into a safe mode, is an alternative, where a physical barrier is impractical. Two hands are better than one in a system where both have to be employed to activate a function, ensuring that one hand does not get caught in the machine. Reaction devices such as trip-cords, kick boards and emergency stop pushbuttons are necessary, of course, but unfortunately more often indicate that the safety hazard has occurred rather than preventing it. Instruction manuals may contain much useful advice regarding safety, but the operator exposed to the danger is frequently the least likely person to have read the manual. Antagonistic devices such as flashing lights, ringing bells, buzzers and horns are all too frequently deactivated.

REFERENCES

1. *Paper, Film and Foil Converter* **56**, (6) (1982).
2. Tadmor, Z. and Klein, I. *Engineering Principles of Plasticating Extrusion*. New York: Van Nostrand Reinhold, 1970.
3. Tadmor, Z. and Gogos, C. G. *Principles of Polymer Processing*. New York: Wiley Interscience, 1979.
4. Agranoff, J. *Modern Plastics Encyclopedia 1981/1982*. New York: McGraw-Hill, 1981.
5. Booth, G. L. *Coating Equipment and Processes*. New York: Lockwood, 1970.
6. Bernhardt, E. C. (ed.) *Processing of Thermoplastics Materials*. New York: Reinhold, 1959.

9
Hot Melt Coaters

Donatas Satas
Satas & Associates
Warwick, Rhode Island

The application of hot melt coating technology has grown at a fast rate in the late 1970's as a result of increased solvent costs and environmental restrictions.

Most of the hot melts consist of a polymer, resin and wax. Commonly used polymers for such blends are polyethylene, ethylene-vinyl acetate copolymers, block copolymers, and atactic polypropylene. Some hot melts might be relatively pure polymers, not blended with resins and waxes. Polyamides and polyesters are used this way.

Hot melts are to be distinguished from waxes and modified wax coating and adhesives, which are applied at lower viscosities, and also from extrudable thermoplastic materials, which are applied at viscosities higher than those of hot melts. Typical application temperatures and viscosities for these coating classes are given below.

Waxes, 0.02 Pa·s at 100°C
Hot melts, 10 Pa·s at 160°C
Extrudable and calenderable coatings, 200 Pa·s at 120°C

These limits for various coating types are not rigid. Any coating applied on hot melt equipment will be called a hot melt. Coatings applied on waxing machines might also be called hot melts, unless they are clearly waxes by composition. Some hot melts might be applied by extrusion coating.

Hot melt equipment is basically an adaptation of machines used for other types of coatings, especially of waxers on the low-viscosity side, and of high viscosity coating equipment, such as extruders and calenders. The hot melt equipment can be subdivided into several classes:

- Curtain coaters
- Roll coaters
- Slotted orifice coaters.

MELTERS

Hot melt is delivered fluid to the coating head and various melting equipment is employed for this purpose. The following equipment types are used for melting:

- Bulk tanks
- Double arm mixers
- Grid melters
- Drum melters
- Roll melters
- Continuous melters.

Bulk Tank Melters

Bulk tank melters are agitated jacketed tanks (Fig. 9-1). The melting rate is slow, but the construction is simple and such melters can be useful for large-volume application. The hot melt is maintained fluid in the tank and cold material is added periodically into the tank. The hot melt is kept at elevated temperature for a long period of time and this method is not suitable for heat-sensitive hot melts.

Double Arm Melters

Double arm mixers are used for melting and the equipment is capable of delivering molten material at a faster rate than is possible in bulk tanks. The temperature is raised gradually to assure that the whole batch is mixed. Formation of two phases, molten liquid and solid pieces, must be avoided in order to have a proper kneading of the batch. The batch when molten is transferred to a holding tank where it might be slowly agitated.

Grid Melters

Grid melter is a convenient way to melt efficiently at a higher temperature and to store the molten material at a lower temperature. A schematic diagram of a grid melter is shown in Fig. 9-2. The grid consists of heated tubing and the storage tank is a jacketed agitated tank. Pieces of hot melt are dropped over the grid where it melts and the molten material flows into the

Fig. 9-1. Melting tank with an agitator and top-mounted immersion heating elements, 200-gal capacity. (*Courtesy D. C. Cooper Co., Chicago, Ill.*)

holding tank. The continuous removal of molten material helps to maintain a good contact between the solid pieces and the hot grid surface.

Drum Melters

Drum melters are constructed on the principle of grid melters. They are suitable for 55-gal drums or 5-gal pails and allow controlled melting on demand. The follower plate of a drum melter is placed over a drum with its cover removed. The hot plate melts the adhesive, which flows along the grooves to a gear pump and is delivered to the coating head. Only the top 5-15 cm are melted at any given time.[1] Figure 9-3 shows a drum melter.

Roll Melters

Roll melters consist of a heated two-roll mill and a reservoir for the molten material from which it is pumped to the coating head. A schematic diagram of such a coater is shown in Fig. 9-4. Pieces of hot melt are placed in the roll

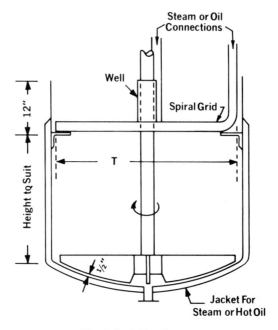

Fig. 9-2. Grid melter.

nip and the molten layer is removed by a scraper from the roll surface. Machines as large as 400 kg/hr capacity are in use.

Continuous Melters

The exposure time of hot melt to elevated temperature is reduced to 60-120 seconds by use of continuous extrusion type melters. Such extruders are also used for compounding of hot melts. Figure 9-5 shows the effect of thermal history on the degradation of block copolymer often used for pressure-sensitive hot melts. The superiority of continuous melters is obvious.[2]

CURTAIN COATERS

A curtain coater head allows formation of a free falling film (curtain), which is deposited on a web or sheets moving underneath the curtain on a conveyor. Curtain coating is especially suitable for coating sheets, or irregularly shaped flat objects. The surface gets coated as the piece moves under the curtain and the excess is collected in a catch basin underneath and recirculated.

Curtain coaters are used for various types of coatings including hot melts. It is required that the coating forms a film which is sufficiently cohesive to

HOT MELT COATERS 151

Fig. 9-3. Drum melter. (*Courtesy Nordson Corp.*)

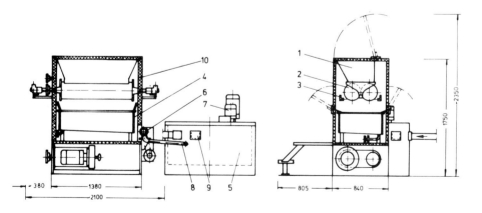

Fig. 9-4. Roll melter. (*Courtesy Maschinenfabrik Max Kroenert, Hamburg, Germany.*)

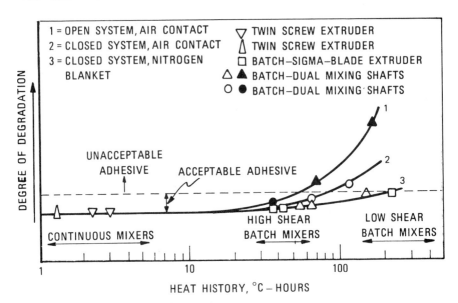

Fig. 9-5. Effect of thermal history on the degradation of block-copolymer. [Coker, G. T., Jr., Lauck, J. E. and St. Clair, D. J. *Adhesives Age* **20**(8):32 (1977).]

maintain a stable curtain. Very high-viscosity hot melts are not suitable for curtain coating. Hot melts of viscosities from 0.025–15 Pa·s have been used with these coaters and formed stable curtains. A schematic diagram of this process is shown in Fig. 9-6. Curtain coaters are relatively simple and inexpensive machines.

The coating head consists of a box with a gap of adjustable width between two knives. One knife is stationary and the other is movable to adjust the gap width, which is generally between 0.1 and 2 mm. The edges of the curtain are maintained by wires or by chains attached to the bottom of the coating head. This prevents necking down of the curtain as it is falling down. The coating is somewhat heavier at the edges and the curtain should be wider than the web to be coated. This does not extend more than 4 cm on each edge.

Two types of curtain coater heads are available: weir type and pressure type. In weir type coating heads the flow of the coating is determined by the fluid head over the weir. This head is the difference between the metering and overflow weirs. This height difference can be adjusted and allows regulation of the flow. The coating flows between the knives (lips) and the gap helps to dampen any flow nonuniformities across the width. Weir type curtain coaters are suitable for low-viscosity coatings of 0.1–1 Pa·s. Higher viscosity coatings are better handled by a pressure head.

A less frequently used variation of a weir curtain coater is the angular

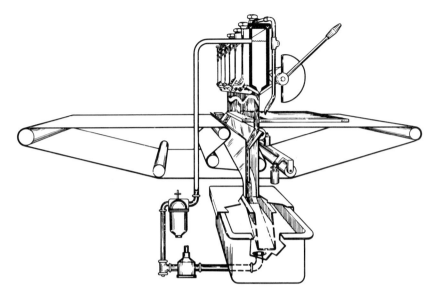

Fig. 9-6. Schematic diagram of a curtain coater. (*Courtesy Ashdee Div., George Koch Sons Inc.*)

plane curtain. The fluid metered by a weir arrangement flows down an inclined plane, instead of falling down as a free curtain. This allows the use of coatings which do not form stable curtains.

The pressure head consists of a chamber with an adjustable gap on the bottom. Coating is continuously pumped by a positive displacement pump. There are two variations of pressure head design. One type utilizes pressurized air above the liquid, which acts as a cushion and helps to maintain a uniform pressure in the chamber. A schematic diagram of such a head is shown in Fig. 9-7. Another variation of a pressure head is completely filled with the coating. The flow is regulated by a variable speed pump or by varying the amount of flow in the bypass, if a single speed pump is used. A completely filled pressure head is more difficult to control than an air cushion pressure head. Pressure heads are suitable for curtain coating of liquids up to 15 Pa·s viscosity.

A curtain coater has the following adjustments:[3]

- Coating head opening
- Coating head height
- Web speed
- Pump speed
- Temperature.

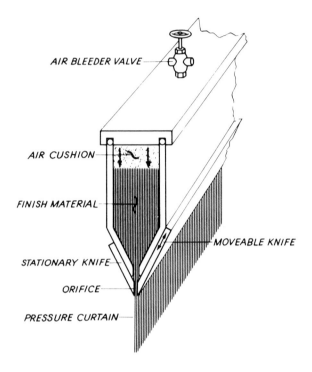

Fig. 9-7. Schematic diagram of a curtain coater pressure head with an air cushion. (*Courtesy Ashdee Div., George Koch Sons Inc.*)

Coating head opening might vary from 0.1-2 mm. Lower viscosity coatings require a narrower gap. A gap of 0.4 mm might be used for a 0.1-Pa·s viscosity coating and is opened for higher viscosity materials.

The gap uniformity effects the coating weight uniformity across the web. The coating head knife edges that form the gap must be parallel. Knives form a 120° internal angle and are extended about 5 cm beyond the width of the web on each side. The stationary knife can be bowed across the width by applying pressure on the screws which are spaced 7-10 cm apart along the knife's back edge. This helps to adjust the gap width more accurately. Variation of the coating deposit across the web width of as little as ± 5% is possible. Variation in the machine direction is negligible.

HOT MELT COATERS 155

Fig. 9-8. Curtain coater. (*Courtesy Ashdee Div., George Koch Sons Inc.*)

The coating curtain accelerates and stretches as it falls down. The gap width determines the initial curtain velocity. The narrower the gap, the faster the curtain velocity for a given pump speed. The gap width, however, has little effect on the terminal curtain velocity. The terminal curtain velocity is determined by the height of the coating head in addition to the initial fluid volume (width of the gap) and the properties of the fluid. The higher the viscosity, the higher should be the location of the coating head. The coating head height might vary between 10 and 30 cm. The speed of the web, or of the conveyor, should correspond to the terminal curtain velocity, or be somewhat higher, since some stretching of the curtain may take place. The web speed should not be slower than the terminal curtain speed; slower speed results in a poor coating surface.

The web speed is the main factor used to vary the coating weight. Curtain coaters usually operate at speeds of 60-130 m/min, but speeds as high as 400 m/min have been reported. The coating width might vary between 30 and 400 cm.

Pump speed determines the amount of fluid discharged through the gap. Continuously variable speed gear pumps are preferred.

Temperature determines the hot melt viscosity and thus its flow characteristics. Low-viscosity (0.025-1 Pa·s measured at 120°C) hot melts are run at 90-120°C, medium-viscosity (1-15 Pa·s) at 130-140°C, and high-viscosity coatings (25-85 Pa·s) are run at 180°-190°C.[3] To maintain the temperature, the coating head might be heated by circulating hot oil or by other means.

The excess coating drops into the pan underneath the conveyor. Air might get trapped, especially at the point where the coating hits the pan. Larger air bubbles in the coating might cause the rupture of the curtain; small air bubbles affect the appearance and barrier properties of the coating. This is especially a problem with high-viscosity hot melts which do not allow the air bubbles to escape easily. In extreme cases, a deaerator might be required to remove the entrapped air from the return hot melt stream. Air might also get trapped between the substrate and the curtain when impermeable webs are coated at high speeds. In case of high-viscosity coatings, this might happen at speeds above 40 m/min.[4]

The adhesion of hot melt to the substrate can be improved by preheating the substrate over a hot cylinder just prior to coating. The surface gloss is improved by shock cooling over a chrome plated cylinder.

SLOT ORIFICE COATERS

Slot orifice, or fountain, coaters have an important place in hot melt coating. The coating flows under pressure through a slot, forming a thin sheet which is deposited onto the web. Earlier coaters have been designed to coat an unsupported web, but such an operation was too tension-sensitive and it was found that a backing roll helps to improve the coating consistency. The clearance between the die lips and the back-up roll, and the web speed are the main factors which determine the magnitude of the shear force acting on the hot melt. The pump speed determines the amount of hot melt discharged through the coating head. The pump speed is synchronized with the web speed, so that a constant coating weight is delivered regardless of the change in the line speed. A general description of the slot orifice coaters can be found in References 5, 6, 7, 8 and 9.

Slot orifice coaters have wide operating characteristics. The hot melt viscosity range is 0.5-250 Pa·s, and coating of hot melts as viscous as 400 Pa·s has been reported. Viscosities below 0.5 Pa·s may cause pressure fluctuations in the coating head and yield a non-uniform coating deposit. Coating weights varying from 6-600 g/m^2 can be applied and deposits as high as 1800 g/m^2 have been reported.[9] Naturally, low-viscosity hot melts are most suitable for low coating weights and high-viscosity hot melts for heavy coating weights. Coating speeds up to 350 m/min are possible.

HOT MELT COATERS 157

The orifice gap is adjustable, but the same gap can be used for a wide range of coating conditions. The gap does not determine the amount of coating deposited. In case of low-viscosity hot melts, a gap that is too wide might not allow the development of adequate pressure inside the coating head and this might cause coating weight variation across the width. The trailing die lip helps to distribute the hot melt uniformly across the width and acts a smoothing bar. It should be parallel to the web surface and a separate adjustment independent of the gap setting might be provided to maintain this parallelism. The gap between the web and the die lip depends on the required coating thickness. For very thin coatings the lip might be set at a pressure against the rubber covered back-up roll.

Angle of the die relative to the backing roll can be adjusted. The slot width can be varied to avoid the application of the hot melt over the rubber back-up roll.

Uniform temperature must be maintained across the die. Temperature variation effects melt viscosity and therefore the coating weight. Circulation of a hot fluid through the coating head gives a more uniform temperature than externally mounted electrical heaters.

The first slot orifice coater was Genpac coater and a schematic diagram of the coating line is shown in Fig. 9-9. Unlike the later slot orifice coaters, Genpac coater deposited an excess of hot melt, which was removed by scrapers and smoothing bars. The coater was able to handle hot melts of 1-20 Pa·s in viscosity at speeds of 300 m/min and to deposit a coating weight of 4-30 g/m^2. Several other slot orifice coater designs have been described in the patent literature.[10,11]

Slot orifice coaters of several designs are on the market. The Park[6] coating

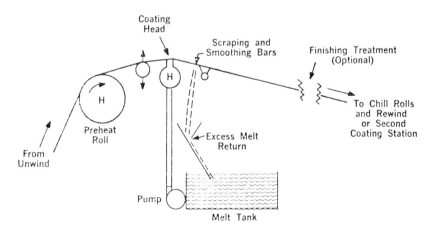

Fig. 9-9. A schematic diagram of a Genpac coater.

head consists of a supply tube, bed tube and coating tube. The cross-sectional area of this slot orifice head is shown in Fig. 9-10 and a schematic diagram of the coater in Fig. 9-11. The rubber coated backing roll supports the web and is held against the die lips by pneumatic cylinders. Another slot orifice coater is shown in Fig. 9-12, and it has been described by Mainstone.[9] The die is movable against the stationary back-up roll.

ROLL COATERS

Roll coaters can be subdivided into two main categories: direct coaters and reverse roll coaters. In the direct coaters the substrate and both rolls travel in the same direction and the hot melt splits between the substrate and the melt roll. Hot melt roll coating is basically different from calendering in this respect. The coating does not split in the calendering operation. In reverse

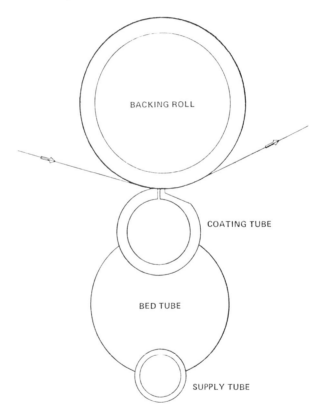

Fig. 9-10. Park coating head. (*Courtesy Bolton Emerson, Inc.*)

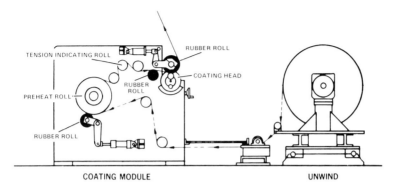

Fig. 9-11. Schematic diagram of a Park coater. (*Courtesy Bolton Emerson, Inc.*)

roll coating the melt roll is rotating in the opposite direction to the web. The schematic diagrams of these operations are shown in Fig. 9-13 and 9-14.

Hot melts are applied by many different variations of roll coaters. Figure 9-15 shows a kiss roller followed by a scraper to remove the excess coating. Figure 9-16 shows a roll coater where the amount of hot melt applied is determined by the gap setting between the doctor knife and the smooth steel applicator roll. It is important that the adhesive reservoir and the applicator roll are heated in order to maintain the hot melt sufficiently fluid.

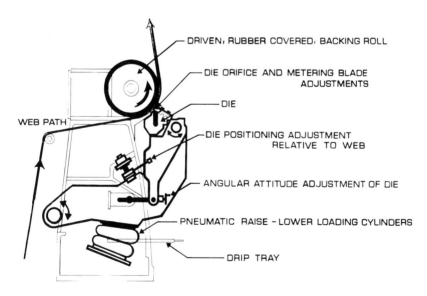

Fig. 9-12. Hot melt slot orifice coater. (*Courtesy Black Clawson Co.*)

160 WEB PROCESSING AND CONVERTING TECHNOLOGY AND EQUIPMENT

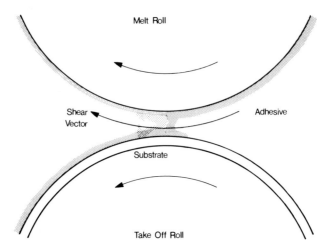

Fig. 9-13. Interface of melt roll with the substrate in direct coating.

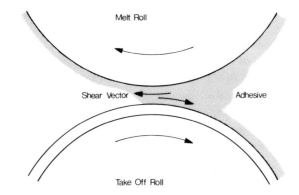

Fig. 9-14. Interface of melt roll with substrate in reverse roll coating.

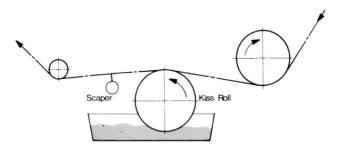

Fig. 9-15. Kiss roll hot melt application.

HOT MELT COATERS 161

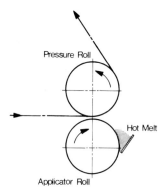

Fig. 9-16. Doctor knife hot melt application.

GRAVURE COATERS

Lightweight coatings, in the lower viscosity range of hot melts and especially waxes, can be applied by gravure coating. Both direct and offset gravure coating is used. Offset gravure gives a smoother coating which is more desirable for barrier type coatings. The gravure pattern obtained in direct gravure application can be smoothed out by passing the coating over heated smoothing bars. A schematic diagram of direct gravure coating is shown in Fig. 9-17. Hot melt gravure coating is suitable for viscosities up to 5 Pa·s. The coating weight range is 3-100 g/m², and the temperature range is 80°-200°C, with some operations reportedly run as high as 250°C.

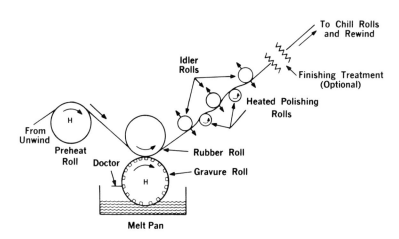

Fig. 9-17. Hot melt coating by direct gravure.

162 WEB PROCESSING AND CONVERTING TECHNOLOGY AND EQUIPMENT

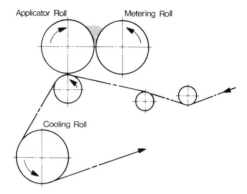

Fig. 9-18. Direct roll coating.

HIGH-VISCOSITY MELT COATERS

Hot melt coaters capable of handling hot melts up to 250-Pa·s viscosity have been constructed. Such coaters can attain speeds of 350 m/min and coating weights up to 300 g/m². Roll bending might become a problem in case of melt viscosities of over 100 Pa·s. Crowning and roll-crossing are required to compensate for the bending. Yet the coating weight within ±5% accuracy across the web is possible with well-designed high-viscosity roll coaters.[12]

A schematic diagram of a direct roll coating arrangement is shown in Fig. 9-18 and a reverse roll coating technique is shown in Fig. 9-19. A coating machine utilizing this coating method is shown in Fig. 9-20. This machine is specifically designed for coating of pressure-sensitive adhesives. A similar, but more versatile, machine is shown in Fig. 9-21. It has as many as 50 different web and roll arrangements for various film splitting, reverse roll coating, impregnation and lamination operations. Figure 9-22a, b and c shows

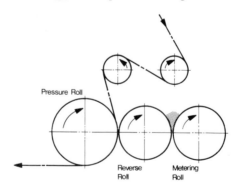

Fig. 9-19. Reverse roll coating arrangement.

HOT MELT COATERS 163

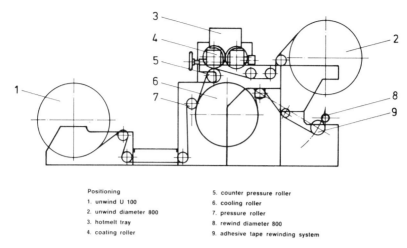

Positioning
1. unwind U 100
2. unwind diameter 800
3. hotmelt tray
4. coating roller
5. counter pressure roller
6. cooling roller
7. pressure roller
8. rewind diameter 800
9. adhesive tape rewinding system

Fig. 9-20. A schematic diagram of a direct roll coater. (*Courtesy Maschinenfabrik Max Kroenert.*)

some of the possible configurations. Figure 9-22a shows dipping of the web into a wax bath followed by squeezing in a nip. If a glossy coating is desired, the web is then run around the cooling roll located in the center of the assembly. Figure 9-22b shows a reverse roll coating arrangement, where the coated web contacts the cooling roll immediately after coating. Figure 9-20c shows a similar arrangement, except the rolls are set up to rotate in the direction of the web travel, giving a direct coating arrangement.

A reverse roll coater-laminator is shown schematically in Fig. 9-23. A photograph of such a machine is shown in Fig. 9-24.

High-viscosity hot melt coating machines are shown in Figs. 9-25 and 9-26. Figure 9-25 shows a Zimmer machine, which features cast iron precision melt rolls, hollow cored to accommodate electric heating rods and water. Water serves as a heat transfer medium. The rolls can be heated to temperatures up to 220°C. The hot melt is introduced directly into the nip, either cold or preheated on compounding equipment. The range of coating

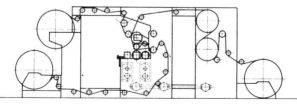

Fig. 9-21. Hot melt coating machine. (*PAK 600, Courtesy of Maschinenfabrik Max Kroenert.*)

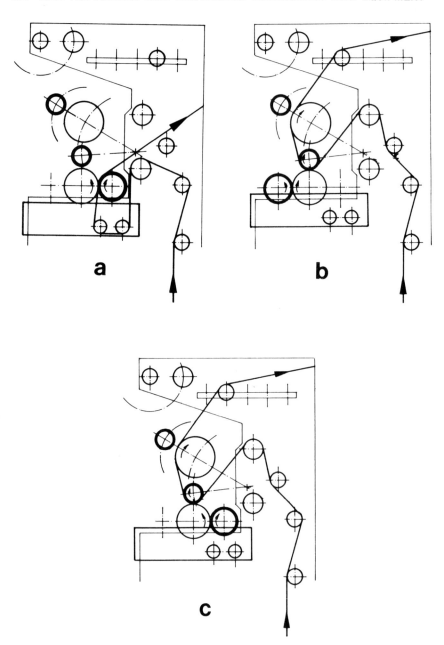

Fig. 9-22. Various arrangements of a hot melt coating machine. (*PAK 600, Courtesy Maschinenfabrik Max Kroenert.*) a—Two-side wax coating b—Reverse roll coating c—Direct roll coating

HOT MELT COATERS 165

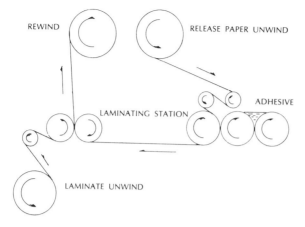

Fig. 9-23. Schematic diagram of a reverse roll coater. (*Courtesy The Wallace Co.*)

weight is 10-250 g/m², speed 5-50 m/min. The machine can handle high-viscosity melts. Another machine also suitable for high-viscosity hot melts and plasticized polyvinyl chloride is shown in Fig. 9-26. This machine resembles a calender, but its price is considerably lower.[15]

Fig. 9-24. Reverse roll hot melt coater-laminator. (*Courtesy The Wallace Co.*)

166 WEB PROCESSING AND CONVERTING TECHNOLOGY AND EQUIPMENT

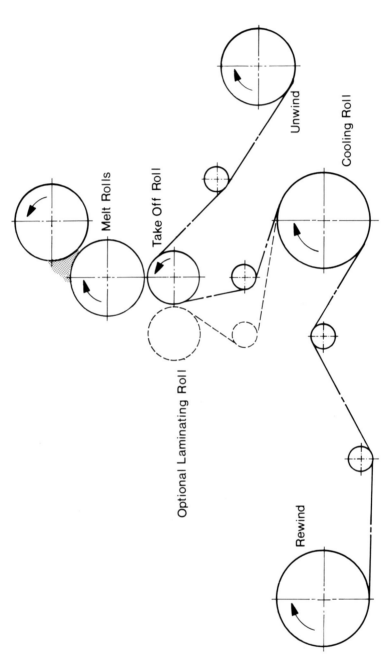

Fig. 9-25. Schematic diagram of a Zimmer coating machine.

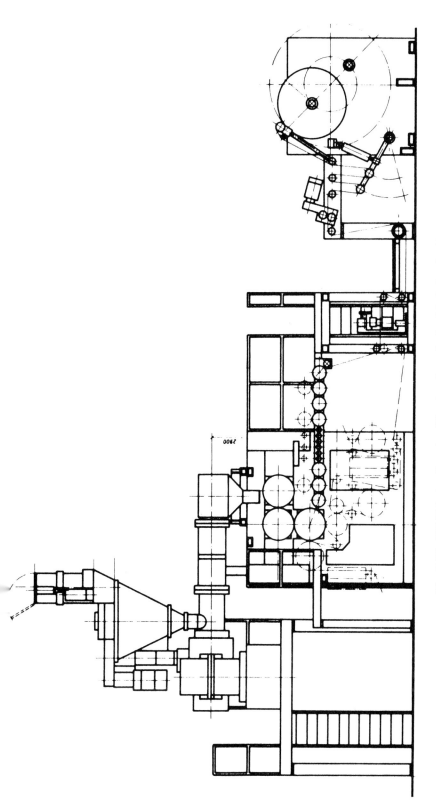

Fig. 9-26. BEMA hot melt coater. (*Courtesy Interplastica SA, Switzerland.*)

Narrow Hot Melt Coating Machines

Initially intended as pilot plant machines, these 150–450-mm wide coaters have found production applications with many pressure-sensitive label manufacturers.[7,13] Most of these machines feature slot orifice application heads. Pressure-sensitive label manufacturers normally purchase coated label stock, slit, print and die-cut to produce labels. The narrow coating machines offered a possibility to integrate vertically by producing the label stock in-house. Figure 9-27 shows a narrow coater of this type.

WAX COATERS

Coating of molten waxes is an old technology and hot melt coatings have been developed from coating and impregnating of packaging materials with wax. Wax coatings are still used for packaging applications as moisture barrier coating over paper and board, for lamination of foil to paper, heat sealing and coating of single-use carbon paper. Paraffin wax is the most commonly used material for these coatings. It has a poor gloss retention and therefore it is often modified to improve its physical properties. Modifiers

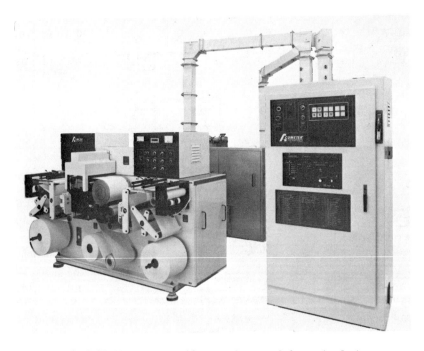

Fig. 9-27. Narrow coater. (*Courtesy Acumeter Laboratories, Inc.*)

are usually microcrystalline wax, ethylene-vinyl acetate copolymer, polyethylene, butyl rubber or polyisobutylene.

Wax may be applied over the paper so that it saturates the sheet, leaving little coating on the surface. Lower viscosity application attainable at higher temperature favors this condition. Such paper is called "dry-waxed."

Higher viscosity wax applied on the surface and cooled quickly by chilling on the rolls or by water immersion gives a glossy coating on the surface of paper or board. Such coatings are called "wet waxed." The viscosity of such waxes is 10–25 mPa·s.

Wax coatings have lost their importance to hot melt coatings and especially to extruded coatings and plastic films. The demarcation line between wax and hot melt coatings is not very clear. It is sometimes considered that wax coatings with less than 5% of polymeric additives such as EVA or LDPE are fortified waxes and above that hot melts.

Wax coatings are of low viscosity and are applied at lower temperature than hot melts. The requirements on the equipment are easier. Melt viscosity of 1 Pa·s at 120°–165°C constitutes the upper performance limits of a wax coating machine. Wax melts at 50°–67°C. Waxers generally run at 100–150 m/min, although some of the modern waxers can coat as fast as 600 m/min.

Waxing machines are direct adaptations of regular coating machines. Molten wax might be applied by a kiss roll, a heated gravure roll, pressure rolls and reverse roll coaters. Wire wound rods might be used as metering devices. Dip coating is often used for two-side application, and squeeze rolls are employed to remove the excess of the coating. Coating weight may vary from 3–15 g/m^2 for glassine to 15–30 g/m^2 for kraft paper.

A large application for wax coatings is precut carton blanks. Such coatings might be continuous, or they might be applied in a pattern, leaving uncoated areas for glue application or printing. Gravure and smooth roll coaters are used for such blank coaters. The roll coaters are especially suitable for pattern coating.

Standard gravure pattern is a dot structure of 1.5 mm in diameter and 0.75 mm deep set on a 45° helix. Other patterns are also used. Such coaters can handle sheets from 15 x 15 cm to 112 x 56 cm in size. The sheets are usually fed by a top combing mechanism. The gravure coating roll is 60 cm in circumference; it is heated and may be equipped with a heated doctor knife.

It is difficult to use higher viscosity hot melts on blank coaters. The blank tends to stick to the roll and may wrap around. The lighter the sheet, the more difficult this problem becomes, although machines capable of handling hot melts up to 17 Pa·s in viscosity have been designed. Some of the designs employ small (up to 8 cm) applicator rolls which reduce the tendency to wrap.[4] Another design employs a grooved roll and a rubber covered roll nip to decrease the tendency to wrap. This self-stripping principle is shown in

170 WEB PROCESSING AND CONVERTING TECHNOLOGY AND EQUIPMENT

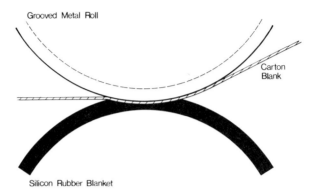

Fig. 9-28. Self-stripping device.

Fig. 9-28. The pressure applied by the grooved roll causes deformation in the rubber coating and prevents the board from wrapping. Hot melts of viscosity up to 30 Pa·s have been used on the blanks by a machine employing this principle. A blank coater is shown in Fig. 9-29.

FINISHING

The applied coating may be leveled to improve the surface gloss by a heated reverse driven roll or a nip. Lower viscosity coatings can be satisfactorily leveled by reheating and chilling. Infrared radiators or flame polishing devices are used for reheating. Reheating and especially burnishing also helps to improve the barrier properties of the coating.

Fig. 9-29. Blank coater, Model 88B (*Courtesy International Paper Box Machine Co.*)

HOT MELT COATERS   171

Web might be preheated in order to improve adhesion of the hot melt coating. Preheating increases the flow time of the hot melt and allows the adhesive to penetrate or wet the surface better.

Air entrapment might become a serious problem in fast coating operations of impermeable webs. Boundary air layer travels along with the web and gets trapped between the hot melt and impervious web. The entrapped air bubbles become visible and detract from the appearance and barrier properties of the coating. This might become a problem at speeds above 40 m/min and coating weights of over 6 g/m² for high-viscosity hot melts. Decreasing the cooling rate helps to remove the air bubbles and also to improve the adhesion.

FOAMING

Application of foamed hot melt adhesives has been developed.[14] The gas is introduced into the hot melt between the stages of a two-stage gear pump (Fig. 9–30). The gas dissolves in the adhesive and is released again after the

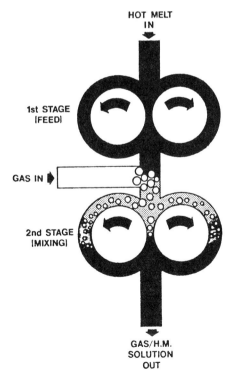

Fig. 9-30. Foaming of hot melt adhesives. (*Courtesy TAPPI.*[14])

adhesive is applied over the substrate, leaving a foamed adhesive coating. Carbon dioxide or nitrogen are generally used to foam the adhesive. The coating density is decreased to 50-20% that of the original. This decreases the amount of adhesive required, decreases the set time, and has some other processing and produce advantages.

REFERENCES

1. Plasschaert, J. *Course Notes, 1978 International Hot Melt Short Course.* TAPPI, 1978, pp. 68-84.
2. Coker, G. T., Jr., Lauck, J. E. and St. Clair, D. J. *Adhesives Age* **20** (8):30-35 (1977).
3. Cox, E. R. *TAPPI* **51**(7):39A-42A (1968).
4. *Equipment for Applying Hot-Melt Coatings.* PL 15-1068. E.I. du Pont de Nemours & Co., Inc., Wilmington, Del. 1968.
5. Weiss, H. L. *Coating and Laminating Machines.* Milwaukee: Converting Technology, 1977, pp. 116-123.
6. Park, G. C. *Proceedings Pressure-Sensitive Tape Council,* June 23-24, 1981, Rosemont, Ill., pp. 156-160.
7. Watson, C. and Satas, D. *Handbook of Pressure-Sensitive Adhesive Technology.* D. Satas (Ed.). New York: Van Nostrand Reinhold, 1982, pp. 558-573.
8. Booth, G. L. *Coating Equipment and Processes.* New York: Lockwood, 1970, pp. 197-209.
9. Mainstone, K. A. *1981 Hot Melt Adhesives and Coatings.* Short course notes of the Technical Association of the Pulp and Paper Industry. TAPPI, 1981, pp. 75-79.
10. Diescher, A. J. U.S. Patent 2,243,333 (May 27, 1941).
11. Van Guelpen, L. J. U.S. Patent 2,464,771 (March 15, 1949) (assigned to The Interstate Folding Box Co.).
12. Womack, H. G. and Wallace, R. C. *Course Notes, 1978 International Hot Melt Short Course.* TAPPI, 1978, pp. 94-114.
13. McIntyre, F. S. *1981 Hot Melt Adhesives and Coatings.* Short course notes of the Technical Association of the Pulp and Paper Industry. TAPPI, 1981, pp. 7-12.
14. Crosby, D. *Ibid.,* pp. 69-74.
15. Zickler, D. *Kunststoffe* **70**(1):14-17 (1980).

10
Powder Coating

Donatas Satas
Satas & Associates
Warwick, Rhode Island

Powder coating is a well-established technique to coat various irregular metallic objects, but it is not much used for coating of continuous webs. There are many other more accurate and simpler techniques in coating of flat surfaces. Nevertheless, powder coating has found uses for metal coils, fusible textile interlinings, shoe materials, and other products where heavy thermoplastic coatings are needed.

Powder coating consists of two steps: distribution of the powder over the surface and sintering of the coating to impart integrity and adhesion to the substrate. Sintering might be accompanied by compression in a roll nip. Powder when heated coalesces, spreads over the surface, penetrates the surface and levels. These processes in powder coating have been discussed by Wolpert and Wojtkowiak.[1] The heating is accomplished by infrared heating or by convection heat transfer in drying ovens. Since there is no vapor to remove, the heating ovens for powder sintering are of simple construction.

There are several methods used to distribute the powder:

- Scatter
- Gravure
- Rotary screen
- Spray
- Fluidized bed.

Random scatter and also gravure and rotary screen powder coating methods have been developed for manufacturing of fusible interlinings.

Fusible interlinings are used to reinforce textile materials used in shirt collars, coughs and suit fronts. Woven or non-woven fabrics with a sintered thermoplastic powder coating are heat laminated to such textile materials needing reinforcement. Dry cleaning and laundering resistance is required. High-density polyethylene, polyamide and polyester powders are used for this application. Such coatings are applied as randomly scattered coatings, or are printed in a regular dot pattern. Fusible interlinings have been developed in Europe and therefore European equipment manufacturers have been more active in development of powder coating equipment for use in continuous web applications.

SCATTER COATING

A schematic diagram of a scatter coating is shown in Fig. 10-1. The depressions in an engraved roll are filled with free-flowing powder from a hopper. The powder is removed from the depressions by a counter-rotating brush. Powder distribution may be improved by placing an oscillating sieve over the fabric. Powder coatings for shirt collar interliners weighing 12-15 g/m^2 are deposited by this method. Figure 10-2 shows a photograph of a scatter coating head which can be mounted on a web coating line. Coating weights up to 1000 g/m^2 are possible. Scatter heads are available up to 5000 mm in width. The scatter coating is followed by an infrared heating to fuse the powder.[2]

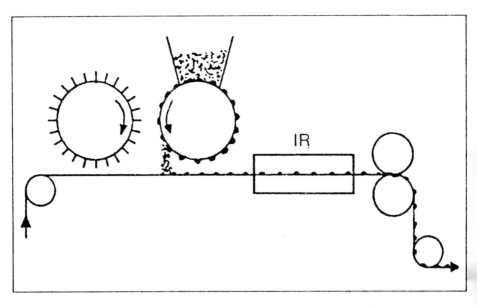

Fig. 10-1. Schematic diagram of powder scatter coating process.

Fig. 10-2. Scatter coating head. (*Courtesy Saladin AG.*)

In an alternate process, the powder is retained in the interstices between pins which are inserted into the roll surface. The powder is removed from the interstices between tightly packed pins by vibrating steel bristles and falls through a perforated oscillating plate onto the fabric. The oscillating plate is driven by compressed air or by electromagnetic vibrating motors. Uniform powder coatings at speeds up to 50 m/min are obtained.

POWDER SPOT COATING

Interlining fabric may be printed with uniformly distributed spots by a gravure process.[2] Coating weights of 10–20 g/m^2 are obtained at speeds of 10–30 m/min. Several variations of this system are available. The roller is engraved with a triangular pattern of indentations, each about 0.5 mm deep up to 10 indentations per linear cm.[3] The powder spot coating is not very suitable for synthetic fiber fabrics because of heat shrinkage on hot rolls.

Figure 10-3 shows a schematic diagram of the Caratsch process. The powder enters the cavities in the roll from a hopper and the excess is removed by a doctor blade. The hopper walls and the doctor blade mounting are cooled to prevent heating the powder in the hopper. The temperature of the engraved roll is maintained at 50–60°C for the polyamide powder and at a lower temperature for other lower melting powders. The engraved roll forms a nip with a heated mating roll where the fabric contacts the powder. Fabric is well-wrapped around the heated roll for a maximum heat transfer. The

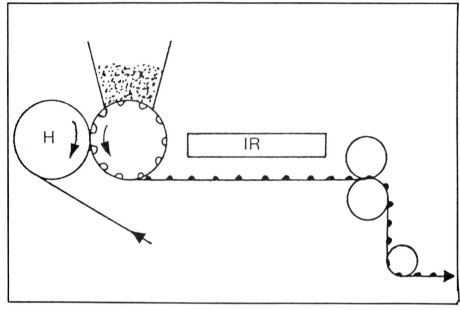

Fig. 10-3. Caratsch powder spot coating system.

heated roll is 50% larger in diameter than the engraved roll. After the transfer the powder is fused in an infrared radiation heated oven.

The Saladin process is shown in Fig. 10-4. The temperature of the heated roll is maintained at 170°-260°C and the powder is fused in the first nip. A photograph of such a machine is shown in Fig. 10-5.

A machine capable of powder spot and scatter coating is shown in Fig. 10-6.

ROTARY SCREEN PRINTING

Powder of grain size up to 80 μm has been printed for fusible interliners by the rotary screen printing process. The schematic diagram of the process is shown in Fig. 10-7. The web passes between the pattern perforated roll and a supporting roll. Powder is fed into the interior of the perforated drum and is forced by a squeegee through the screen openings.

The same process is also used for printing of powder paste and polyvinyl chloride plastisol for fusible interliners. Aqueous dispersions containing of up to 40% fine powder compounded with thickening and flow agents are used. Powder paste process has some advantages over dry powder printing: it is easier to introduce additives, such as plasticizers, and it is easier to fuse the coating without causing shrinkage of the substrate. Nevertheless, powder

POWDER COATING 177

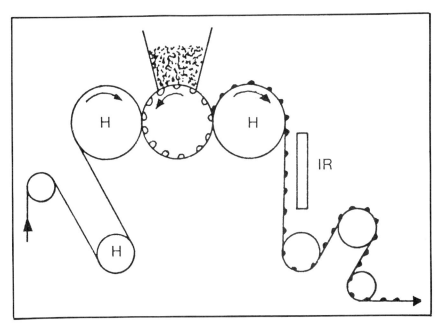

Fig. 10-4. Saladin powder spot coating system.

Fig. 10-5. Powder point machine. (*Courtesy Saladin AG.*)

Fig. 10-6. Combined powder-point and powder-sprinkling machine. (*Courtesy Schaetti & Co.*)

coating by gravure methods has mostly replaced rotary screen printing for manufacturing of fusible interliners.

FLUIDIZED BED COATING

Fluidized powder exhibits some of the properties of a liquid and therefore can be coated by the techniques developed for liquid materials. The fluidized bed apparatus consists of a tank subdivided into plenum chamber and fluid bed chamber separated by a diffuser plate. When pressurized air is introduced into the plenum chamber it passes through the diffuser plate and percolates through the powder separating the particles, lowering the apparent density of the material and causing the powder-air blend to behave like a liquid.

Fluidized bed coating is the most accepted technique of powder coating various irregular metal parts, but it is not used much for coating of continuous webs. Most of the webs, except metal coil or wire cloth, cannot be heated to a sufficiently high temperature to fuse the powder upon contact with the web. The method is not suitable for deposition of uniform coatings below 0.1 mm thick, unless the fluidized bed is combined with use of electrostatically charged powder. In such a case, thinner coatings of 50 μm are possible.

Powder particles can be electrostatically charged in the fluid bed by ionizing the surrounding air or by recirculating the powder through a charging grid. The web is charged with an opposite charge. If the substrate surface is nonconductive, it might require a treatment with a conductive solution.

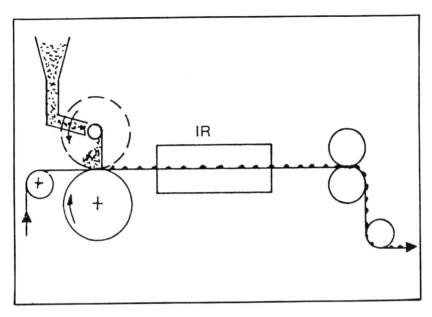

Fig. 10-7. Rotary screen coating of paste.

Powder covering the substrate with an opposite charge gradually eliminates the surface charge. This has a self-leveling effect on the coating, it also limits the maximum coating thickness. Very thick coatings (above 500 μm) are not possible because of this so-called back-ionization phenomenon.[4]

Fluidized particles can be fed at a controlled rate by an apparatus made for such purpose. A metering screw is mounted on a powder container which in turn is mounted on a vibrating unit. A special carburator mixes the powder with the carrier gas and the mixture is delivered by the screw to its point of use. Powder size ranges up to 1 μm in diameter. A photograph of such a feeding machine is shown in Fig. 10-8.

Fluidized particles retain some of their free-flowing characteristics even when the air flow is interrupted. This temporary condition, static aerate, is different from a fluid bed: the particles are stationary and not in a turbulent flow; they have a certain angle of repose. Such static aerate can be fed in a thin uniform layer over a moving substrate.[5] Static aerate can be coated by such methods as knife-over-roll.

SPRAY COATING

Powder can be sprayed over a heated part so that it melts upon contacting the surface. The uniformity of such coatings is poor.

The electrostatic powder spray method is a considerable improvement

180 WEB PROCESSING AND CONVERTING TECHNOLOGY AND EQUIPMENT

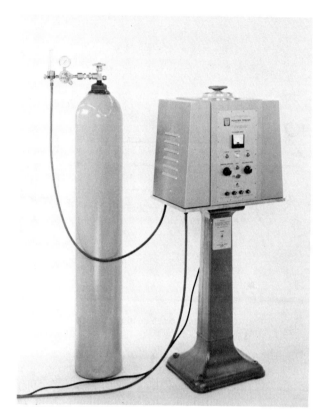

Fig. 10-8. Powder feeding machine. (*Courtesy Sylvester & Co.*)

over the simple spray. Powder leaving the gun is electrostatically charged and the web to be coated is electrically grounded and carries an opposite charge. The web can be at room temperature, or it can be heated. In case of an unheated web the powder particles are held in place by their electric charge. At some thickness the accumulated powder prevents any further deposition of more powder with the like charge. The charge accumulation limits the coating thickness and also promotes the coating uniformity.

Steel coil cleaned through three to five cycles of detergent washes and rinses and iron phosphate treated for improved adhesion can be coated by such spray-coated method.[6]

Powder spray guns or fluid bed applicators may be used. Spray guns may have the following charging systems:

- Annular charging zone
- Point charging (needle electrode)
- Internal charging system.

Mixture of powder with air can be explosive. Lower explosive limit (LEL) for most plastic powder is about 20 g/m^3

KNIFE COATING

Heavy powder coating can be applied by a knife over channel or over blanket coating. The powder is placed in front of a knife from a vibrating hopper or fed from a fluidized bed. The thickness of powder coating is determined by the gap setting. The coating is fused by infrared heaters and calendered to obtain a compressed coating of uniform thickness. This technique is used to coat some shoe components.

REFERENCES

1. Wolpert, S. M. and Wojtkowiak, J. J. Flow at the surfaces of powder coatings. *Nonpolluting Coatings and Coating Processes*. Gardon, J. L. and Prane, J. W. (Eds.) New York: Plenum, 1973, pp. 251-269.
2. Schaaf, S. and Luethi, H. Bonding of garments with fusible interlinings. Zürich: Emser Werke AG, May 1981.
3. Smith, L. M. Fusible adhesives for textiles. *Developments in Adhesives, 1*. Wake, W. C. (Ed.). London: Applied Science Publishers, 1977, pp. 199-221.
4. Bright, A. W. Plant and equipment for the application of powder coatings. *Powder Coatings* 1 (3):3-16 (1978).
5. Brooks, D. H. U.S. Patent 3,167,442 (January 26, 1965) (assigned to International Protected Metals, Inc.).
6. Miller, E. P. Electrostatic powder coating—potentially pollution-free finishing method. *Nonpolluting Coatings and Coating Processes*. Gardon, J. L. and Prane, J. W. (Eds.). New York: Plenum, 1973, pp. 225-234.

11
High-Vacuum Roll Coating

Ernst K. Hartwig
Leybold-Heraeus GMBH
Hanau, West Germany

It is possible to build up a coat by vaporizing the coating material, then condensing it on a substrate surface. If there is no reaction of the vapor with gases or other contaminants, nor any entrapment of foreign particles on the substrate during the condensation process, the coat obtained will retain its physical properties and remain identical to the material in its solid form.

This is only possible in an environment where the number of particles, other than those of the coating material, is comparatively low in respect to the coating material. This means that the coating process has to take place in a vacuum. The optimum vacuum level applied depends on the kind of coating material and the coating process used. As a rule, coating quality increases with decreasing pressure. For high-quality coating in modern equipment, the working pressure ranges from as high as 5×10^{-4} mbar to as low as 5×10^{-5} mbar. Some special products need working pressures below 5×10^{-5} mbar. These working pressures apply to thermal evaporation processes. In a sputtering process, the working pressure is in the range of 5×10^{-3} mbar. The coating processes described above are called high-vacuum coating or vacuum coating.

There is one important modification of the process of vacuum coating which expands its scope of applications. This modification is referred to as reactive coating. In a reactive coating process, a controlled amount of gas—e.g., oxygen or nitrogen—is introduced into the area between the source and substrate. The reaction of these gases with the coating material results in an oxide or nitride formation which is then deposited on the substrate.

THIN FILM TECHNOLOGY

The main step in high-vacuum coating is to evaporate a coating material and to transfer it as vapor. By controlling the evaporation rate and the time the substrate is exposed to the source, well-defined very thin films of the coating material can be deposited. For instance, it is usual to coat aluminum in a thickness range of up to 0.1 μm. This is about one hundredth of the thickness of aluminum foil used in commercial applications.

Building thinner and thinner coats from a certain thickness down, the film may have properties which are different from those of the solid material used for coating. The technology around this range of thickness is called "thin film technology" and opens new possibilities for product applications.

Among the numerous reasons why the thin film vacuum coating process continues to gain such popularity, two are of significant importance.

- The same or almost identical properties of a material in its solid form can be obtained by vacuum coating with much lower material consumption.
- Thin film vacuum coated materials have useful properties which the same materials in their solid form do not possess.

These two factors are the main economical and technical reasons why a large number of vacuum coated products are now entering the market.

The thin film vacuum coating applied to rolled flexible web materials is called "high-vacuum roll coating." High-vacuum roll coating is a method to produce coatings for industrial high-volume applications.

THIN FILM PROPERTIES

Vacuum coated thin films offer a number of properties which are the basis of a large variety of applications. Characteristic properties are:

- Purely decorative effects. Metallic appearance generally, highest "silver" brilliancy or satin gloss, any desired metallic effect color.
- Selective transparency or reflection of visible light, ultraviolet light, short wavelength and long wavelength infrared light.
- Selective electrical conductivity or resistivity.
- Transparent electrical conductivity (heatable transparent films).
- Partially or totally releasable coatings.
- Storage properties.
- Magnetic properties.

COATING MATERIALS AND SUBSTRATES

Practically all metals, many of their oxides and alloys and a number of other elements and compounds are suitable for deposition as thin layers on substrates in a vacuum coating process. The criteria for the selection of a coating material are its thin film properties, cost, availability and demands on the equipment.

The majority of all roll coating plants built until now operate with aluminum as the coating material. Vacuum deposited aluminum has good electrical and reflection properties, as well as a decorative appearance. It adheres to the substrate well and has good corrosion resistance. Aluminum has a relatively low melting point and is available in wire form, which makes it well-suited for low-energy evaporation and for continuous feeding. Its price is relatively low, and the world's reserves of aluminum are larger than those of other metals. The total coating costs are lower than for any other metal, with the exception of zinc. Aluminum can be expected to remain the most widely used coating material for many years. The shares of other materials already used for roll coating will, however, increase in the future, as the field of application is growing steadily. Metals in question are nickel, cobalt, chromium, copper, silver, steel, tin, indium, zinc and titanium, as well as their alloys and oxides.

As a substrate, the industry offers a number of plastic films which are suitable for high-vacuum roll coating. Particularly common are polyester, polypropylene, polyethylene, nylon and polycarbonate films. Polyester is the material most appropriate for high-vacuum roll coating. It gives no problems to equipment and process and is used for higher value products; e.g., capacitors, window films, barrier packaging, metallic yarn and cardboard laminates.

Polypropylene is next to polyester. It needs to be pretreated before vacuum coating. It can only be coated on one side because of roll blocking, if the pretreatment is done on both sides.

Selected papers and cellophane usually require pre-coating and make greater demands on the pump set of a roll coating plant.

PROCESS AND EQUIPMENT

Basic Equipment Design

High-vacuum roll coating equipment today is highly diversified and tailored to different applications for different products. Despite the constantly changing state-of-the-art, there exists an unchangeable basic design for all vacuum roll coaters which consists of the following three main construction elements:

- A source which holds the coating material and causes its transfer to a substrate.
- A holding and transport system for the substrate.
- A vacuum to accomplish the process.

Processing Sequence. Production of coated film in a high-vacuum roll coating plant proceeds in the following stages:

1. Fitting the roll of substrate into the transportation and winding system.
2. Evacuating the chamber in which the coating process takes place.
3. Starting the winding system which exposes the running substrate to the coating source.
4. Activating the coating source.
5. Coating the substrate.
6. Breaking the vacuum and cooling.
7. Changing the roll of substrate and maintenance of the plant.
8. Starting next cycle.

The detailed specification of a plant is determined by:

- The coating material, its properties and the thickness of the coat to be applied.
- The number of coats applied in one cycle.
- One-side or two-side coating.
- The kind of substrate, its thickness, width and length.
- The required output.
- Economical considerations.

Coating Sources

Depending on the way solid material is transformed into vapor, one differentiates between thermal evaporation and cathode sputtering.

Evaporation is the change of the state of a material achieved by supplying it with heat until it transforms from a solid state into a liquid state and then into vapor, which is transferred and deposited.

Sputtering is the displacement of particles from a solid surface by ion bombardment and the precipitation and deposition of such particles onto a substrate.

The various sources used are shown in Fig. 11-1.

Radiation Heated Coating Sources (Oven). In this form of thermal evaporation the energy is supplied by heat radiation. The coating material is

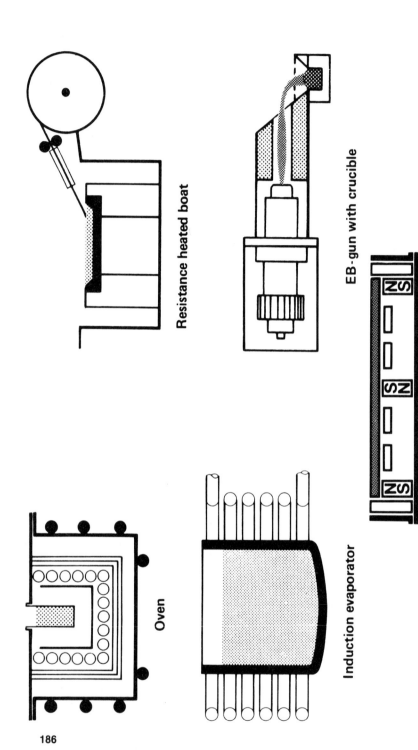

Fig. 11-1. Evaporation sources.

situated in boats made of a material with a considerably higher melting point than the coating material itself. (Fig. 11-2). The evaporation source contains one or more boats. The material is fed into it in batches. After conclusion of one or several working cycles, the boats are filled up again, or exchanged for new boats containing pre-melted material.

A radiation-heated evaporation source requires an extensive heating-up period before gaining thermal stability which then, however, remains very stable. Such a source is not suitable when fast control is needed.

This oven is the classic evaporation source for zinc and other materials with a low melting point. Its employment is limited in temperature by the thermal stability of the boats and shields in comparison to coating materials with higher melting points. With regard to other evaporation sources, the efficiency rate of a well-designed oven is high.

Induction Heated Coating Sources. In this process, metal in a nonconductive crucible is heated up by an induction coil surrounding the crucible. The evaporation system consists of a number of induction-heated crucibles, arranged in one row covering the web width.

Fig. 11-2. Radiation heated source (oven).

The material is usually fed batchwise in the shape of granules or small pieces. In semi-continuous operation, the size of the roll of substrate is limited. For air-to-air plants, not continuously fed evaporation sources are unsuitable.

The induction-heated evaporator can be employed for a wide range of melting temperatures, which makes it suitable for many coating materials. For many years it was the most popular evaporation source, particularly in the U.S.

The induction-heated evaporator is thermally slower than evaporation boats, because its crucible contains a considerably larger quantity of metal.

For the evaporation of aluminum, the induction-heated evaporator is inferior to the more modern resistance-heated boats. This is especially true with regard to uniformity of film thickness, maintenance and the cleanliness obtainable for the system.

Resistance Heated Sources. These sources are made from combinations of titanium diboride (TiB_2), boron nitride (BN) and sometimes aluminum nitride (AlN). The heating of the source is done by electrical current passage through the material itself. The coating material is fed continuously as a wire (Fig. 11-3).

The entire evaporation system consists of a number of equidistantly arranged boats. In modern roll coaters the power is supplied to each boat by individually controlled power supplies. All power supplies are activated and controlled by an integrated master system (Fig. 11-4).

The wire feed system supplies the same amount of wire to all boats. This is the most important requirement for a constant and uniform evaporation rate in the entire system.

Fig. 11-3. Resistance heated source (boat).

Fig. 11-4. Resistance heated source (complete evaporator system).

For aluminum, this evaporation system presently offers the most advanced technology adaptable for high thickness and high-uniformity coating.

The same type of boat may also be used for different metals. The range of application of these boats is limited by the thermal stability of the semiconductive material and the surface-wetting properties of the coating material.

Electron Beam Sources. Evaporation by electron beam is a thermal process, like evaporation by induction-heated systems or resistance-heated boats. The energy is directly supplied to the surface of the coating material by electron beam. The material is evaporated from ceramic or metal crucibles, the latter being water-cooled (Fig. 11-5).

By scanning the focused beam, the heated area on the surface of the coating material may be of different size and shape adapted according to the requirements of the process. Depending on the width of the substrate, an elec-

190 WEB PROCESSING AND CONVERTING TECHNOLOGY AND EQUIPMENT

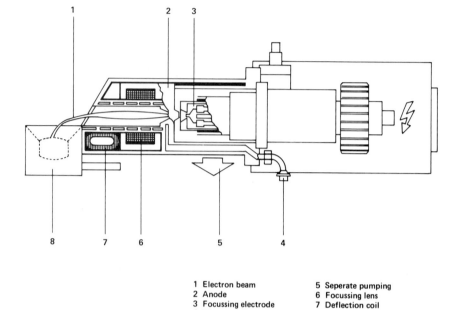

1 Electron beam
2 Anode
3 Focussing electrode
4 Water - cooling
5 Seperate pumping
6 Focussing lens
7 Deflection coil
8 Water - cooled copper crucible

Fig. 11-5. Electron beam gun—schematic.

tron beam evaporator system in a roll coater will be equipped with one or more guns (Fig. 11-6). The coating material may either be supplied in batches or continuously by wire feeding or by vertical insertion of a metal rod (rod feed system).

The electron beam gun can be used for the evaporation of a wide range of metals, especially those with a relatively high melting point and for a number of non-metallic materials; e.g. SiO or SiO_2.

By using several crucibles with different coating materials supplied either by one electron beam gun or by separate guns the deposition of alloys in a selectable ratio is possible. Rod feed systems allow the evaporation of molten alloys from one crucible, if their components have melting points very close to each other. Nickel/cobalt alloys represent one example.

Important advantages of electron beam evaporator systems in roll coaters are:

- Direct and fast supply of energy to the coating material evaporation area.
- Fast control of the system with regard to evaporation rate, deposition thickness and thickness uniformity, via power supply, scanning system, beam deflection system.
- Long life span of water-cooled crucibles.

HIGH-VACUUM ROLL COATING 191

Fig. 11-6. Electron beam sources—six guns.

Evaporation of materials by electron beam is the most versatile and most modern method of all thermal evaporation processes.

Sputter Coating Sources. Vacuum sputter coating is the displacement of particles from a solid surface by ion bombardment and the precipitation and deposition of such particles onto a second surface. Depending on whether the ion bombardment is induced by direct voltage, as in the case of metals, or by radio frequency voltage, as in the case of many semiconductors or insulators, the process is called d.c. sputtering, or r.f. sputtering.

It was found that in a suitably dimensioned and aligned magnetic field the number of particles transferred is increased and the accompanying loss is heat reduced. This means that the source becomes stronger and more efficient by means of a magnetic field (Fig. 11-7). A magnetic field may be applied both to the d.c.-energized cathodes and the r.f.-energized cathodes. This technique is called "high-rate sputtering" or "magnetron sputtering."

The elements of this source are sputter cathodes covering the entire coating width. The coating material as a target is fastened to the cathode by gluing, soldering or clamping, depending on the type of cathode. The entire sputter source consists of a number of cathodes arranged along the web path. They may be arranged around a coating drum. The power is supplied to each cathode by its own individual power supply. The specific rates of the sputter cathodes are much lower than those of all evaporation sources. Of all sputter cathodes, the specific rate of the magnetron cathode is the largest.

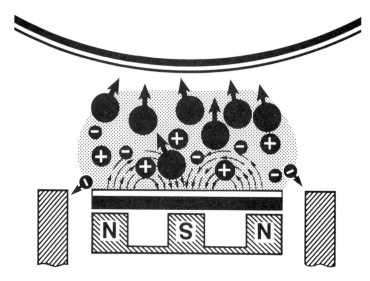

Fig. 11-7. High-rate sputtering—principle.

Sputtered films have a finer, more compact structure; i.e., smoother surface, and therefore higher reflectivity and higher barrier properties for comparably the same thickness of coating. Sputtering offers a highly extensive application concerning the coating of various materials.

The material efficiency and the operational reliability of the source is higher in sputtering than in all other techniques.

Roll Arrangement

The overall design of a plant, in respect to its roller arrangement, film guidance and winding system, is decisively influenced by the question of whether the plant is to be operated continuously or semi-continuously. Which form of operation is chosen depends not only on technical considerations, but is also influenced to a great extent by economical considerations.

Semi-Continuous Operation. In a roll coating plant designed for semi-continuous operation, the roll of substrate is situated inside the vacuum chamber (Fig. 11-8). Here the roll is unwound, exposed to a source and rewound again under vacuum. The working cycle consists of four steps:

1. Pumpdown
2. Evaporation
3. Venting and cooling
4. Recharging.

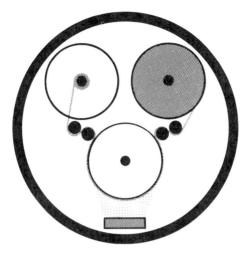

Fig. 11-8. Semi-continuous system.

The working cycle chart, Fig. 11-9, illustrates the relationship between these four different stages. The higher the percentage of time spent on evaporation, the greater its efficiency.

Semi-continuous roll coating is the preferred process at the moment, as it has a great number of important advantages in comparison to other processes:

- High flexibility of the system for processing various base materials and products.
- Very favorable winding conditions; no air-telescoping at high-speed, low-tension rewinding.
- Ideal dust-free and scatch-free winding conditions for freshly applied sensitive coats.
- Technical and economical superiority with all films of less than 90 μm thickness.

Continuous Operation. In continuous roll coating plants, the unwind and rewind rolls are situated outside the vacuum chamber. The substrate must therefore enter and leave the vacuum chamber through locks. The air-to-air plants (Fig. 11-10) closely reach the ideal of 100% utilization of the operating time for coating purposes. They are designed in such a way that the working cycle is interrupted only when maintenance is required. However, there are some difficulties still being experienced with this process that have prevented the full exploitation of the concept. The latest plants running at high speeds require sophisticated lock systems and a flying splice winding

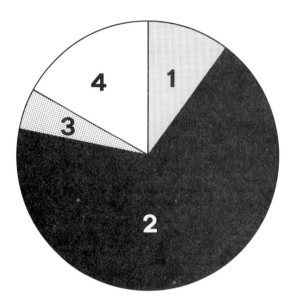

1 Pumpdown time 10%
2 Evaporation 68%
3 Cooling and venting 5%
4 Re-charging 17%

Operating data:
Material thickness 12 μm
Roll O.D. 600 mm
Film length 22 000 m
Web speed 6 m's^{-1}

Fig. 11-9. Working cycle chart.

system which allows roll changing without stopping. The design of such plants is considerably more complicated. The gain in production output is much lower than its price increase justifies in comparison to semi-continuous plants.

The key to the technical success of air-to-air plants can be found in the lock system, which must be capable of transporting the film material into the vacuum chamber. A further source of problems is the maintenance of the

Fig. 11-10. Air-to-air system.

thermal evaporation source during long-term continuous operation. Extensive efforts are being made to solve these problems.

Winding System

There are some significant differences between winding systems of continuous plants. Since the winding system of a semi-continuous plant is within the vacuum during operation and exposed to the vapors of the coating material, special demands are made on its design:

- The winding system must be as compact as possible to keep the size of the vacuum chamber small.
- The winding system must not release any gases, vapors or fluids which may have any adverse effects on the vacuum. This requirement presents particular problems of lubrication.
- The winding system has an air side and a vacuum side. The drives, gears and in some cases also the belt or chain transmission systems are situated outside the vacuum chamber connected by vacuum-tight lead-throughs.

For air-to-air and for semi-continuous roll coating, the following is of importance.

During the coating process, the substrate is exposed to heat. In order to prevent thermal damage, the substrate is passed over a coating drum while being coated. This drum is chilled. In many cases, water is a sufficient cooling agent. For the processing of very thin films, refrigeration equipment needs to be employed for constantly cooling the coating drum down to -30°C or even lower. When such refrigeration equipment is being used, the chill roll must be brought to room temperature before flooding the vacuum chamber to prevent condensation of water vapors.

When working with extremely thin films, special devices are engaged to ensure good contact with the coating roll. For example, an additional drive is essential between the coating drum and the take-up roll to ensure good thermal contact between the substrate and the coating drum without influencing the rewinding tension.

Drive and Control of the Winding System. It depends on the given speed range and the range of tension to be supplied to the web which particular drive system is to be selected. For plastic films of 12 microns or less, a three-motor calculator controlled system is preferred which operates as shown in Fig. 11-11.

The tension force, which is exerted on the web, is dependent on the mate-

196 WEB PROCESSING AND CONVERTING TECHNOLOGY AND EQUIPMENT

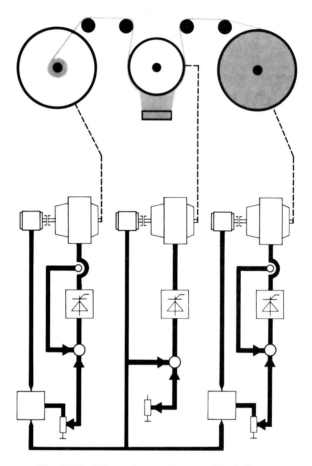

Fig. 11-11. Drive and control system—block diagram.

rial and can thus be controlled. This force should remain constant during the total winding process. The applied torques at the feed shaft, coating drum and take-up shaft are controlled toward this end.

Each one of the drives is controlled by a thyristor control element. Special low-inertia d.c. motors are used which have excellent dynamic properties. The web is unrolled from the feed roll with constant tension. This is produced by the braking force of the unwind motor, which is proportional to the feedstock roll diameter. The instantaneous diameter is calculated from the rpm of the unwind roll and coating drum. The chill roll diameter is a fixed constant.

The motor of the chill roll accelerates the web to a preselected speed. A starting integrator makes it possible to increase the rpm at a constant rate up to the selected value. During that operation, a special device controls the winding and prevents possible loops and tears in the band. All guide rolls and bowed rolls maintain the same speed as the web by synchronization.

The drive of the take-up shaft is constructed similarly to that of the feed shaft. The web tension remains constant, as the rpm decreases with the increasing roll diameter. The winding electronics make it possible to automatically decrease the web tension continually with increasing roll diameter if desired.

The drive system can be switched to "loading" or "coating" operation. When defects occur; e.g., tears in the band, the system is stopped automatically; the feed roll is stopped with maximum current. An essential characteristic of this system is the need to balance the static and dynamic frictions of the system.

Should an additional drive be needed to the ones mentioned above in order to separate web tension between coating drum and rewinder, it can be integrated into the described system.

Two-Side Coating—Two-Source System. Coating both sides of a substrate is necessary for certain applications. Examples of these are plastic films and papers for various types of capacitors or polyester film for high-quality metallic yarn.

Two-side coating is possible when the web is passed twice through a plant with one evaporation source. In this case, the process has to be interrupted for turning the roll.

In the case of thermal evaporation a roll coating plant for one path two-side coating has two evaporation sources and two coating drums (Fig. 11-12). The winding system of such a plant may be built to allow two modes of operation:

- One coat on either side of the web
- Two coats on one side of the web.

The possibility of applying two coats to one side of the substrate can be utilized to produce particularly thick coats in an economical manner. If two different coating materials are used and the web transport system is made reversible, one will already have a general purpose plant for the production of multi-layer coats (sandwich-coating). For materials requiring different vacuum conditions, the coating chambers must be separated from each other. In addition to this, it is essential that a suitably arranged pumpset is installed to evacuate both chambers.

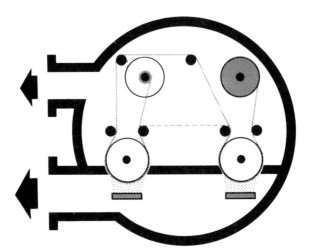

Fig. 11-12. Two-source system.

Multi-Source Systems—Sputter Roll Coaters. If more than two different coating materials are to be applied in one run, the two-source system could be extended in principle.

Multi-source systems for material over 1 m wide could present special handling problems. Until now the demand for such arrangements has been low. This is different in the field of sputter roll coaters. Here multi-source multi-layer systems are a must.

The design of a sputter roll coater differs significantly from a conventional roll coater working with thermal evaporation sources. The unique design of web sputtering plants is due to two distinct characteristics of the sputter cathodes:

1. They do not operate with liquid metals. Their surface, therefore, does not need to be arranged horizontally.
2. Sputter cathodes have a relatively low specific sputtering rate. To improve throughput rates, the material target area must be as large as possible.

In a typical sputtering plant, the web is allowed to pass over a chilled coating drum of large diameter, where it is exposed to as many sputter cathodes as possible to arrange around the drum, as shown in Fig. 11-13.

This design must be modified if different materials are to be coated in one run, especially if reactive and non-reactive coating will be done in one run. Chamber subdivision may also be essential.

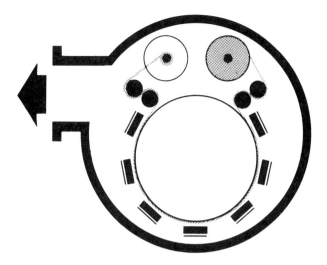

Fig. 11-13. Basic design, sputter roll coater.

Chamber Subdivisions

A further possibility of varying plant design is to subdivide the vacuum chamber into two or more separate compartments. Different vacuum conditions can prevail in each of them. They are frequently evacuated by separate pumping sets. There are a number of reasons for a subdivision of the vacuum chamber, including:

- Two or more coating stations working at different pressure levels.
- Two or more coating stations without any cross-contamination being allowed.
- Generating and holding the more expensive high-vacuum level in the coating area, only.

During the many years of development of web coating plants, models with one, two, three and four chambers have been built for one-layer, one-station coating. Advances in process development, however, have created a situation in which more than two chambers no longer appear to be necessary for general-purpose use. The two-chamber systems, Fig. 11-14, is now the preferred design for the production of evaporated high-quality coatings on plastic films. It is essential for technical as well as economical reasons when coating aluminum on paper.

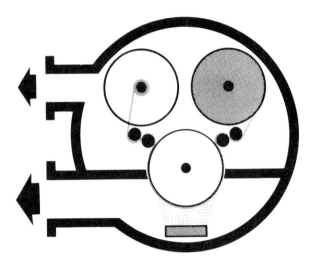

Fig. 11-14. Semi-continuous, two-chamber system.

Vacuum Systems

Today's vacuum technology offers a large variety of different types of pumps for different purposes, different pressure ranges and different pump capacities.

Pump sets have to be combined from pumps of different types, each of them designed for a certain pressure range to pump down in two or three steps from atmospheric pressure which is 1000 mbar to a high-vacuum level which is as low as 10^{-5} mbar. A vacuum control system commands each of the pumps. Such control systems automatically run the full program, including pumpdown, holding and watching the working pressure and finally breaking the vacuum and venting the chamber to end a working cycle.

If a multi-chamber system is employed as described above, one pump set has to be installed for each sub-chamber. Figure 11-15 shows the principle of a two-chamber system pump set as it is applied for aluminum coating by evaporation on film and paper. The coating chamber is equipped with an oil diffusion pump supplying a working pressure in the lower 10^{-4} mbar range. The winding chamber is pumped down and operated more economically in the 10^{-3} mbar and lower 10^{-2} mbar ranges. Pumps suitable for these pressure ranges are oil booster pumps. The forepumps working to atmospheric level are connected to each other to balance the load. Figure 11-16 shows typical pumpdown curves of a two-chamber system.

If the substrate to be coated is emitting a notable amount of water vapor, cold-trap systems are used. The water vapor is pumped off by freezing on the surface of a deep-cooled panel. The most successful pumpdown of water

HIGH-VACUUM ROLL COATING 201

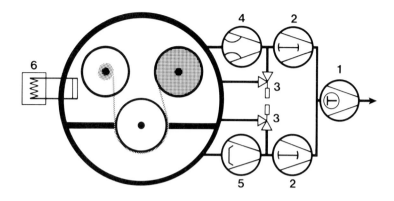

1 Trochoid pump
2 Roots pump
3 Pneumatic angle
4 Oil booster pump
5 Oil diffusion pump
6 Cryogenic trap

Fig. 11-15. Two-chamber system, pumpset principle.

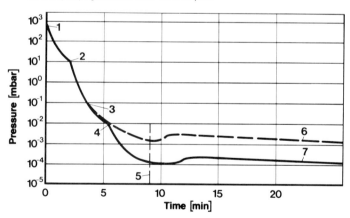

1 Prevacuum pump
2 Mechanical booster pump
3 Oil booster pump
4 Diffusion pump
5 Start of evaporation
6 Winding chamber
7 Evaporation chamber

Fig. 11-16. Two-chamber system, pumpdown diagram.

vapor in coating paper is a closed loop cryogenic system installed in the winding chamber of the roll coater. The temperature required is about -70°C at a pressure in the middle of the 10^{-3} mbar decade in the winding chamber.

In a coating process, demanding a higher level of cleanliness, like sputtering, cryogenic pumps are used successfully in some cases. Cryogenic pumps work at the temperature of liquid helium.

A third type of high-vacuum pump sometimes employed is the turbo-molecular pump, an ultra-high-speed turbine system of a special design. Both the cryogenic pumps and the turbo-molecular pumps, in comparison to oil diffusion pumps, have the advantage of generating an oil-free vacuum. Considering all advantages and disadvantages, the oil diffusion pumps are preferred in vacuum roll coating with aluminum.

Layer Thickness Monitoring

In vacuum roll coating the thickness of the coat has to be measured directly or indirectly. The thickness of vacuum thin film coating ranges between 10 and 100 nm. The amount of material used to deposit an aluminum layer of 50-nm thickness on a substrate is O.15 g/m².

A number of methods are used for measuring the amount of coating material deposited. In the case of aluminum, the more common techniques use the measuring of layer resistivity. After coating, the web passes over two metallic rolls which touch the coated layer. The resistance of the layer just between the two rolls is measured continuously. A well-known relation between thickness and resistivity allows calculation of thickness.

The disadvantage of this method is the inevitable integration over the total web width. The measurement delivers no indication for partial deviations within the full web width.

With other techniques there is a choice between several optical methods of measuring transmission (optical density) or reflectivity over the whole spectrum of visible light or parts of it. Advantages of this method are that no physical contact is required between monitor and film that measurement is differentiated in sections, for instance for each boat of an evaporator system. Disadvantages are low accuracy on high thickness and no signal if the coating is opaque.

For multi-layer systems, more sophisticated optical methods are available with which, under certain conditions, one layer among three can be selected for measuring.

A further method is to change the electrical conditions of a frequency-selective circuit with the layer to be measured. This method requires a minimum layer thickness of about 50 nm.

When all methods mentioned fail, the amount of material emitted from the source is usually measured with a probe as it is practiced in coating processes where single parts are coated.

STANDARD TYPES OF ROLL COATING EQUIPMENT

High-vacuum roll coating of plastic films and papers is a highly specialized field with diversified applications.

Aluminum roll coating equipment can be subdivided into two separate groups:

1. *Decorative plants.* These first-generation plants are predominantly suitable to metallize for aesthetic purposes. This today still constitutes the majority of all vacuum roll coaters in the U.S.
2. *Functional roll coaters.* These are plants suitable to coat functional layers to be applied to products, for their light transparency, oxygen barrier or resistivity. There is an increasing demand in coating capacity to meet the requirements of modern functional products.

High-Vacuum Capacitor Roll Coaters

Capacitor roll coaters hold the most records.

- The thinnest films are metallized: 1.5 μm polyester.
- The highest coating speed is applied: 12 m/sec for coating zinc on paper; 10 m/sec for coating aluminum on film and paper.
- The longest rolls are coated: more than 30,000 m for films in the range from 3–9 μm
- The shortest pumpdown times are practiced: about 4 min down to 4 x 10^{-4} mbar under normal working conditions.

Currently five different series of standard plants are available. Each of these series consists of two or three different types for different maximum web widths between 500 and 800 mm (Fig. 11-17). The five series are:

- Aluminum one-side coating of ultra-thin capacitor films.
- Aluminum one-side coating of regular thin film.
- Aluminum two-side coating of regular capacitor films and papers.
- Zinc one-side coating of regular capacitor films and papers.
- Zinc two-side coating of mainly capacitor paper.

The 650-mm web width type has an annual output of 150 metric tons of 5 μm polyester film which is equivalent to 30 × 10^6 m^2.

Fig. 11-17. Capacitor vacuum roll coater.

Converter Type Vacuum Roll Coaters

These are roll coaters for modern packaging application and related fields where the functional properties of the coatings must be superior. In its standard version this type is for aluminum coating only. The most modern design includes high accuracy, high thickness and high uniformity of thickness as known from capacitor plants. Plastic films of 10 μm polyester and higher thickness and papers as used for cigarette packaging, labels or various packaging applications are accepted.

Because of the wide variety of substrates and gauges it is ensured that the plants' high output can also be utilized for manufacturing a number of different products in smaller volumes. This type of plant is built in different models varying in maximum web width and output.

The preferred converter type in 1982 has an annual output of 4000 metric tons of 40 g/m^2 paper or 1500 metric tons of 12 μm polyester, which is equivalent to 100 $\times$ 10^6 m^2 (Fig. 11-18).

Vacuum Sputter Roll Coaters

Lack of versatility and, as a result, lack of standardization are complicating the introduction of sputter roll coaters to the market.

For a number of sputter roll coaters, built mainly for the production of window film, the basic design shown in Fig. 11-13 was used. Five or seven sputter cathodes are arranged around a chilled coating drum. The coating material—target material—may be the same or different, providing the sput-

HIGH-VACUUM ROLL COATING 205

Fig. 11-18. Converter-type vacuum roll coater.

tering process is non-reactive on all cathodes. A certain versatility is given by the choice of coating material.

The design becomes more complicated if non-reactive and reactive sputtering is to be applied in one plant or in one run. Generally, sub-chambers will need to be used to allow separate pumping and to avoid cross-contamination. Figure 11-19 symbolizes a subdivision into three sputter

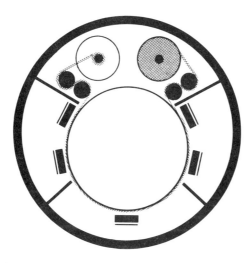

Fig. 11-19. Sputter roll coater subdivision—schematics.

206 WEB PROCESSING AND CONVERTING TECHNOLOGY AND EQUIPMENT

chambers for reactive-non-reactive-reactive coating. Here again the sputter target (that means the coating material) can be changed. This plant also could be operated reactive or non-reactive on all stations. This design is for d.c. magnetron sputtering, d.c. magnetron sputtering and r.f. sputtering were not compatible in one plant until recently.

Sputter coaters of the type described are operated at a web speed between 1 and 10 m/min. The majority of plants are designed for web widths between 1.5 and 2 m. The preferred pumps are diffusion pumps.

Vacuum Roll Coating Lab Plants

A somewhat universal type of lab plant has been developed (Fig. 11-20). It is a semi-continuous type coater with two main coating stations, each of which may be equipped with interchangeable sources. Electron beam guns, resis-

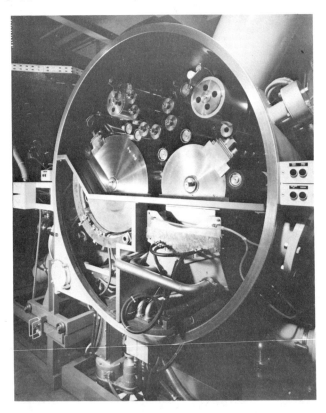

Fig. 11-20. Lab coater—front view.

tance heated boats or sputter cathodes may be used. A three-chamber system allows differential pumping. The winding direction is reversible to enable multilayer coating.

The web width is limited to 300 mm and in some cases has to be smaller. The flexibility is paid for by low web speed and efficiency, which are not the most important features of lab plants. Allowances are made for additional equipment and measurements.

PRODUCT APPLICATION

Coating Paper

Paper was the subject of metallizing trials from the early beginnings of high-vacuum roll coating. Today the problems of this process have been solved. The primary demand for metallized paper is to replace metal foil with paper laminates in products, like cigarette bundle wrapping, labels and confectionary outer wraps.

There are two important points to be considered due to the nature of paper:

1. Most papers have a surface too rough for vacuum coating to produce high brilliancy.
2. Paper has a high residual moisture content which cannot be accepted in a vacuum process.

These problems may be solved by the following:

- Prior to vacuum coating, the paper is pre-coated with a lacquer to smooth the surface.
- The final step in pre-coating the paper is drying it until its water content is acceptable to the vacuum process and to the vacuum equipment.
- The high-vacuum roll coater is equipped with an extra-strong pump set designed to handle the increased pumping load caused by the moisture in the paper.

When pre-coating paper, particular attention must be paid to:

- Selecting a suitable kind of paper.
- Processing the know-how of the proper pretreatment procedure peculiar to aluminum materials.
- Optimizing the correct combination of paper quality, pretreatment materials and process parameters of pretreatment.

The optimizing is especially important for economical reasons. The optimum residual moisture in the paper is less than 4.0%. Deviations are possible, but each situation must be examined individually. There are disadvantages caused by the moisture, such as reduced metal layer quality, reduced machine troughput and higher operating costs. A typical high-vacuum roll coater designed for plastic film will not economically process paper, but a modern vacuum roll coater designed for paper being equipped with highly sensitive winding systems will coat plastic film successfully.

The Transfer Process in High-Vacuum Roll Coaters

In this process, plastic film, like polyester or polypropylene, is coated in the usual way in a high-vacuum roll coater. Through special preparation of the film the coating is made releasable. The transfer from the vacuum coated film to another web material is a non-vacuum process and will be done in a special coating and laminating plant as a second stage (Fig. 11-21).

There are some advantages for this kind of indirect vacuum coating. It may be applied:

- To materials which are unsuitable for coating under vacuum.
- To materials too thick to be coated economically in a high-vacuum roll coater.

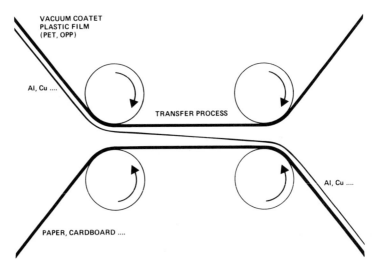

Fig. 11-21. Transfer process—principle.

- If the demand on the web material is higher with regard to vacuum coating than for the final application.
- For extra-high requirements of brilliancy for the metal coat.

The following points are to be taken into consideration when comparing direct with indirect vacuum coating:

- The transfer is an additional process stage done on separate equipment.
- Coating plastic film as used for transfer is cheaper than coating paper directly.

For the transfer process described it is anticipated that the transfer film is used multiple times.

Barrier Coating

Flexible food packaging now represents one of the major markets for metallized films. After years of using the aesthetic properties of metallized films only, the most important step was to improve the functional properties, like gas, moisture and ultraviolet radiation barrier, to a level high enough to be suitable for this application. The thickness of a vacuum coated layer performing barrier functions is less than 3% of the thickness of aluminum foil for packaging application. The main product currently used for high-barrier flexible packaging is a laminate of metallized PET-film on a heat-sealable film, generally LDPE.

Studies conducted in different places show that the barrier quality of metallized film is dependent on:

- The substrate, material and surface structure.
- The vacuum coating, deposition process, thickness and thickness uniformity.
- The protection of the metal layer against mechanical and corrosion damage.
- All production steps to finalize the packaging material, like laminating, protective lacquering, printing, passing through packaging machines and heat sealing.

Polyester is by far the best substrate for building up barriers by vacuum coating with aluminum. The oxygen barrier properties of polyester are superior by a factor of ten or even more to OPP and PE. Oriented polyamide and cellophane are closer to PET, but have disadvantages in production steps following vacuum coating.

Regarding the metal layer itself, the most important points are the microstructure of the layer, its thickness and thickness uniformity as well as the grade of microstructural damages caused mechanically or by heat.

In the future barrier properties may be improved by changing the process or the coating material. Sputtered aluminum layers, for instance, exhibit significantly higher oxygen barriers with same layer thickness (Fig. 11-22), but this process is much more expensive.

Window Films

Window films turn windows into transparent insulators. They act as selective filters. The summer film selectively rejects short as well as long infrared wavelengths, while exhibiting high transmission to visible light. This means that in warm climates window films reduce the transfer of heat from the sun, thus keeping the room cool and reducing air conditioning costs.

The winter film reflects long wave infrared radiation from interior room surfaces and passes the solar spectrum, including visible light. This means that in cold climates window films reduce the loss of heat from inside the building and also permit a substantial amount of solar heat and the majority of the daylight to enter the building.

Window films are produced by vacuum roll coating plastic film, mainly high-quality polyester with high optical properties.

Using different coating materials and several layer construction, window films with different performance characteristics can be produced.

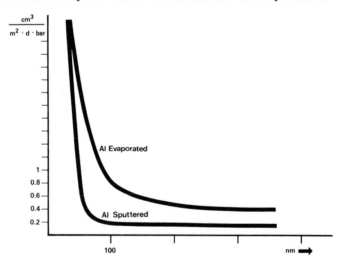

Fig. 11-22. Barrier properties—evaporation versus sputtering. Coating thickness versus oxygen permeability.

- Metals, chiefly Al, Cr, Ni, Ti, stainless steel or alloys of these metals may either be deposited as single layers, or as multi-layer coats.
- Semi-conductive materials, preferably oxides like TiO_2, SnO_2, IO_3 and BiO_3, sometimes with small amounts of nitrides or similar compounds as additives, are deposited as single or multi-layer coats.
- Combinations of both groups are built up for multi-layer constructions, for instance three-layer systems comprising oxide-metal-oxide.

The cheapest, the most developed and best-sold window film is produced by thermal evaporation of aluminum.

Aluminum sputtered films instead of evaporated ones give a higher transparency for a given reflectivity which makes the window film more attractive for some application.

Metals other than aluminum are evaporated in the modern window film technology by electron beam guns or they are sputtered with d.c. magnetron cathodes. The semi-conductive layers may be deposited either by reactive d.c. magnetron sputtering or by r.f. sputtering.

Multi-layer combinations of metallic layers and oxides are deposited in d.c. magnetron sputter technology. In this case, the sputter roll coater is subdivided into zones of reactive and metallic sputtering, as described previously.

Magnetic Tape

Magnetic tape is one of the very high-volume products where conventional coating will be successfully substituted by high-vacuum roll coating. Two important reasons for this are that:

1. There are no pollution problems in vacuum coating, as caused by the large quantities of solvents used in the conventional coating of magnetic tape.
2. There are important advantages in the magnetic properties and consequently the performance of the vacuum coated product, for instance a significantly tighter packing density.

Today the production of high-vacuum coated magnetic tape is only in the very early stages. In lab and pilot plants, magnetic material currently is evaporated by electron beam. Parallel activities are being applied to sputtered magnetic layers. Figure 11-20 shows a lab plant for magnetic tape development.

REFERENCES

1. Wilder, H. *Application of Intermetallic Evaporation Sources.* SVC Conference, April 21-23, 1982.
2. Hartwig, E. K. and Jaran, J. R. *Paper, Film and Foil Converter* **54**(2): 68-72 (1980).
3. Spring, W. C. *Paper, Film and Foil Converter* **54** (2):79-81 (1980).
4. Hodge, M. *Windows for Energy Efficient Buildings* **1**(1) (1979).
5. Hartwig, E. *Coating* **5**:143-146; **7**:207-210 (1981).
6. Levin, B. P. *Windows for Energy Efficient Buildings* **1**(1) (1979).
7. Kelly, R. S. A. *Performance of High Barrier Metallised Laminates—An Update.* Pira Seminar, March 26, 1982.

12
Saturators

David R. Hardt

American Tool and Machine Company
Fitchburg, Massachusetts

The coating methods used for saturating, impregnating or treating of webs involve the simultaneous coating of the web on both sides with an excessive amount of coating, which is then metered by various types of thickness control devices. The saturated product might be a completely saturated web without any internal voids, or it might be a fibrous product which has been only lightly impregnated with a polymer to improve its physical properties.

The coating heads used for this process have three distinct sections which are designed in accordance with the substrate characteristics and type of coating used.

The first section is a pre-wet assembly which applies the coating to one side prior to its overall saturation. On porous webs, this allows the coating to be pushed through the material from one side, driving out the air prior to immersion in the coating bath.

The immersion or dip section is where the material picks up sufficient coating to allow the final metering process to provide the desired thickness or application coating weight.

The metering section provides the proper weight or thickness control, and also the finish characteristics of the coating. The product might be also finished after drying at the end of the saturating line. Calendering is usually employed to impart the desired finish at that point.

PRE-WET SECTION

The pre-wet section usually consists of a large-diameter fountain roller immersed in a coating pan in which a coating level is maintained to allow a suitable pickup of the coating. The fountain roll size is dependent on the substrate, coating used and required line speed, but usually varies in size between 20 and 40 cm. Because of its size, this roll is normally driven. For the pre-wetting application, the coating is applied to the inside of the substrate and the applied web tension forces the coating through the web to eliminate the air between the fibers. To maximize the effect, the web wrap on this roll is approximately 180°, with a narrow dip pan being contained between the exit and entering webs. To increase the coating carry-up, the roll is furnished with a matte finish. This also improves the wettability on the roll and helps to avoid coating skips in the pre-wet process. Smooth finish chrome rolls should not be used as pre-wet rolls because of the wetting problem.

New coating material from the circulation system is usually introduced at the pre-wet unit, which is normally at a higher plane than the main coating fountain, and the excess is allowed to cascade down a spill plate to the main fountain area. The saturation level of the surrounding air keeps the pre-wetted sheet from drying before it is further processed, and the distance from the pre-wet roll to the fluid surface of the fountain allows sufficient time for the pre-wetting action to take place until one side, and most of the internal structure, is thoroughly coated. Since the cascading process tends to increase the solvent evaporation rate, is it important that the pre-wet roll and cascade area be furnished with tightly fitting covers.

The distance from the pre-wet roll to the fountain fluid surface ranges from 30–90 cm, and, in the more versatile coaters, this distance would be adjustable to accommodate a variety of operating conditions. Although roll pre-wet units are the most common, another method used is a die or applicator style design. These systems normally rely on a fixed displacement pump with a variable speed drive to meter a suitable amount of fluid for pre-wetting. These systems have the advantage of being able to use the pressure to force the coating material into the web. With the pumping arrangement used, the pre-wet coating is accurately pre-metered resulting in more uniform pre-wetting. As product quality demands increase, greater use can be expected for this type of pre-wet system.

The higher precision die units are sometimes used to provide total saturation of a porous substrate. Used in conjunction with a suction box mounted above the unit, controlled voids can be opened up in the sheet to provide specialized products such as filter media or coated scrims.

IMMERSION SECTION

The immersion, or dip, section provides contact with the liquid coating for a sufficient amount of time to allow the web to absorb the desired amount of material. The design parameters of this section are mainly dictated by the characteristics of the web being run, and the requirements of the finished products. The variables in this section are the immersion time, and the mechanical and hydraulic action applied to the web. The time will vary with the openness and absorbancy of the sheet, and with the degree of saturation desired. The hydraulic and mechanical action on the web will tend to increase the rate at which saturation occurs and can be effective in lowering the immersion time for a desired end result.

Ideally the web path arrangement through the saturating area should be designed in a way that allows mechanical and hydraulic action to be applied to the same side of the web on which the pre-wet coating is applied. This has the effect of introducing any new coating material from the same side of the sheet and thereby pushing out any air which tends to be entrained in the sheet. If the coating fountain rollers first contact one side and then the other, any residual air bubbles left after pre-wetting will tend to be trapped in the saturated web. This can be a severe problem in webs used for circuit boards and other electrical products. Various saturation techniques are shown in Fig. 12-1 through 12-4.

The wet strength of the substrate or web determines the tension at which the material can be processed and the mechanical and hydraulic forces that can be applied during saturation.

Each idler used requires a certain applied tension to make it turn, and, as the number of non-driven idlers increases, a higher initial tension is required to turn them. If this required tension approaches the wet strength of the material, the user will experience an increasing amount of web breaks. With weak substrates or tension-sensitive materials, it may be necessary to drive

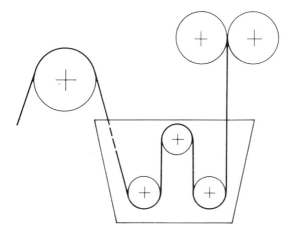

Fig. 12-1. Conventional sawtooth saturation.

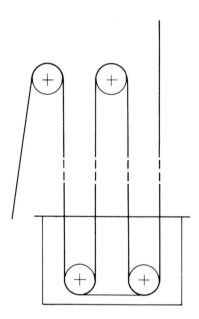

Fig. 12-2. Waterfall saturator arrangement.

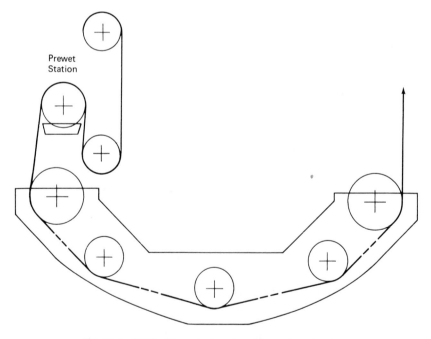

Fig. 12-3. Single side contact enclosed fountain saturation.

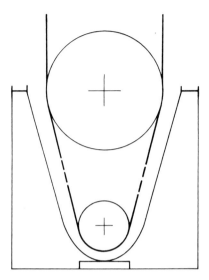

Fig. 12-4. Double roll dip.

the coating fountain rolls to avoid this problem. If the fountain rollers are larger than normal idler sizes, they are generally driven to maintain uniform tension during the coating process. With non-driven idler rolls in the coating fountain, the tension is usually controlled between the pre-wet roll and a dryer head roll using a vertical drying tower, or between the pre-wet roll and a pull roll unit in a horizontal machine. In this tension loop, the saturated web goes from wet pliable material to semi-rigid and elastic. In this web section it is desirable to sense the tension and use a feedback control method to maintain constant processing tension. This is somewhat difficult to do in practice because the material is wet at this point, so contact is difficult.

Additionally, the tension in the loop is affected by the type of metering system used, as there is usually a tension difference on either side of the metering unit. For webs with low wet strength, the fountain tension should be kept low to avoid breaking the sheet in the fountain.

The fountain pan design is important to obtain a good saturating result. A properly designed pan will increase the hydraulic action on the material at the points where it is properly supported by rollers. If the pan is furnished with a curvature slightly larger than that of the roll, but adjacent to it, the turning idler roller will force the liquid through the restrictive gap created, resulting in hydraulic action which will speed up the saturation. The pan design is also important in maintaining thorough circulation of the coating so that the liquid coating maintains its homogeneity. The better pan designs also provide surface skimming features to remove any foam generated either by processing or from the pumping system.

Another necessary feature of the fountain area is the capability to either remove the web from the fountain or the fountain from the web. Mechanically it is easier to remove the web from the fountain, but this disturbs the machine tension which can cause product variations until it is restabilized. Also, some web take-up means are provided to keep the web taut through the line to avoid side guiding or tracking problems. When the fountain rollers are driven, it is generally easier to move the fountain pan. Since most saturating pans hold a good amount of liquid coating, they must be adequately constructed. For this reason they are not as easy to handle as might be expected, and the better systems contain motorized pan raising and lowering equipment. Since some saturating materials tend to be expensive, the pan is designed to require a minimum amount of coating to fill the pan.

The coating pan design should allow fluid inlets at the front and rear of the pan, beyond the material width, so that incoming fluid velocity does not disturb the saturating process. The pan should have a drain at its lowest point which can fully evacuate all of the coating material at the end of a run. In normal operation the coating overflows into the skimming style drains at each side so that surface foam is carried from the fountain area. For large fountains and easily foaming materials, an additional foam skimmer at the center may be necessary.

The coating circulation system usually has an external tank which, in some cases, has a settling chamber to allow any foam to dissipate. Sometimes mixing is also required. In these cases provisions are made for temporarily or permanently mounting a mixer. Predictable pumping rates require some kind of a positive displacement pump, usually a diaphragm or gear style unit. These pumps also have less of a tendency to shear the materials and, therefore, do not tend to induce foam. Since the homogeneity of the saturating material, free of any inclusions, is important, a filter is usually inserted between the pump and the coating fountain. The best designs are plate style filters which are not susceptible to clogging and are easily cleaned. Quick-change hose connections are usually used on the filter and pump to allow fast flushing for cleaning purposes.

METERING SYSTEMS

After the material is saturated to the degree desired, the excess is then metered from the sheet. The metering system used also provides the thickness control and the appropriate finish required. Various metering means are in common use today, and include the following devices:

SATURATORS 219

- Reverse roll metering sets
- Inflatable bars
- Mayer bars
- Air-knives
- Bar scrapers
- Doctor blades
- Squeeze rolls
- Dip and flow units.

Some of these metering systems are shown in Figs. 12-5 through 12-9.

The reverse roll metering sets are the most widely used and generally provide the best control of surface finish and accuracy, but this is also one of the more expensive systems to purchase. The rolls and bearings are usually constructed to the same tolerances that are used for reverse roll coaters. The runout of the roll in the bearings is usually held from 0.0001-0.0002 TIR. Some saturated products require very stringent tolerances on straightness (for computer applications) and taper, as some materials must be held within 1% thickness variation. In these cases, the rolls might have to be specially honed to meet the requirements.

The rolls used for metering are usually mirror finished to a range of 0.1-0.2 µm, which is done by a super-finish process. Because of the accuracy and finish, these rollers are quite expensive to manufacture. To obtain a clean metering surface as the roll is rotated, each metering roll is doctored using a blade which wipes any residual coating from the roll prior to the

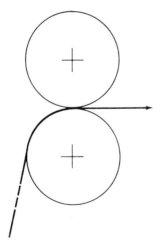

Fig. 12-5. Squeeze nip metering.

220 WEB PROCESSING AND CONVERTING TECHNOLOGY AND EQUIPMENT

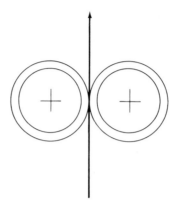

Fig. 12-6. Inflatable membrane metering.

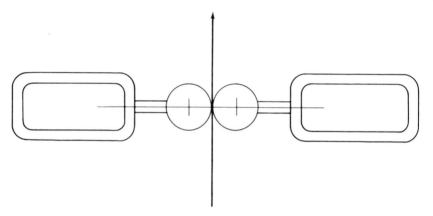

Fig. 12-7. Bar metering.

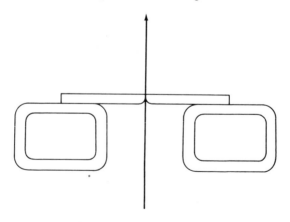

Fig. 12-8. Opposed blade metering.

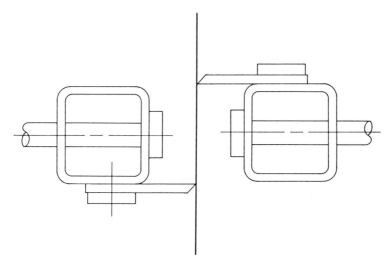

Fig. 12-9. Offset opposed blade metering.

metering nip point. Since this surface is highly finished, standard steel doctor blades can cause wear damage. For that reason, either phosphorous bronze, composite materials or plastics, such as nylon or Teflon, are used for cleaning.

The speed of rotation of the metering nip set is a major operating variable. The finish of the saturant or coating is substantially affected by the settings. The coating weight applied is also changed by this variable, but coating weight can be held constant by varying the gap between the two metering rolls. The other variable which affects the coating is the speed of the web. In metering systems the speed element that is important is the relative speed between the metering nip rollers and the web. For low-speed operation, the metering rolls are run at a low speed in the reverse direction to the web. Low-speed lines generally run from 2—20 m/min for good metering. As line speed increases, the metering roller speed should decrease to zero and then start turning in the web direction, and then as web speed is further increased, the metering roll speed should increase as well. In practice, the coating nip experiences greater coating carry-up as the web speed increases, so, in many cases, it is possible to set a fixed speed for the metering rolls which is used for varying line speeds.

The best finish is usually obtained by operating the reverse roll metering set at 2-3 m/min surface speed in the reverse direction for lines running up to 20 m/min. For these conditions only, reverse running metering rolls are required. For speeds above this, it is common practice to provide a reversing drive so that the metering rolls can also be rotated in the same direction as the infeeding web. The standard design practice for metering roll drives is to provide a surface speed of the roll which is half of maximum line speed. Some customers require that the metering rolls be able to run to full line speed, and in some cases, to 50% over the line speed. Under these design conditions, the metering roll drive is used at its minimum speed where it is most susceptible to cogging, which, if it occurs, can have disastrous results for the coating. At these low speeds it is important to have good motor cooling and a good-quality drive regulator. Tachometer feedback, inertia and friction compensation drive loops also improve performance. The ideal situation is to run the motor at a high speed and obtain the necessary gear reduction through a high-ratio gearbox. If the wide speed range is required, a multi-ratio gearbox can be used. Since metering roll accuracy is the key design criterion to maintain product thickness uniformity, these rollers are usually manufactured from materials which provide rigidity and dimensional stability.

The metering nip is usually fitted with a fixed mechanical stop to keep the two rolls from contacting each other, and thereby damaging the mirror finish of the rolls. If the metering rolls are reground, they should be ground in pairs to the same diameter, and these fixed stops must also be ground to assure the minimum settings.

The gap for the metering roll settings is usually controlled by a pair of precision wedge blocks. These are usually ground in pairs to assure the same taper on each block so that simultaneous adjustment results in a uniform closing or opening of metering roll nip. To obtain this result, it is also necessary that adjusting screws be uniform in manufacture. These screws are usually fitted with backlash eliminators or ballscrews, which have minimum clearances. For servo actuated wedge adjustments, it is even desirable for ballscrews to have backlash elimination, and also pre-loading features. For adjustments on saturators, the equipment should be designed to allow single-side adjustment, from one or either side, so that both wedges can be moved simultaneously. It should also be possible to move only one wedge, with the other remaining fixed, to eliminate any gap variation or to compensate for non-uniform materials.

Since most saturated materials are lapped or sewn together, the additional thickness at the splice requires that the gap be opened to allow the splice to pass. If the gap is opened appreciably, the coating for that section will be excessive. The excess coating can drip back on the coater station, causing a mess which must be cleaned before the coater will function properly. This problem is usually handled by equipping the metering nip air loading system with a three-way, three position valve. In the center operating position, the loading pressure is relieved, but the roll is not moved. When the splice goes through the metering nip, it then pushes the two rolls apart, only enough to allow the splice to pass. After the splice goes through, the pressure is immediately reapplied to prevent any excess coating buildup. In this way the web is fully coated and no excess is allowed to accumulate. The valve used for this system can be a manual type which the operator sets, or a solenoid type, operated by a splice detector.

13
Laminating

Lee A. Mushel
Faustel, Inc.
Butler, Wisconsin

Films, papers, glassines, metal foils and nearly every material which can be manufactured in roll or sheet form are combined in the process we call laminating. The resulting multiple laminates have physical and chemical properties that would be unobtainable from any single component. Moisture vapor and gas transmission rate, abrasion resistance, stiffness, gloss and tensile strength are only a few of the properties which can be manipulated through the proper choice of material and adhesive system. Flexible packaging, health care, electrical and energy conservation industries have long made use of the laminating process.

There are three laminating techniques:

- Wet bond or combining before solvent removal or curing.
- Dry bond or combining after solvent removal.
- Thermal or combining by heat and pressure only.

All three share the basic mechanical configuration shown in Fig. 13-1. A steel backing roller is mounted on fixed position bearings in suitable side frames. A movable elastomer covered pressure roller is brought into contact with the backing roller. The nip formed brings the webs into intimate contact with each other. The controllable variables of the process are pressure, temperature and time (dwell). Table 13-1 summarizes the process conditions for the three types of laminating.

The steel cylinder will be constructed with provisions for heating if intended for dry bond or thermal laminating. Uniform temperature across the entire working face of this roll is important and its inner construction will re-

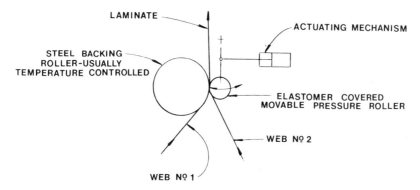

Fig. 13-1. Elements of a typical film roll-to-roll laminator.

flect this need. While electrical resistance heating is sometimes used, steam, hot water or oil is more common. The heat transfer medium is pumped through rotary unions to the drum. Internal construction should provide positive rapid flow over the entire inner surface for maximum efficiency. A single-wall hollow cylinder is not suitable for most applications. As the roll speed increases, stratification of the transfer medium is difficult to avoid; even with baffles and siphons a 10°C variation across one meter of width is not unusual.

Double-wall construction utilizes two concentric cylinders with the medium circulating between them. Higher and more uniform fluid movement improves heat transfer. The addition of spiral wires or baffles is a further improvement. These baffles also provide support between the inner and outer shells, which reduces deflection. Mechanical deflection or bowing, either because of roller weight or as a result of roller to roller pressure, will alter the elastomer footprint. Footprint is a common term used to describe the actual pressure roll contact area. As the elastomer is increasingly deformed by pressure, this area will increase in width. Simultaneously, deflection will produce a narrowing toward the center. This effect is illustrated in

Table 13-1. Laminating Process Conditions

	WET BOND	DRY BOND	THERMAL
Temperature	Low	From ambient to distortion of web	70°C to distortion of web
Pressure	Low (3 kg/cm or less) to medium (20 kg/cm)	9–45 kg/cm	20–60 kg/cm

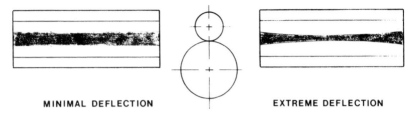

Fig. 13-2. Effect of deflection on pressure roll contact area.

Fig. 13-2. Minor contact distortion of this sort can be tolerated by most webs with at least moderately low coefficients of friction. But light-gauge, high-friction webs are intolerant of the instantaneous velocity variations across the elastomer surface. The result is web wrinkling and stress patterns. Three methods of correcting this problem exist:

1. The rollers are designed for maximum stiffness using solid forgings or thick wall tubing. Several patented and highly effective journal support techniques also exist.
2. The rubber roll may be ground with a raised or crowned center (see Fig. 13-3).
3. Provisions are made for slightly skewing the pressure roll so that its axis is not parallel to the heated steel roll.

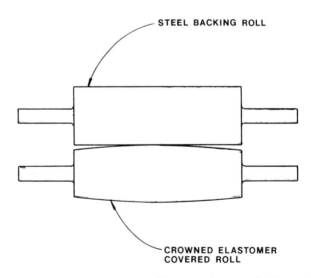

Fig. 13-3. An elastomer covered pressure roll is crowned to control effects of deflection.

Contact area may be measured by placing carbon paper-tissue strips in the nip and momentarily bringing the system to desired pressure. If this is done several times across the width of the laminator a clear picture of deflection, if any, will appear. Special pressure indicating papers are also available for this purpose. Unless the laminator has been designed to high-pressure low-deflection standards, the contact width in the center could well be only half that found at the edges. For most applications this will not be troublesome.

A variety of elastomer coverings are used on the pressure roll. Neoprene or Buna N are entirely satisfactory for most wet bond applications where high pressures and temperatures are not encountered. EPDM* has excellent toughness and can be used in higher temperature situations.

Any surface roughness or contamination on either roll will be reflected in the web. The elastomer is ground during manufacturing using an independently driven wheel mounted on a lathe tool holder. This is followed by a polishing operation using a fine abrasive paper. A Class II[1] finish is adequate for most work in converting although a Class I may be required for certain critical applications. No numerical specifications exist for surface finish. Suitability, in the last analysis, will depend upon agreement between the supplier and the converter.

Concentricity, as measured with a dial indicator, should be within 0.05 mm total indicated runout (TIR). This standard is more difficult to achieve with softer rolls (less than 50 Shore A) but since most laminating is done with harder rubbers (75 Shore A and higher) it should not be compromised.

Several rubber hardness measuring devices exist. These include those by Pusey & Jones, Shore Instrument & Manufacturing, and the International Rubber Hardness Degree Instrument. There is no industry consensus for the ideal laminating rubber hardness. Unless experience or preference indicates otherwise, 80 ± 5 Shore A will constitute a satisfactory specification. A softer rubber will give a wider footprint for a given pressure resulting in longer exposure or dwell at the process temperature. But for thick webs with high specific heats and fast line speeds this will be insignificant. If we assume a speed of 200 m/min with a contact width of 2 cm the dwell is only 0.006 *seconds*. This is not much time in which to raise adhesive interface temperature or accomplish plastic flow. Nor is a harder pressure roll a guarantee of greater optical clarity.[3] As hardness is increased, perhaps to 100 Shore A or more, the adjustment of the laminator becomes more difficult. Minor defects in the pressure roll become apparent and objectionable. It is not unusual to find local irregular spots or areas across the face of a roll that

---

*Terpolymer of ethylene, propylene, and diene side chain. See ASTM D1418–79 giving the nomenclature for various elastomers.

deviate from true concentricity. A local variation in hardness will alter the effects of the grinding and polishing operation frequently resulting in a low spot. Such a defect is quite difficult to detect and will all too frequently go undetected during casual inspection. A very hard roller, 100 Shore A or above, will have little tolerance of this and the spot will appear as nonuniform contact reflected in the laminate as haze or tiny included air bubbles. Such a defect is easily identified since it will repeat and the distance between identical points will equal the circumference of the pressure roll.

Mechanical deflection tends to limit combining pressures. Dry bond and thermal laminating are commonly done at pressures as high as 230 kg/cm of width. A very hard roll with its inherently narrow contact width will display all variations in concentricity that happen to be present.

Accidental contamination of elastomer covers is inevitable. Adhesive from wet or dry bond laminating will foul the roller and unless promptly and completely removed with a suitable solvent or cleaner, it will cure, making a grind and polish operation necessary.

Nearly all common covering elastomers are affected appreciably by the common organic solvents used during adhesive cleanup. Since exposure during cleaning is limited, do not hesitate to use them. Solvent damage will be of little consequence when compared to that done by cured adhesive.

Eventually the point will be reached where recovering is needed. Always refer to an original manufacturer's print or specification when choosing the replacement. Rubbers tend to harden with age and a minimum thickness must be maintained for accurate gauging. After several grinds, the covering might be too thin to permit measurement. A thick covering makes possible the maximum number of grinding operations between recoverings. However, the thickness must be limited to prevent excessive heat build-up due to hysteresis, which would needlessly shorten roller life.

WEB PATH CONTROL

Web path control idlers are an important part of any machine laminating section. They are used to control pre- and post-nip contact. Pre-nip contact or dwell on the hot roll is for either adhesive/sealant activation or for temperature conditioning to avoid the effects of thermal stress on the laminate in the nip. Post-nip control will affect curl and undesirable heat-induced distortion. Adjustment should not exceed the limits imposed by the general principles of web tractioning.

It is often helpful to include some type of spreading or smoothing roll immediately before the combining point. Some indication of web tension before combining is also mandatory for reliable curl control. Curl, the tendency for a web or laminate to spontaneously coil itself when not under tension, is nearly always unwanted. It can result from unresolved tension imbalance between webs. This does not mean that web tensions should be equal before combining. But the correct tension values, once obtained experimentally, will remain constant. Curl may also be induced by a coating or adhesive. This could be the result of either coating/adhesive application that has a directed application pattern, or because the coating/adhesive shrinks upon drying, or because it exhibits a markedly different coefficient of thermal expansion and moisture sensitivity from that of the substrate. These properties tend to produce a diagonal curl which is especially detrimental in applications requiring sheeting.

It is also helpful to include some web aligning or guiding capability immediately before combining. The edge of one web is sensed immediately before the nip and this signal is used to align the secondary web. In wet or dry bonding, the substrate receiving adhesive is referred to as the primary web, and the web being combined is called secondary.

Blocking or wrap-to-wrap adhesion will occur during dry bond laminating if a strip of adhesive is exposed by web misalignment. For this reason it may become necessary to make the secondary web slightly wider than the primary. Small sideways movement will then be unimportant.

WET BOND LAMINATING

If an adhesive is applied to a substrate and immediately combined with a second web or sheet, we refer to the process as wet laminating. The name is highly appropriate since the description dates from the time when most adhesives were water-based modified caseins, hide glues and silicates. After laminating, the web passed through a drying system where the water was evaporated. This definition has now been expanded to include any adhesive product where combining occurs before solvent removal or curing. One hundred percent solids radiation curable adhesives as well as one- and two-component moisture and self-initiated types are included. If a solvent based adhesive is used, one of the webs must be permeable to the solvent so that removal in the drier is not impeded. When planning sequences of converting operations, an all too common error is lacquering the permeable web before laminating. Of course, 100% solids materials can be assembled in any order desired. If the permeable web is absorbant and light adhesive application is used, forced drying may not even be necessary. The moisture present will be absorbed without adverse effect. This is by no means a universally applicable practice. An incompletely dried foil laminate, for example, will seriously corrode in a short time.

Since most adhesives have relatively low viscosity in order to facilitate application to the substrate the combining nip conditions must:

1. Avoid forcing the adhesives through a permeable or porous web (strike-through).
2. Avoid squeezing the fluid from between the plies, thus contaminating the laminator and web backside.

This is done with minimal nip pressures, proper choice of substrate and careful adhesive selection.

The laminator is located close to the adhesive coater. This is done to avoid reducing the efficiency of the adhesive through premature solvent loss, for improved web control and because a drier or curing device is often located very near the coater giving very little room for the wet bonder.

Unless the adhesive is an actively functional part of the laminant such as a fire retardant, wet bond laminating is done with low-adhesive coat weights. The typical range is 0.75–2.0 g/m^2. Under no circumstances can a bead of fluid adhesive be tolerated in the combining nip. If close examination shows that one exists the coater must be adjusted to supply less adhesive. A wet nip represents an uncontrolled situation which will inevitably lead to edge bleed even though short-term appearance considerations may make such an operation seem desirable.

WET BOND PROBLEM ANALYSIS

Strike-Through

Adhesive may be applied to an impermeable substrate (treated polyethylene in the case of a nonwoven/adhesive/polyethylene laminate), but the combining pressure of the nip will drive it through the permeable web. Contamination of the laminator rollers and following support idlers may foul within seconds or it may be a gradual situation with adhesive accumulating over a longer period. It may even manifest itself only by the appearance on rollers of small clumps of fibers from a paper or tissue. Four areas should be explored in order to correct the problem.

1. Reduce the amount of adhesive.
2. Improve the holdout properties of the porous web.
3. Reduce combining pressure.
4. Alter the adhesive by increasing viscosity, solids or both.

Blistering or Delamination

These can be large or small areas of delamination which occur immediately after combining. Large irregular patches which occur principally on one side of the laminate are good indicators of inadequate combining pressure or insufficient adhesive across the entire face. Excessive differential tensions between the plies will also cause this. A tendency to wrinkle or pull in local areas is also a good indicator on non-uniform pressures. Many small areas of delamination are more likely to be caused by a general failure of the adhesive. Either it is not adequate or perhaps the nip isn't completely closed.

Edge Contamination

This is certainly due to the presence of a bead or wet nip flooding past the web edges. This may occur at slow speeds and disappear entirely when operating faster. The quantity of adhesive or nip pressure must be reduced.

DRY BOND LAMINATING

In dry bond laminating, adhesive is applied to a substrate, the solvent is removed in a drier and the exposed adhesive brought into contact with the receptive surface of the secondary web in a combining nip (Fig. 13-4). In contrast to wet bond laminating, most dry bond combining is done at elevated pressures and temperatures. Adhesive coat weights may be very low (0.6 g/m^2) to very high (15 g/m^2 or more.) After solvent removal the viscosities are so high, at least 50 Pa·s, that flow from between the plies is virtually nonexistent. In fact, heat and pressure are used to intentionally level any residual application irregularities that might degrade laminate clarity. But this is not always successful because thermal web distortion may occur before any appreciable flow takes place.

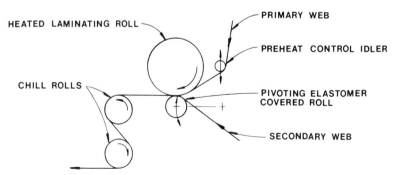

Fig. 13-4. A dry bond adhesive laminator with adjustable web contact provisions.

DRY BOND PROBLEM ANALYSIS

Inadequate Bond Strength

Since few dry bond adhesives require high pressures and temperatures for normal performance it's actually quite unusual for bond strength problems to be primarily caused by the dry bond laminating section. Bonding difficulties originate elsewhere. However, low immediate or green bond strength can definitely cause or accentuate a variety of other problems as indicated in the following paragraphs. Common sources of inadequate bond immediately off the laminator include:

1. Adhesive associated problems including diluent contamination, improper compounding and decomposed or defective adhesive.
2. Application related problems such as unequal coat weight.
3. Substrate related sources such as poor or absent corona discharge treatment and migrating jaw release or slip additives.

Lack of clarity is a common complaint when dealing with transparent laminates. These problems may be grouped by specific defect.

Haze

With the naked eye there is no evidence of discrete bubbles or delamination. The haze may occur in repeating, arcuate or feathered patterns. The cause is most often a defective rubber pressure roll although the effects are more pronounced under conditions of high pressure and temperature. A web path that gives minimum contact with the heated cylinder is often best.

Haze (Minute Bubbles)

Careful examination under magnification shows the tiny delaminated areas. If the pattern is related to the cell count of a knurled or etched application cylinder it is probably safe to conclude that the problem is related to insufficient flow or leveling of the adhesive. A more random pattern could be indicative of inadequate bonding because of surface preparation. Hot jaw release additives and slip agents are frequent offenders.

Delaminating (Irregularly Spaced Bubbles)

Examination under a microscope will reveal an included dirt particle near the center of each bubble. The sources include air contamination of convection driers or environmental sources aggravated by static electricity. Air filters and static eliminators will control the problem. Premature contact of primary and secondary webs before the actual combining nip can also produce a random bubble pattern.

Delaminating (Ridges Across the Web)

Ridges extending inward from the web edges are almost certainly caused by improper web tensions at the moment of combining. An edge view examination will show that one ply is raised and the other is flat, indicating that the flat web was under excessive tension. If inspection fails to show that the ridges are actually an open tube, the situation should be treated as one of inadequate bond.

Delaminating can often be controlled by careful rewind tensions. A tightly wound roll will help. Extremely improper tension patterns will result in delamination in the machine direction as well as across the web.

THERMAL LAMINATING

Thermal laminating is more complex than either dry or wet bonding. The high temperatures and pressures normally encountered produce extreme web stresses which make outstanding controls for all variable factors absolutely mandatory. Small changes in pressure and temperature will affect bonding dramatically. Tiny tension variations produce serious wrinkles. Thermal laminating requires a heat activatable element which can serve as the adhesive. This adhesive component may take three forms:

1. A coating preapplied to either or both webs to be combined.
2. A coating applied in-line on the machine which is heat activatable as contrasted to most dry bond adhesives.
3. A thermoplastic web such as an ethylene-vinyl acetate modified low-density polyethylene. Adhesives of several types are available in web form.

Thermal laminating done with a single nip (Fig. 13-5) will be a slow process with top speed not much above 70 m/min. Making the heated roll very large will increase the top speed in some cases, since this will permit web contact (preheat) before the combining nip. Such an approach presents some practical problems. If the machine is stopped for any reason, plastic webs will melt or adhere to the hot roll. The machine must be rethreaded before starting.

The situation is not as serious if the roll is fluorocarbon coated (as all hot rolls for thermal laminating should be), but automatic web retractors are

234 WEB PROCESSING AND CONVERTING TECHNOLOGY AND EQUIPMENT

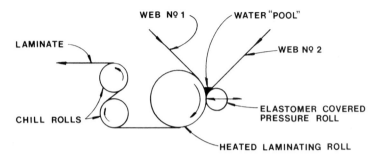

Fig. 13-5. Thermal or dry bond laminator with moisture assist in the form of a nip held pool.

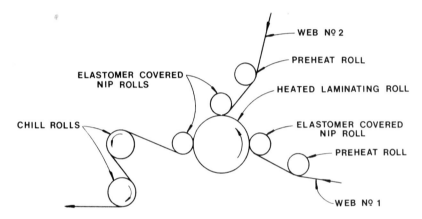

Fig. 13-6. Single hot roll thermal laminator with multiple combining points.

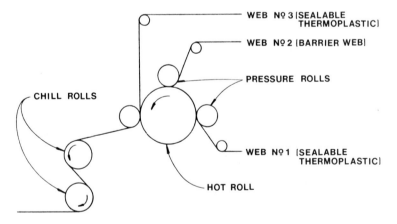

Fig. 13-7. Tri-plex thermal laminator without provisions for web preheating.

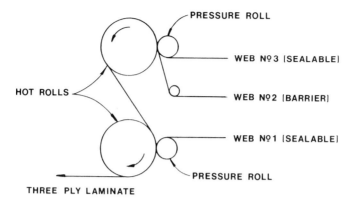

Fig. 13-8. Dual roll tri-plex laminator without web preheat provisions.

almost mandatory. Simply raising the hot roll temperature will cause thermally induced wrinkles, since most plastic webs have very large coefficients of thermal expansion. This effect can be eliminated through the use of preheat rolls at some elevated intermediate temperature to "precondition" the web. The system in Fig. 13-6 shows a single large-diameter hot roll with preheat capability.

It is not necessary to have an adhesive web buried between two others. A three-ply structure with thermoplastic webs on either side of a core of high strength or barrier material as shown in Fig. 13-7 is perfectly practical. Two smaller hot rolls (lower cost, more easily fabricated with equal accuracy) will also permit higher speed (Fig. 13-8). Corona discharge treatment is required as an adhesion promoter and its use under certain conditions is covered by U.S. patents.

A thermal laminator must have easy operator access for threading and clearing the inevitably accidental wraps around the backing and pressure rolls. It is especially important to minimize the possibility of accidental operator contact with the hot components. At the same time the structural rigidity of the mechanical components must be maintained. Because of the potential operator hazard, all nips should be equipped with safety sensors. These photoelectric devices will detect the presence of a carelessly placed hand or article of clothing and automatically open the nip and stop the machine.

THERMAL LAMINATING PROBLEM ANALYSIS
Inadequate Bond Strength

Untreated or poorly treated surfaces, bleed of film additives, contamination from contact with silicones or similar inhibitors and low laminating temperatures are the most common causes of unexpectedly low bonds. The cause of failure should be investigated carefully before beginning corrective action. For example, does the failure occur at the interface between the plies of the laminate, rather than between a film substrate and a previously applied coating. The latter would imply a defect in or inadequate coating anchorage while the former would more strongly implicate the laminating operation.

Haze

Haze is usually caused by incomplete bonding. Microscopic random points remain unfused. Although the maximum bond is not reached under these conditions, the product may be entirely serviceable. If clarity must be improved, review of the suitability of a water or steam nip. Increased pressure or temperature may help. Investigate the pressure roll contact width while increasing actuation pressure. If appreciable roll deflection occurs, it is entirely possible that a decrease in actuation pressure will result in more efficient pressure roll action. If harder elastomer coverings are considered, deflection will be especially critical. Temperature increases may result in web distortion and curl problems. Improper storage of hygroscopic webs and retained solvent from coating or printing operations may also cause haze.

Film Adhesion to Pressure Roll

Polyethylenes, unoriented polypropylenes and their co-polymers exhibit fairly good adhesion to rubber pressure rolls. Most thermal laminating applications involve one web that is essentially unaffected by the process temperature. If this web is made slightly wider than the more plastic or adhesively active web, adhesion will take place preferentially to it rather than the pressure roll. An overlapped covering of fluorocarbon pressure-sensitive tape around the pressure roll at the very edge of the web will also minimize the downtime otherwise required for cleaning or roller grinding. Polymer coated films tend to adhere to the pressure roll in the contact area if the nip remains closed during machine stop periods. While removing the film is not difficult the coating will be very troublesome. Furthermore, it cannot be ignored because every revolution of the pressure roll will leave a permanent mark in the laminate. Automatic nip opening is necessary.

ANCILLARY EQUIPMENT

Moisture Assist

Experience has shown that moisture added to polymer coated webs immediately before combining will dramatically aid clarity. This can be done with a flooded nip as shown in Fig. 13-5, or with a steam source as indicated in Fig. 13-9. Steam is far less troublesome from a practical standpoint although a clean steam source will probably make such a system more expensive. The flooded nip approach, while inexpensive and relatively easy to test, has definite practical problems. A dam system must be fabricated to contain the water to form the pool. The water will be released every time the nip is opened. While the volume of the pool will be small, this can cause waste, if moisture sensitive materials are used. Liquid level control is another important aspect. Manual supervision would detract from operator attention to more important detail. A simple overflow orifice in one end dam is one possibility and definitely preferable to elaborate electronic control. Moisture addition is also used as a cure accelerator for single-component 100% solids urethane adhesives.

Heating Systems

Direct electrical resistance heating is used for smaller laminating rollers. But it is difficult to design a large hot roll, heated this way, with acceptably uniform face temperatures. A fluid medium (oil, water or steam) is easily controlled and existing designs give very uniform surface temperatures over wide widths. Compact heaters, complete with temperature controllers and pumping system, are available which require only simple plumbing to rotary unions on the hot roll to complete the installation.

Pressurized water systems have the advantage of low-cost heaters and inexpensive transfer medium. The disadvantage is limited maximum tempera-

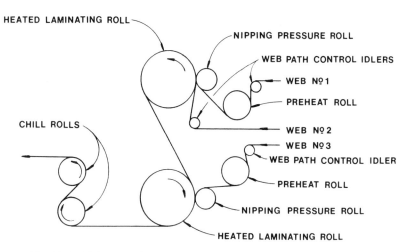

Fig. 13-9. High speed tri-plex dual roll thermal laminator with preheat rolls.

ture. Water-heated systems are used for operating temperatures below 120°C. Steam systems react quickly to desired temperature changes but will require licensed operating personnel for higher temperatures.

Oil type transfer media are quite expensive. Oil systems are slow to respond to temperature changes but do offer the advantage of high temperatures without the need for highly trained personnel.

Web Chilling

Webs exiting thermal laminators and most dry bond laminators are at highly elevated temperatures and must be cooled before rewinding. A coated,

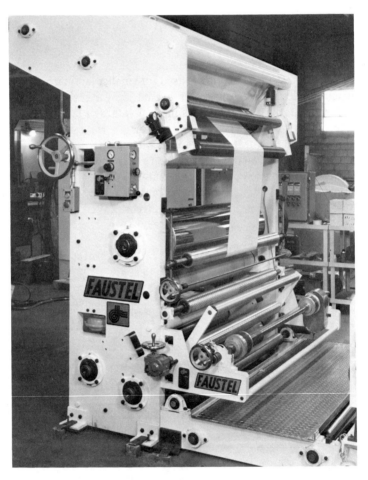

Fig. 13-10. Dry bond laminator with metal foil unwind.

LAMINATING 239

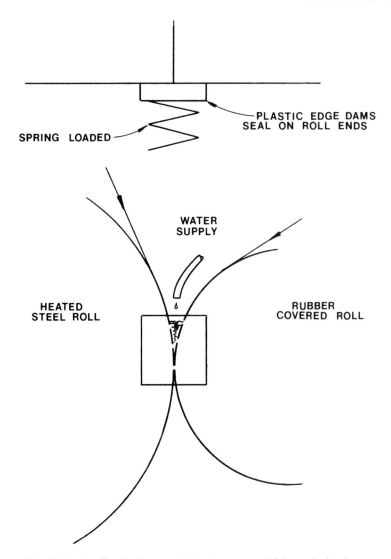

Fig. 13-11. Details of moisture assit showing water addition and edge dams.

tightly wound, very warm roll will otherwise tend to block. It is not necessary to chill far below ambient. In fact, if catalyzed adhesives are involved, it may actually be desirable to rewind the laminate slightly warm to accelerate cure. Only very specialized applications require cooling below 25°C.

An extremely important factor in thermal laminating is control of web tension variations induced by thermal expansion and contraction of the

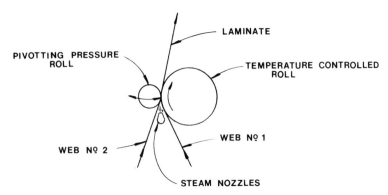

Fig. 13-12. Dry bond or thermal laminator with moisture assist from steam nozzle or a steam chest.

backing and chill rollers. In the interest of drive system economy it is common practice to have a fixed speed ratio between these elements. The tension changes between these points can be appreciable, but still unimportant, if highly thermoplastic materials only are involved. However, a metal foil or other nonextensible web will exhibit relatively short randomly located machine direction wrinkles. A new machine intended for high-temperature operation should include provisions for compensation. If wrinkles are observed on an existing machine and tension is a suspected cause, simply bypass the chill or other potentially offending component temporarily. Addition of a mechanical speed variator is not difficult and will permit operation over any process temperature range that does not include web distortion. Automatic temperature control of cooling rolls is desirable for several reasons:

1. Excessive chilling is a needless waste of energy and may actually be detrimental to adhesive cure.
2. Cooling far below the dew point will produce condensation which accelerates rusting of unprotected roll journals.
3. Accumulated condensate may create waste product if moisture-sensitive webs are involved.

REFERENCES

1. *Roll Covering Handbook*. Rubber Manufacturers Association, Inc., 1980.
2. Allen, Lindley and Payne (Eds.). *Use of Rubber in Engineering*. London: Maclaren & Sons, 1967.
3. Middleman. *The Flow of High Polymers*. New York: Interscience, 1968.
4. Ritchie, P. D. *Physics of Plastics*. New York: D. Van Nostrand, 1965.

14
Surface Treatment

Wayne M. Collins*

Enercon Industries Corporation
Menomonee Falls, Wisconsin

Plastics in general have chemically inert and non-porous surfaces with low surface tensions that cause them to be non-receptive to bonding with printing inks, coatings and adhesives. Polyethylene and polypropylene have one of the lowest surface tensions of the various plastics and are the two materials most often subjected to surface treatment to improve their bonding characteristics. Surface treatment, however, is not limited to these two materials and can be used to improve the bonding ability of virtually all plastic materials, as well as some non-plastic materials, such as foil and paper. The three methods by which surface treatment are accomplished are (1) corona discharge; (2) etching by means of acid or plasma; and (3) flame treatment. Flame and etching are used entirely for molded or blow molded parts and are of no direct interest to the converter. Flame treatment of film and sheet has been completely supplanted by corona discharge surface treaters. This chapter will deal with techniques and equipment employed in surface treatment by corona discharge.

The object of corona treating is to improve the wettability of the surface, therefore improving its ability to bond to solvents and adhesives. In order for a surface to be properly wet by a liquid, the surface tension of the plastic must be higher than the surface tension of the liquid.

Ideally the surface tension of the plastic should be 10 dynes/cm² higher than the surface tension of the solvent or liquid. For example, a printing ink having a surface tension of 30 dynes/cm² would not adequately bond to a material having a surface tension less than 40 dynes/cm². A method has been

---
*Presently with Pillar Corporation, Milwaukee, Wisconsin.

Table 14-1. Concentration of Formamide-Cellosolve Mixtures Used in Measuring Wetting Tension

Formamide (volume %)	Cellosolve, (volume %)	Wetting Tension (dynes/cm^2)
0	100.0	30
2.5	97.5	31
10.5	89.5	32
19.0	81.0	33
26.5	73.5	34
35.0	65.0	35
42.5	57.5	36
48.5	51.5	37
59.0	41.0	39
63.5	36.5	40
67.5	32.5	41
71.5	28.5	42
74.7	25.3	43
78.0	22.0	44
80.3	19.7	45
83.0	17.0	46
87.0	13.0	48
90.7	9.3	50
93.7	6.3	52
96.5	3.5	54
99.0	1.0	56

devised for measuring surface tension known as the "wetting tension test" (ASTM D2578). With this test, a series of mixed liquids with gradually increasing surface tensions are applied to a treated plastic surface until one is found that just wets the surface. The surface tension of the plastic is approximately equal to the surface tension of that particular mixture. Test solutions are available from various manufacturers of corona treating equipment. Table 14-1 gives the ratio of formamide and Cellosolve for various surface tensions.

The wetting tension test method is by far the most prevalent measurement used to determine treatment level. Several other methods are in limited use, however, and the following chart briefly describes these methods.

METHODS OF CORONA GENERATION

Figure 14-1 depicts a basic corona treating arrangement. Essentially all corona treating installations conform to this basic scheme. Low voltage 60-Hertz electrical power is fed into an electrical device that raises the frequency to some higher level. This high-frequency electrical power is applied

Table 14-2. Other Treat Level Measurements

TEST METHOD	PROCEDURE	MEASUREMENT
1. Water spreading test	A specified volume of distilled water is dropped on the material before and after corona treating.	The area covered by the water is measured and compared to determine the treat level.
2. Contact angle test	A drop of distilled water is placed on the web.	The relative contact angle between the droplet and the film surface is measured and compared.
3. Tilting platform test	A drop of distilled water is placed on the material which is secured in a horizontal plane. The plane is gradually tilted.	The angle at which the water begins to move down the plane is measured and compared.
4. Dye stain test	Material specimens are dipped into a special dye and dried in a vertical position.	The dye stains a treated surface but not an untreated surface.
5. Adhesion ratio test (based on ASTM tentative D2141-63R)	Pressure-sensitive tape is applied with equal force to treated and untreated material.	The degree of force required to peel the tape from the material is measured and compared.
6. Ink retention test	Ink is spread on the surface of the material, allowed to dry and covered with pressure-sensitive tape.	The relative area of ink that is lifted from the treated surface when the tape is removed is measured and compared.

to a step-up transformer that increases the voltage to some higher level. The high-voltage, high-frequency electricity is discharged from an electrode through the web being treated to a metallic ground roll.

Although the basic principles remain the same, many improvements have been made over the past 20 years which greatly increase the capability and performance of these systems. The earliest power supplies were motor generator types that proved unreliable for long-term continuous operation due to mechanical breakdowns. These were replaced by power supplies using a tesla coil and spark gap to generate the high-frequency and high-voltage electrical power. This was an improvement over the motor generator systems, but was not reliable due to the erosion of spark gaps. Next came solid state power supplies using transistors as the output device. Transistors have

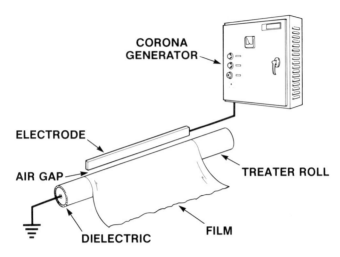

Fig. 14-1. The conventional corona treating system.

a limited output power capacity, however, and these units required as many as 16 transistors connected in parallel to achieve the required power levels. Although transistors are generally very reliable solid state devices, the larger number of components required caused a proportional increase in the chance of a random failure. The most recent step in the evolution of power supplies has been the development of an inverter using silicon controlled rectifiers (SCR) as the output device. The SCR type inverters have been in widespread use for the past 15 years and have proven themselves to be reliable in every respect.

Power supply reliability problems have historically been insignificant when compared with failures in the mechanical portion of the treating system, specifically the roll covering.

An electrode and roll arrangement can be drawn schematically as a large capacitor (Fig. 14-2). The electrode represents one plate of the capacitor, the grounded roll represents the other plate of the capacitor, and a combination of the web, roll covering and air make up the dielectric. The air in the gap has a relatively high dielectric strength and effectively insulates the electrode from the roll until voltage and frequency levels reach a point high enough to cause ionization (corona is generated). At the point of ionization, the resistance of the air in the gap is reduced drastically due to a greater number of free electrons. The wattage required to ionize this air determines the size of the treater power supply. The greater the ionization (or corona), the higher the level of treatment applied to the film.

Since air loses its capacity to serve as an insulator when ionized, it is necessary to introduce another dielectric such as a roll covering to prevent a

SURFACE TREATMENT 245

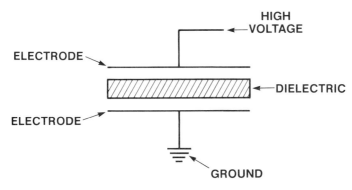

Fig. 14-2. Capacitor.

direct arc from the electrode to the roll. A direct arc is nearly equal to a short circuit and would cause an overload which trips the power supply off before corona can be established. Unfortunately, a dielectric roll covering material has never been found that can withstand exposure to heat and corona over a long period of time. Materials such as polyester, chlorinated polyethylene, ceramic, silicone rubber, epoxy and even glass have been tried and met with varying degrees of success. In spite of many years of experimentation with different roll covering materials, the problem of burn-out has been only slightly alleviated and periodic roll covering replacement continues to be a necessity. The search for a truly reliable roll covering material is still going on, but without much promise of success.

Fortunately for converters, a couple of important breakthroughs were made that have eliminated this troublesome source of down time. The first discovery was that a properly designed power supply could generate corona at a much lower voltage level. A lower voltage level means that a much thinner dielectric can be utilized and in many cases the film itself can serve as the dielectric and no roll covering is necessary. The limitation to this method is the fact that a minimum film thickness of approximately 0.05 mm for low-density polyethylene and 0.025 mm for polyester or polypropylene is required. This system fits well on many blown film applications where a layflat tube is being treated, but still leaves a large percentage of corona treating applications with troublesome dielectric covered rolls.

The remedy for these applications was found by a German company which began using quartz as the dielectric and placed the dielectric on the electrode rather than on the roll (Fig. 14-3). The electrode is in the form of a quartz tube filled with conductive material. Quartz has the advantage that it is impervious to damage by heat or corona and can operate continuously for an infinite length of time without replacement. Placing the dielectric on the electrode, rather than the roll, allows the quartz to be easily cooled by the

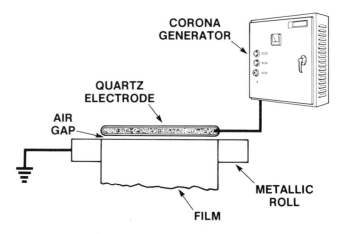

Fig. 14-3. Quartz electrode system.

same air that normally removes ozone from the system and permits quick replacement in case of failure.

An incidental, but important, advantage is that the system can treat conductive as well as non-conductive webs with equal ease. The quartz electrode and dielectricless treating systems have been in use for several years and have proven themselves to be extremely reliable.

CORONA TREATING APPLICATIONS

The majority of corona treating equipment is found on extruders. Normal practice in the past has been to treat film at the time of extrusion and ship it to the converter for printing and laminating without further treatment. Since film is easier to treat immediately after extrusion, this has always been considered the optimum location for the treater. Recent years have brought a distinct trend toward in-line treatment on the laminating or printing machine. One of the factors contributing to this trend is the fact that treatment levels decay somewhat with time and can be obliterated by contact with idler rolls during subsequent machine operations. Also there is a practical limit on the treatment level that can be applied to a film prior to winding if blocking is to be avoided. Film suppliers have traditionally provided treatment levels in the 36-42 dynes/cm^2 range. Treatment levels in this range are adequate for flexographic printing using solvent based inks and are just acceptable for bonding with solvent based adhesives.

Use of water based inks and adhesives and 100% solids, or high solids adhesives, have necessitated the placement of corona treating equipment on converting machines. This is partly because the inks and adhesives have a

SURFACE TREATMENT 247

higher surface tension and don't wet the film as well as solvents, and partly because of fatty acid slip additives present on the surface of the film. Initial treatment on the extruder is accomplished before much of the slip additive has migrated to the surface. The slip additive has relatively poor wetting characteristics and can cause problems in bonding unless subjected to a second in-line corona treatment. Even before the advent of new types of adhesives and printing inks, a high percentage of converting machines were equipped with corona treaters simply as a quality assurance tool for avoiding problems caused by inadequately treated film from a supplier.

Constant development of plastic materials, printing inks and adhesives as well as a wide diversity of applications for the corona treating process dictate that individual treat level requirements be determined according to the product application and end use requirements. Specifications for treatment level can often be obtained from the raw materials supplier or the machinery manufacturer. If treatment level data is not available from these sources, actual trials using a corona treater will establish the minimum required treatment level. Most manufacturers of corona treating equipment have laboratory set-ups that are available for this purpose.

Experience with converting machines has indicated that there is no upper limit on treatment level. The higher the treatment level, the better the expected bond. Treaters with output ratings of 20 kW are available, so there is no application that is beyond the power range of present-day corona treating equipment. It is however, unwise from a cost standpoint to purchase treating equipment with a capacity beyond that which is necessary.

Thermal bonds such as those made by an extrusion coater or by a thermal laminator are also greatly improved by corona treatment. In extrusion coating, the paper, foil or film substrate is treated just prior to the extrusion nip. Precisely why the treated surface of paper is more receptive to bonding by extrusion coating is not known, but it does promote stronger bonds at higher line speeds and the treatment is commonly used in practice. Foil surfaces are thought to be improved by chemical change of rolling oils left on the surface from previous forming operations. In the case of annealed foil, treatment has an effect on the residues of the oils. It has been found that the introduction of ozone in proximity to the extrudate also assists in the bonding process. Specialized ozone generators have been developed just for this purpose. Thermal laminating, in which two plastic webs are bonded by heat and pressure, requires treatment of both webs to a level above 40 dynes/cm^2.

CORONA TREATER SELECTION

After determining the proper treatment level for a given application, a treater of adequate power output capacity should be selected in order to achieve that level. Sizing is accomplished by determining the number of watts per unit area required to achieve the desired treatment level. The number of watts/m^2 can vary from as low as 5 watts for low-density, low-slip polyethylene, to 100 watts or more for high-slip polyethylene or polypropylene that has been in storage for a long period of time.

Much has been said about various electrode configurations and different frequencies having the ability to treat film more efficiently. Extensive field experience and laboratory trials at the author's place of employment, Enercon Industries, have indicated that regardless of the shape of the electrode or the frequency of the electric current, it is the power applied to the film that determines the treatment level. The main consideration in an electrode design is that it should provide a uniform distribution of corona and be of adequate size to permit full rated output from the power supply. The frequency should be high enough to eliminate objectionable audible noise from the electrode and low enough to avoid television and radio interference. An ideal frequency range is between 9 and 10 kHz. Nine kHz is high enough to suppress audible noise, and staying below 10 kHz avoids the necessity of obtaining an FCC transmitter license.

In keeping with the modern trend to eliminate operator involvement to the greatest degree possible, the power supply should be of a type that automatically matches to varying web widths. Some power supplies are very sensitive to changes in load such as different film thicknesses or electrode lengths. These power supplies are adequate for fixed width and fixed thickness films, but require tuning by the operator when variables are introduced. Also the power supply should be equipped with a loss of treatment sensor that sounds an alarm if power output falls below the desired level. This alarm will immediately alert the operator and avoid production of scrap due to the treater not putting out adequate power.

Two optional features that may be beneficial are a remote control station which places the treater controls at a location convenient to the operator, and a zero speed switch which shuts the treater off when the line stops.

Another important consideration is whether or not conductive webs such as metallized films or foils are being processed on the machine now, or will be in the future. Conductive webs require that the insulation be placed on the electrode. The two basic types of treater stations for accomplishing this are the covered roll type and the quartz electrode type. The covered roll type stations use multiple dielectric covered rolls mounted around a bare metal ground roller. Earlier in this chapter, the difficulties in maintaining dielectric covered rolls were pointed out. Since the foil treating station uses multiple rolls, the problems are compounded and stations of this type are rapidly becoming obsolete. The quartz electrode treating station is the only sensible method in current use for treating conductive webs.

If a treater station is to be located in an area where solvent fumes are present, it will be necessary to purge the system to prevent explosion hazards. Generally non-explosion-proof treating stations are much more convenient to thread-up and to maintain than are the purged stations, so if possible the treater station should be located outside of the hazardous area. If this is not possible, however, the station should be equipped with gasketed doors, interlocks and timers for proper purging.

The normal air gap setting between electrode and roll for most corona treating systems is 0.15 cm. This air gap distance can be varied to permit treatment of thicker materials such as foam and sheet. Air gaps up to 0.5 cm are often used. A well-designed treater station should have air gap adjustments external to the treater station. This permits air gap adjustment while the equipment is in operation and eliminates potential harm to plant personnel from electrical shock. The electrode should also be mounted in a way that permits it to pivot out of the way to clear an obstruction on the web such as splice.

Since generators used in corona treatment produce potentials as high as 10,000 volts, extreme care should be taken in operating and installing the equipment. The high-voltage transformer should be located as close as possible to the electrode to avoid unnecessary personnel exposure. The optimum arrangement is to have the high-voltage transformer be an integral part of the treater station. With stations of this type, all high-voltage wiring and connections are internal to the treater station and no high-voltage wiring is done at the time of installation. If for some reason it is not possible to mount the high-voltage transformer directly on the treater station, the high-voltage wire should be run through a metal pipe via plastic spacers providing about 5 cm of clearance between the high-voltage wiring and the inside wall of the pipe. Cautions taken with high-voltage wiring will not only provide operator safety, but will eliminate one of the most common sources of failure in corona treating systems: high-voltage shorts to ground.

Corona discharge generates ozone which is highly corrosive and toxic in concentrated amounts. A good treater station should be constructed so that all parts exposed to ozone are made with materials that resist corrosion such as aluminum, stainless steel, or protective plating. An exhaust system with sufficient capacity to remove ozone from the plant environment should be provided.

There is a great deal of speculation and theory about what exactly happens physically and chemically when a film is subjected to corona discharge. The prevailing theory is that polar groups are formed on the film surface. Regardless of the exact mechanism, there is no question that corona treatment improves the bonding properties of plastic material, thus facilitating their use in printing, laminating and coating operations. With the current trend toward lower solvent emissions, the future should bring even more extensive use of corona treaters in converting operations.

15
Drying

Donatas Satas
Satas & Associates
Warwick, Rhode Island

Web drying is a process during which volatile material is removed from the web leaving non-volatile solids behind. These solids can be concentrated on the web surface, such as in coating, can be distributed throughout a fibrous web, as in saturating, or it can be located at the interface between two webs providing adhesion, as in laminating.

In addition to drying, curing or crosslinking might take place at elevated temperature. Curing involves a chemical reaction between adjacent polymer molecules and it changes the coating properties by making it stronger, stiffer, insoluble. Curing quite often follows the drying step and is carried out at temperatures above those used for drying.

Fusion is another operation that might be carried out on similar equipment. Fusion of powder particles into a continuous coating, or fusion of a blend of solid polymer particles and liquid plasticizer as in vinyl plastisol, are the processes for which drying equipment might be used.

Web drying is a process in which heat and mass transfer take place simultaneously. Consequently, a drier must provide a source of heat, means of heat transfer to the web, means of vapor removal away from the web and means of conveying the web through the drier. The driers are classified either according to the method of heat transfer; i.e., convection driers, infrared driers, or according to the method of web handling: conveyor driers, catenary driers, etc.

DRYING PROCESS

The drying process is a complex phenomenon, because simultaneous heat and mass transfer can take place by several different mechanisms or their combinations. The overall drying rate depends on the rate of the slowest step and can be accelerated by increasing the rate of the slowest step.

At the initial drying stages, the liquid is evaporating from the web surface, which is saturated with the liquid. Equilibrium between the heat transfer rate into the web and the mass transfer rate of evaporation and removal of vapor is established. The evaporation rate can be simply increased by increasing the heat transfer rate. At such conditions the drying rate can be expressed by the following equation:

$$\frac{dw}{d\theta} = \frac{hA\Delta t}{\lambda} = k_g A \Delta p, \qquad (1)$$

where:

$\frac{dw}{d\theta}$ = drying rate (w is weight of liquid, θ is time)

h = heat transfer coefficient
A = area of heat transfer
Δt = $t - t_s$, where t = gas temperature and t_s = temperature of the evaporation surface
k_g = mass transfer coefficient
Δp = $p_s - p$, where p_s = vapor pressure of liquid at the surface temperature t_s and p = partial pressure of liquid in the gas stream.
λ = latent heat of evaporation at t_s

The drying rate is constant as long as there is sufficient liquid at the surface. This constant rate period might take up most of the drying time for porous materials, such as paper. In drying of solid coatings, this period is short and the drying soon enters into the falling rate period. The drying rate becomes dependent upon the transport of the liquid to the surface. This can take place by diffusion, capillary flow, flow caused by shrinkage and pressure gradients, gravity flow or flow by vaporization-condensation sequence. In drying of coatings, diffusion is the most usual mass transfer mechanism, but, regardless of the particular mechanism, the drying rate continuously decreases during this period. The heat transfer rate is higher than required to evaporate the liquid that reaches the surface and the web surface temperature increases. The diffusion process is described by the general diffusion equation:

$$\frac{\delta w}{A \delta \theta} = -D \frac{\delta c}{\delta x} \qquad (2)$$

where:

- w = the liquid flow by diffusion
- θ = time
- c = concentration of liquid per unit volume
- x = half thicknesss of the solid layer when drying takes place from both surfaces
- D = diffusivity of the liquid
- A = area of the mass transfer.

The diffusion equation has been transformed and solved for various boundary conditions by several authors and has been summarized in numerous publications.[1,2] The diffusion equation fits the experimental data quite well.

While the mechanism of mass transport to the surface is diffusion in most solvent-based coatings, the drying of latex coating is somewhat different. Several drying mechanisms have been proposed including diffusion, capillary flow and others. In drying of latex coatings, the water removal is less difficult and higher drying rates can be sustained than in drying of solvent based coatings.

Drying by conduction takes place by a complex mechanism of vaporization-condensation within the fibrous web. Drying by infrared radiation is similar to drying by convection, except that the constant rate drying period can be extended, since heating takes place from inside the coating: the radiation penetrates somewhat below the surface. This effect of heating the coating uniformly is more pronounced in high-frequency drying.

Drying Curves

The drying data is best represented by drying curves. Such curves are constructed by plotting percent vehicle remaining in the coating against drying time as shown in Fig. 15-1. More informative is to plot drying rate as a function of time as shown in Fig. 15-2. This plot is for a typical solvent based coating of about 0.1 mm wet thickness. Various drying periods are clearly distinguishable in such a plot.

The segment A–B is the cooling down period, which is exhibited by many solvent based coatings. The temperature of the coating decreases, because of fast solvent evaporation rate at the beginning of the drying. The segment B–C represents the constant drying rate period. Although the duration of this period may be short, a large amount of solvent evaporates during this time. The constant rate period is somewhat longer for latex coatings, and very long for simple removal of water from porous webs, as in paper manufacturing. Generally it is desirable to extend this period. Lower temperature

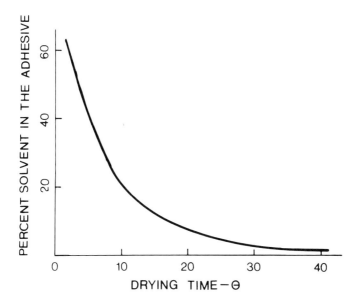

Fig. 15-1. Drying curve. Vehicle remaining in the coating versus drying time.

drying, applying heat from the backside, and other means are used to extend the duration of the constant rate period.[38] The segment C-D represents the transition from the constant rate to the falling rate period represented by the segment D-E. The falling rate period dominates this particular drying process of a solvent based coating.

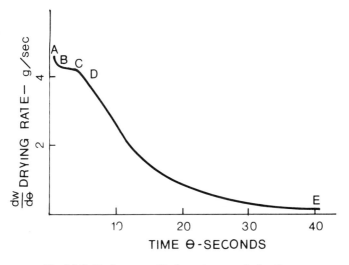

Fig. 15-2. Drying curve. Drying rate versus drying time.

Drying curves can also be constructed in several other variations: plotting the drying rate against percent solvent in the coating, etc.[3]

Heat Transfer

The heat can be transferred to the web by several mechanisms. In hot air ovens the heat is transferred mainly by convection from the higher temperature air to the web. In infrared ovens the heat is transferred by radiation, some by conduction (within the web) and also by convection, since infrared ovens are usually used in conjunction with air flow to remove the vapors. In hot drum drying the heat is transferred mainly by conduction. In hot drum-impingement driers the heat is transferred by convection and by conduction. In high-frequency drying the energy is transferred by an electromagnetic field and converted to heat internally within the polymer structure.

Convective Heat Transfer

The heat transfer by convection is the most important mechanism in drying of coatings. The heat is transferred by a stream of hot air contacting the surface. It must be transferred through a boundary layer of stagnant air and the thickness of this boundary layer is a good measure of the resistance to the heat transfer. Most of the air ovens operate well into the turbulent flow region at high Reynolds numbers $[LV(\varrho/\mu)]$. At Reynolds numbers above 500,000, the Colburn equation can be used.[4]

$$\frac{h}{C_p V \varrho} \left(\frac{C_p \mu}{k} \right)^{2/3} = \frac{0.036}{[LV(\varrho/\mu)]^{0.2}} \qquad (3)$$

where:

h = heat transfer coefficient
C_p = specific heat at constant pressure
μ = viscosity
ϱ = density
k = thermal conductivity
V = average velocity
L = heated length.

For air, Eq. 3 can be simplified to

$$h_m = 0.0128 \, G^{0.8} \qquad (4)$$

where:

h_m = heat transfer coefficient based on length mean Δt
G = mass flow ($V\varrho$).

Equation 4 indicates that the heat flow is related to the 0.8 power of the air velocity.

At Reynolds numbers below 80,000, the Pohlhausen equation becomes valid, which in its simplified form is shown below.

$$h = n(V/L)^{1/2} \qquad (5)$$

where:

n = a constant dependent on the air properties.

The equation shows that the heat transfer coefficient is proportional to the reciprocal square root of L, which is the function of nozzle spacing. The closer the nozzles, the smaller is L.

While most of the driers operate in the turbulent flow region, some have been developed to operate at low Reynolds numbers. In such cases close spacing of nozzles becomes very important. The theory behind the construction of such driers has been discussed by Gardner.[5]

CONVECTION DRIERS

The direction of the air flow is the most important characteristic of a drying system. In the early convection ovens, the air flow was parallel to the web surface, usually counter-current to the web movement. Such driers are still in use for applications where a high heat transfer is not required. The impingement driers was a major improvement with heat transfer rates three to four times those of parallel driers. In the impingement driers the air is discharged through restricted openings and directed perpendicularly to the web surface. Through driers was a further improvement of porous web drying. The air passes through the open web and the heat and mass transfer coefficients are quite high.

Some heat transfer coefficients for comparison purposes are given below.[6]

Quiet air	10 kcal/(hr) (m²) (°C)
15-mph wind	29 kcal/(hr) (m²) (°C)
Normal driers	100–200 kcal/(hr) (m²) (°C)
High-velocity air driers	170–370 kcal/(hr) (m²) (°C)

Air velocities as used in the convection ovens are roughly characterized as follows.

3–25 m/sec velocity is low
25–50 m/sec velocity is medium
50–125 m/sec velocity is high

Drying rates vary depending on the design of the drier and the drying conditions. The approximate range of water drying rates in various driers is given below.

Tunnel parallel air flow drier	7–15 kg/(m^2)(hr)
Impingement drier	35–65 kg/(m^2)(hr)
Air foil drier	20–50 kg/(m^2)(hr)
Infrared drier	50–140 kg/(m^2)(hr)
High-velocity air drier	140–150 kg/(m^2)(hr)

The heat transfer in a forced air convection drier is determined by the following variables:

- Air velocity
- Air mass
- Air temperature
- Air nozzle geometry.

Air temperature is the most convenient variable to adjust and therefore it is the main variable used to vary the drying conditions. Regulation of the air flow is less convenient and there might be other factors than drying which determine the air flow required. A certain minimum air flow must be maintained when drying solvent based coatings in order to ensure that a flammable concentration is not reached. Floater ovens might require a certain minimum air flow to convey the web adequately.

The air nozzle geometry: the distance of nozzles from the web surface, the spacing between the nozzles, shape and type of nozzles employed and other variables of nozzle design and arrangement are not amenable to frequent change or adjustment. The best arrangement of air directing accessories for a given drying application is chosen on the equipment acquisition and only infrequent adjustments are made.

Impingement Nozzles

The impingement nozzles have a high local heat transfer coefficient at the point of impingement, which drops rapidly as the distance from the jet increases. The locat heat transfer coefficient is greater everywhere, even between the jets, than that of turbulent boundary layer, which is greater than the heat transfer coefficient of a laminar boundary layer.[2] This is illustrated schematically in Fig. 15-3.

The heat transfer rate, therefore, is substantially higher in impingement driers than in either parallel or counter-current flow driers. The drying rates are more than three times higher than those obtained in a parallel flow drier.[7]

Two basic types of impingement nozzles are used: slotted and round, the former being preferred by most oven manufacturers. When the nozzles are compared on the basis of equal fan energy and air flow rate, the heat transfer coefficient was found to be greater for round nozzles.[8] Round nozzles, however, are more likely to produce less uniform drying across the web, but they are easier to construct than slotted nozzles. Slotted orifice nozzles must be accurately machined, in order to perform properly, and the gap width must be maintained constant. Change in the gap width will have an effect on the drying rate.

The heat can be applied on one or on both sides. Coated materials are usually dried from the coated side, although additional heating of the web from the bottom might help to bring up the temperature of the coating without causing much evaporation. This helps to extend the constant rate drying period and also to increase the diffusion rate of the vehicle through the coating. Saturated materials are dried from both sides. Two-side drying might not be practical for some oven constructions, i.e. conveyor ovens.

An air cap nozzle is shown in Fig. 15-4. Air tube nozzle with an annular space for return air is shown in Fig. 15-5. Many different designs of slot ori-

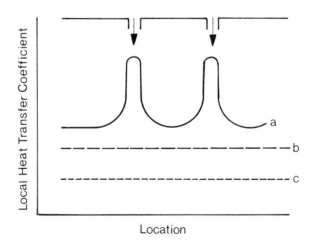

Fig. 15-3. Local heat transfer coefficient for impingement nozzles.

258 WEB PROCESSING AND CONVERTING TECHNOLOGY AND EQUIPMENT

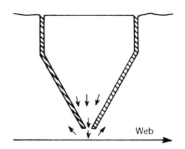

Fig. 15-4. Aircap nozzle.[37]

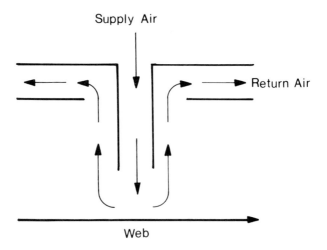

Fig. 15-5. Air tube nozzle.

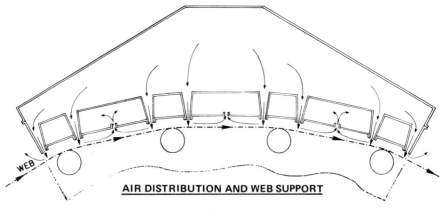

Fig. 15-6. Slot orifice nozzles. (*Courtesy Black Clawson Co.*)

fice nozzles are available. A typical air distribution system in a slot orifice drier is shown in Fig. 15-6.

Interferences

The exhaust of spent air must be accomplished with minimum disturbance of the impinging jet and the proper design of the air return system is quite important for a well-functioning of the drier.

Two types of air flow interferences can occur in a forced convection drier. Interference can be caused by the spent air of the upstream jets (cross-flow interference) and it can be caused by the adjacent jets. The cross-flow interference is not important in well-vented driers. Well-vented design requires that the maximum length of the spent air path is below 30 cm.[8]

On narrow webs, up to 150-cm width, the air return might be sideways along the web. In case of wider webs the return openings must be located between the nozzles in order to avoid overdrying the edges and excessive crossflow interference. Round nozzles might be provided with annular space around each nozzle for the return air. In slotted nozzle driers the air is returned through the slots placed between the nozzles as illustrated in Figs. 15-7 and 15-8.

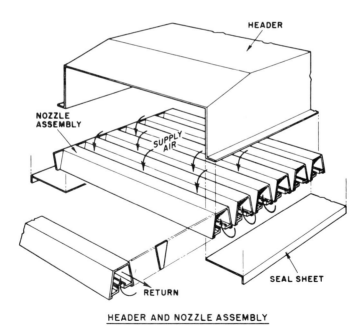

Fig. 15-7. Header and nozzle assembly. (*Courtesy Black Clawson Co.*)

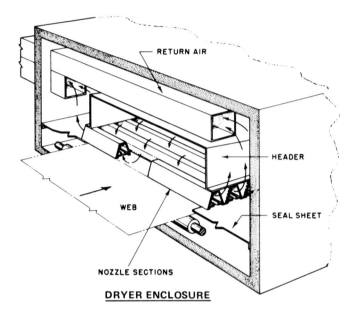

Fig. 15-8. Return air flow. (*Courtesy Black Clawson Co.*)

The variation of heat transfer coefficient with the distance between nozzles is shown in Fig. 15-9. The air velocity and the temperature were kept constant at 16,000 fpm and 400°F. The interference from the adjacent jets becomes important when the spacing between the nozzles is reduced to about 1″. A decrease of heat transfer coefficient is observed at the narrow spac-

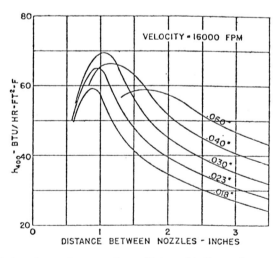

Fig. 15-9. Variation of mean heat transfer coefficient with distance between nozzles for five different orifice sizes.[6]

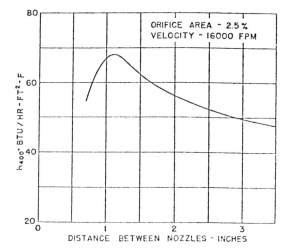

Fig. 15-10. Variation of mean heat transfer coefficient with distance between nozzles where orifice size is 2.5% of nozzle spacing and air energy is constant.[6]

ings. The effect of cross-flow interference is noticeable in the curves for 0.060-0.040" orifice sizes.

Figure 15-10 shows the variation of heat transfer coefficient with distance between nozzles with a constant orifice area of 2.5% and at a velocity of 16,000 fpm. The air horsepower is constant under these conditions, unlike in Fig. 15-9, and gives a good indication of the effect of nozzle spacing. Again the interference from the adjacent jets is noticed at the nozzle spacing below 1".

Nozzle Distance and Width

The heat transfer coefficients of round nozzles are not sensitive to the distance between the jet and the web surface. The heat transfer coefficients in slotted jets are reported to decrease by 15-20% when the jet-to-surface distance was increased from 0.64 to 3.8 cm.[8] It is generally considered that a distance of 1 cm can be maintained without contaminating the nozzles with wet coating.

Janett et al.[9] show the effect of the nozzle width for slot nozzles set on 6" centers and 1" away from the web surface at air velocity of 10,000 fpm (Fig. 15-11). To obtain the heat transfer coefficient value at different air velocities the multiplier shown in Fig. 15-12 is used and to correct for air temperature differences the multiplier shown in Fig. 15-13 is employed.

Gardner[6] shows the effect of percentage orifice area on the heat transfer coefficient keeping the fan horsepower constant at 0.5 HP/ft^2. The air velocity at these conditions varies with the changing percentage orifice area. The

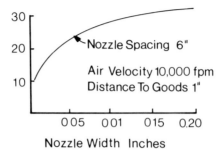

Fig. 15-11. Effect of the nozzle width on the heat transfer coefficient $h = BTU/(hr)(ft^2)(°F)$.[9]

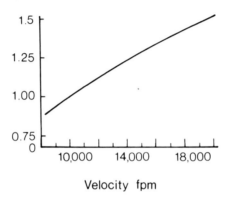

Fig. 15-12. Multiplier for the effect of air velocity on heat transfer coefficient.[9]

data indicates that the highest heat transfer coefficient is obtained when the orifice area ratio to the surface area is 0.03 (Fig. 15-14).

Air Velocity

In a drier with a parallel air flow to the web surface the heat transfer coefficient increases with 0.8th power of the air velocity. The power consumption increases as the cube of the air velocity and therefore the premium paid for higher heat transfer values in the 75-125 m/sec air velocity range is quite

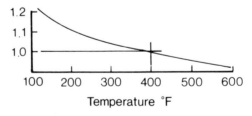

Fig. 15-13. Multiplier for the effect of temperature on heat transfer coefficient.[9]

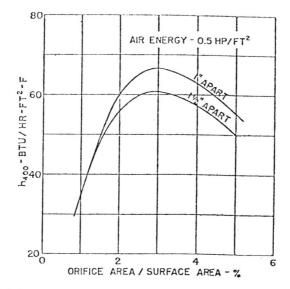

Fig. 15-14. Variation of mean heat transfer coefficient with percent orifice area at constant air energy for nozzles spaced 1″ and 1.5″ apart.[6]

high. It might be more economical to run at lower air velocities and to increase the drying by other means, such as extending the oven length. The maximum air velocity might be also limited for other reasons: a high air velocity might disturb the coating. Also in the diffusion controlled falling rate drying period the heat transfer is not the most important factor which determines the drying rate.

Figure 15-15 shows the variation of mean heat transfer coefficient as a function of air mass velocity at 400°F for four slot orifices of different width. The nozzle to surface distance was ¼″. The data is the result of the tests carried out by Gardner.[6] The curves are exponential and the exponent ranges from 0.55–0.80. The inflection in the 0.100″ orifice curve is due to crossflow interference and it starts when the air flow reaches 520 cfm/ft² of surface. This is a very high flow rate and most of the ovens might have 50% lower air flow rates.

The effect of air velocity and nozzle geometry on heat transfer coefficient is shown in Figs. 15-16 and 15-17. The figures show the plots of Nusselt and Reynolds numbers.

where:

$Nu = hd/kf$ and $Re = \varrho_o v_o d/\mu_f$ for slot orifices
$Re = \varrho_o v_o D/\mu_f$ for round orifices.
h = impingement heat transfer coefficient

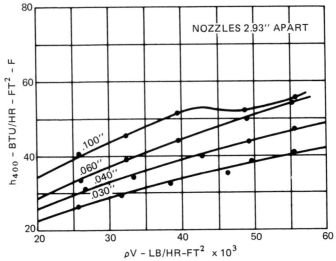

Fig. 15-15. Variation of mean heat transfer coefficient with mass velocity of air at 400°F for four orifices all spaced 2.93" apart.[6]

d = the width of the slot orifices
D = diameter of round orifices
k_f = thermal conductivity at film termperature
ϱ_o = air density in plenum
v_o = air velocity at the nozzle
μ_f = absolute viscosity at film temperature.

Floatation Nozzles

Floatation nozzles perform a dual function: they provide the air flow for drying and also support the web so that it can be conveyed through the drier

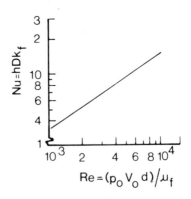

Fig. 15-16. Effect of air velocity and nozzle geometry for slot jet heat transfer.[8]

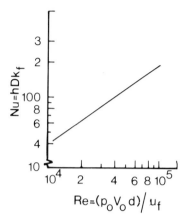

Fig. 15-17. Effect of air velocity and nozzle geometry for round jet heat transfer.[8]

without use of idlers, conveyor belt or other mechanical accessories for support.

There are two basic types of floatation nozzles: direct impingement and air foil. The impingement nozzles support the web by air flow on both sides of the web. The web accepts a sinusoidal contour as it is floated through the oven at a low tension. Such arrangement is shown schematically in Fig. 15-18. Another construction of floatation nozzle is shown in Fig. 15-19. Floatation nozzles exert a flattening effect on the web by keeping it at equal distance from the nozzle. Webs which tend to curl, webs that must be processed at low tension and webs which are coated on both sides are especially suitable for processing by floatation driers. Problems that might be experi-

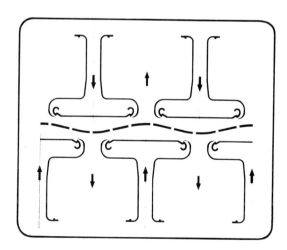

Fig. 15-18. Floatation nozzles. (*Courtesy Pagendarm Apparatebau GmbH.*)

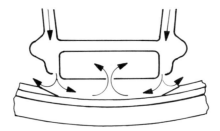

Fig. 15-19. Floatation nozzle, TEC Systems Inc.

enced with a cambered coil in a catenary oven are overcome by using floatation drying. Therefore, one of the first applications of floatation ovens was for coil drying.

Lightweight, low-strength webs may be processed through a floatation oven at tensions as low as 40 g/linear cm.[10] Sometimes, in order to develop a high tension, a suction roll is used at the entrance, providing a nip effect.

Air foil type nozzles utilize the Bernoulli principle to keep the web afloat. When a fluid is discharged through an orifice at a sufficiently high velocity, a restriction of the cross-section of the fluid stream is observed some short distance after the discharge. At this point of *vena contracta* the fluid velocity is higher and the pressure is lower. The web is attracted at this lower pressure area, but it cannot touch the nozzle because of the air cushion (Fig. 15-20). The air balance between top and bottom nozzles is not critical in this type of drying and different air velocities may be used on each side. It is also possible to float the web by using nozzles on top of the web only. The web is held at the low pressure *vena contracta* areas. After discharge the air follows the flat portion of the nozzle. The turbulent air flow at the flat portion of the nozzle is in contact with the web for a longer time than in case of impingement nozzles and this contributes to the high heat transfer rate obtainable in air foil

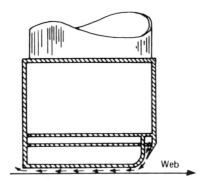

Fig. 15-20. Airfoil nozzle.[37]

Table 15-1. Heat and Mass Transfer Rates of Airfoil Nozzles at Air Temperatures of 470°F[12]

Air Velocity, ft/min	Heat Transfer Coefficient, BTU/ (hr)(ft^2)(°F)	Water Removal Rate, lb/(hr)(ft^2)	Heat Transfer Rate, BTU/(hr)(ft^2)
4000	30.0	6.0	9,600
6000	43.0	8.8	12,700
8000	49.8	11.1	15,200
10,000	53.9	12.9	17,000
12,000	56.2	14.0	18,300
13,000	57.5	14.8	18.900

driers. The extended flat surface of the nozzle is about 60 times the width of the nozzle slot.[11]

Experimental results of drying paper by air foil driers have been reported.[12,13] Some of the results are shown in Table 15-1. Such air foil nozzles are also used for web stabilization and web cleaning.

Another type of a nozzle utilizing the same principle is shown in Fig. 15-21. A single row of eyelid nozzles alternating across the web is used. Every other nozzle is impinging air in the web travel direction, while the alternates impinge air opposed to web direction.

Floatation nozzles are sometimes also used in combination with straight air impingement nozzles, which have a higher local heat transfer coefficient.

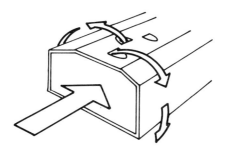

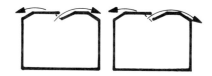

Fig. 15-21. Flakt floatation nozzles.

THROUGH DRYING

Through drying is a very efficient drying method for open webs. The air passes through the open web and high drying rates are obtained because of greatly increased drying surface. The main applications have been in paper making, drying of non-woven fabrics, glass mat and other wet laid products or lightweight saturated webs. Through drying has been reviewed by Villalobos.[14]

Several drier designs are used. The web might be supported by a rotating cylinder which is either under pressure or under suction. The surface of the rotating cylinder might be constructed from metal wire screen, metal fabric (not to be used above 180°C), synthetic fiber fabrics (limited to temperatures below 250°C). Another design employs an open roll for support only and a separate traveling wire belt is carrying the web. A schematic diagram of such a unit is shown in Fig. 15-22.[14] The rolls for such machines are either honeycombed, ribbed, or perforated rolls and might vary in size from 0.5-6 m in diameter. Heated air is introduced through the axis and is either sucked out or forced out through the open web being dried.

Another drier design consists of a flat bed unit. Such a design is more suited for highly permeable webs, because it is more difficult to apply a higher pressure differential than in rotary units.

The air flow used in the through driers is about 150-400 m^3/(min) (m^2). The air temperature is limited by the heat resistance of the carrier and the web. A comparison of the drying rates with other drying methods is given below.[15]

Through drier:	250-600 kg/(hr) (m^2)
Can drier:	7-50 kg/(hr) (m^2)
High-velocity drier:	60-200 kg/(hr) (m^2)

CONDUCTION DRYING

Drying by contacting the web with a hot drum has found many applications for products which are not damaged by such contact. This drying method is very useful in paper making, lamination of films to porous webs, lamination of two porous webs, drying of saturated webs and many other applications. It is not useful to dry coated webs.

The wet web is drawn over a double bank of heated cylinders as shown in Fig. 15-23. The cylinders are arranged to give the maximum wrap. The web is about 80% of the time in contact with heated cylinders and about 20% of the time between the cylinders in open draws. Sometimes felt is used to improve the web contact with the hot roll surface. Instead of a bank of

DRYING 269

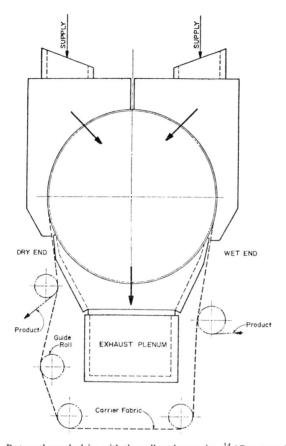

Fig. 15-22. Rotary through-drier with the roll under suction.[14] (*Courtesy AER Corp.*)

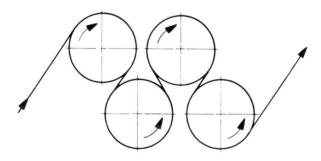

Fig. 15-23. Hot roll drier.

cylinders, a large-diameter heated drum can be used with or without impingement drying. Drying rates can be increased considerably by combining the conduction and convection heating. Air velocity at the nozzle can reach 5000 m/min in such impingement driers and the hot drum serves mainly as a web support, as most of the drying is done by convection.[16] Infrared heaters might also be used in conjunction with the hot drums.

Conduction drying on rotating drums provides means of conveying the web and ironing of the surface against the polished chrome of the drum. Therefore, this drying method has found applications in writing and publication paper manufacturing.

The main resistances to heat transfer in drum drying are the contact resistance between the web and the cylinder and the internal resistance within the web.[17] The drying mechanism is quite complex and depends on the properties of the web and other conditions. Figure 15-24 shows a typical drying curve for hot surface drying.[18] The drying rate $\delta S/\delta \theta$ is expressed as a rate of change of total saturation of the fibrous sheet. Saturation is defined as the fraction of the void space in a porous sheet. The initial portion of the curve AB represents the warm-up period during which the drying rate increases rapidly and approaches a relatively constant level at the point B. The portion BC represents the constant rate drying period, although a slight decrease of the drying rate is noticeable during this period. During this period the water evaporates at the bottom surface and diffuses through the porous body. Some of the water condenses within the fibers and evaporates again. The CD portion of the curve represents the falling rate drying period. This period starts when the water disappears from the outer surface.

Use of the felt further complicates the evaporation process. It enhances the heat transfer from the hot drum to the web because of improved contact,

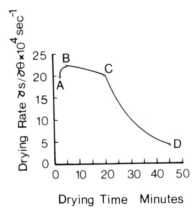

Fig. 15-24. Drying curve for hot surface drying.[18]

DRYING

it absorbs some of the moisture, but it also interrupts the evaporation from the outer surface.

The decrease of the initial water content in the web does not effect the drying rate. The compression of the web and the corresponding decrease of the thickness can increase the drying rate substantially.[17] The surface texture might also have an effect on the drying rate. Smooth surface favors a better contact with the hot drum and improves the heat transfer rate. A 20% difference in the drying rate of paper has been shown between the smoother wire side and more irregular felt side.[17]

Hot drums are constructed from either cast iron, stainless or carbon steel. Single or double wall construction is available. Double wall drums have a thinner outer wall to minimize the temperature differential and a thicker inner wall for structural purposes. The drums can be heated by steam, by hot fluid or by electrical heaters. Steam heated drums are limited by ASME Boiler Code to 165 psi internal pressure (180°C). Hot oil heated drums can be raised to a higher temperature up to 230°C. The drums are of 60-450 cm in diameter and the web wrap might vary between 180-300°.

The drum surface might be chrome plated, which is passivated by a treatment with dilute oxidizing acid in order to improve the release properties. The surface may also be Teflon coated for the same purpose.

The drums are usually driven, because the high inertia of heavy drums does not allow them to be pulled by the web. The drum drive often serves as the master drive for the machine.

INFRARED RADIATION DRYING

Infrared radiation is an important and widely used method to raise the coating temperature for either drying or curing purposes. Infrared heaters are useful in applying a high-intensity heat flux and are especially of interest where a fast temperature rise is required, or where the temperature rise without much evaporation is desired. Infrared drying is usually used in combination with convection drying: air flow is required to remove the vapor from the coating surface.

One of the first infrared industrial uses was baking of automobile finishes. Near infrared radiation lamps were used. First gas infrared burners appeared in the 1920's and were used in the textile industry.

Infrared Radiation

Infrared energy band lies just outside the visible light band in the electromagnetic spectrum as shown in Fig. 15-25. Its wavelength band is 0.7-1000 μm, but the useful band for drying is only in the region of 0.7-11 μm. This useful

272 WEB PROCESSING AND CONVERTING TECHNOLOGY AND EQUIPMENT

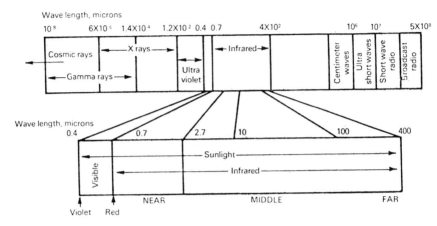

Fig. 15-25. Electromagnetic spectrum.

wavelength band is subdivided into several sections and the radiation equipment is identified according to this classification. Near infrared band is 0.7-3, middle infrared 3-6, and far infrared 6-11 μm.

The Stefan-Boltzmann law is used to calculate the net heat transfer between two bodies.

$$q = \gamma E(T_s^4 - T_t^4), \tag{6}$$

where:

q = net heat transfer flux
γ = Stefan-Boltzmann constant (1.356 x 10^{-12} cal/(cm^2)(sec)(°K)^4
E = emissivity
T_s = temperature of the radiation source
T_t = temperature of the target.

Equation 6 indicates that the heat transfer increases as the fourth power of absolute temperature and therefore the emitter's temperature should be kept as high as possible if a high radiation density is desired.

The radiation density is non-uniformly distributed over a band of wavelengths. The distribution of the radiation energy from a black body is expressed by Planck's law.

$$\frac{E_\lambda}{n^2 T^5} = \frac{C_1(\lambda T)^{-5}}{e^{C_2/\lambda T} - 1} \tag{7}$$

where:

E_λ = radiant flux
n = refractive index of the emitter
T = absolute temperature
λ = wavelength
$\left. \begin{array}{c} C_1 \\ C_2 \end{array} \right\}$ = Planck's law constants: $C_1 = 3.740 \times 10^{-5}$ (erg) (cm^2)/sec; $C_2 = 1.4388$ (cm) (°K).

The wavelength of maximum intensity is inversely proportional to the absolute temperature. This is expressed in Wien's displacement law:

$$\lambda_{max} T = 0.2898 \,(\text{cm})(°K) \qquad (8)$$

The Planck's law equation and Wien's displacement law are represented graphically in Fig. 15-26. The curve in Fig. 15-26 is the maximum limit for an actual infrared radiator. Emitters below 800°C do not act as black body sources and their energy emission must be reduced by the emissivity factor. This lowers the amplitude of the curves in Fig. 15-26. The curves also lose their smoothness in real radiators. Figure 15-26 shows that the radiation energy level is low at long wavelength and that the source must operate at a short wavelength end of the infrared spectrum in order to emit energy of sufficiently high density to be of practical use.

The radiation energy can be absorbed by the target, transmitted or re-

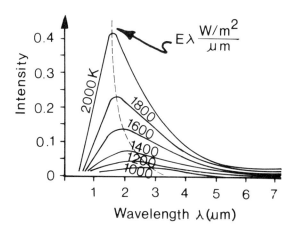

Fig. 15-26. Graphical representation of Planck's equation and Wien's displacement law.

flected. Only the absorbed energy can raise the temperature of the material. Based on the principle of energy conservation

$$\alpha + \beta + t = 1 \qquad (9)$$

where:

α = coefficient of absorption
β = coefficient of reflection
t = coefficient of transmission

If the coating is opaque to infrared radiation, or is sufficiently thick, transmission coefficient becomes equal to 0 and the radiant energy is either absorbed or reflected. The reflected energy can be absorbed by the heater (if the heater is an area source), or reflected by a reflector behind the line or point infrared radiation source. In either case that energy is not lost but reradiated to the target.

Figure 15-27 shows the absorptance and reflectance of energy as a function of wavelength. This material in order to be effectively heated requires that the radiation energy is concentrated between 2 and 3.3 μm and/or between 6 and 8 μm. Thus the selection of a radiator with proper characteristics becomes important in order to raise the temperature of the target fast and effectively. The absorption and radiation curves should be matched as closely as possible. This is not difficult if the radiator is in the near infrared range. A radiator at 800°C spans 2-3.3 μm absorption band of the material to be heated, as shown in Fig. 15-26. Figure 15-28 shows the infrared absorption spectrum of water, which is a common ingredient of latex coatings. The first peak at 2.7-3.2 μm matches the radiation spectrum of near infrared radiation sources. The second peak at 6.2 μm and longer wavelength corresponds to the far infrared radiating at about 700°-840°C.

Infrared radiation is absorbed by the coating and the substrate at various depths. If the material is non-homogeneous, the heating might not be uni-

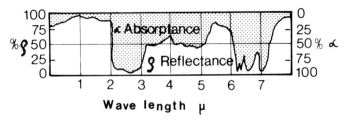

Fig. 15-27. Adsorptance and reflectance of infrared energy as a function of wavelength of a coating.

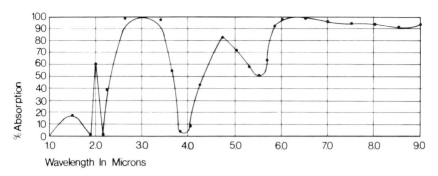

Fig. 15-28. Infrared absorption spectra of water.

form. The depth of penetration varies; the maximum penetration by infrared radiation is about 0.6 mm. Thicker materials are heated by conduction from the section that has absorbed infrared radiation. Far infrared radiation is also absorbed by air reducing the radiation efficiency.

Electrical Emitters

The most important element of the infrared drying system is the emitter and the infrared driers are usually classified according to emitter characteristics. The emitters are subdivided according to the wavelength emitted (near, middle, far, or just near and far), physical configuration (point, line, area source), materials of construction used (tungsten, nichrome) or according to the energy source employed (electrical, gas). Heat transfer efficiency is in the range 40-60% for electrical infrared heaters and 20-30% for gas fired heaters, neglecting heat recovery from combustion gases.

Many materials have been used for electrical heating elements, but tungsten and nickel chromium alloy are the most important metals used. Both are employed as filament wire which is protected by glass, borosilicate glass, quartz, ceramic or metal. Panels are available constructed from borosilicate glass with a thin coating of tin oxide on one surface, which functions as a resistor. The glass has a high emissivity (0.96) and the tin oxide has a low emissivity (0.15). Therefore, very little energy is radiated from the coated side toward the internal reflector.

Infrared bulb with tungsten filament is the well-known near infrared point source. The filament temperature is at 2260°C and the glass of the bulb gets heated to 200-260°C. It also acts as an emitter complicating the emission spectrum of the infrared bulb. Short quartz tubes equipped with reflectors are also used as point sources. Far infrared point sources are made from a nickel chromium alloy wire covered by a ceramic insulator and equipped with a reflector. The temperature of the ceramic might reach 540-650°C.

The near infrared line sources are constructed from a tungsten filament, helically wound and placed in an evacuated quartz tube often refilled with an inert gas such as argon. The wire might be supported by tantalum discs to prevent its touching the wall of the quartz tube and overheating it by conduction. The tungsten filament is at 2260°C. Quartz is 85-95% transparent to the short wavelength infrared energy. Reflectors are used to direct the radiation. The energy from such emitters may be concentrated up to a density of 30 watts/cm^2. The mass of the emitter is low and the response is therefore short. Full output can be reached in 1-2 seconds.

Far infrared line sources are constructed by placing a coiled resistance wire inside an unevacuated quartz tube. The resistance wire radiates at 870°C, the quartz tube at some lower temperature and a compound energy distribution curve characterizes this emitter. Far infrared line sources are also constructed by placing resistance wire inside an opaque ceramic tube. Such sources radiate at 816°C. Fast responding far infrared line sources are constructed by wrapping a ribbon of resistance wire on the outside of a quartz tube. Figure 15-29 shows several types of line sources.

Area sources are all far infrared, there are no suitable materials from which a near infrared area source could be constructed. Various ceramics, quartz, glass, glass cloth are used for the radiating surface. The heating element is embedded into the radiator. Ceramic surfaces are useful up to 540°C, borosilicate glass up to 315°C, quartz plate up to 950°C. Metal surfaces may be also used for area sources. Figures 15-30 and 15-31 show installed infrared area sources.

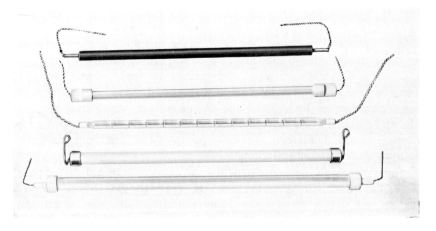

Fig. 15-29. Various infrared line sources. From the top down: metal rod, quartz tube, quartz lamp and two quartz tubes. (*Courtesy Fostoria Industries, Inc.*)

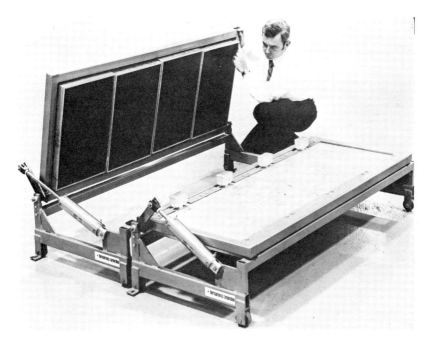

Fig. 15-30. Borosilicate glass panel heaters installed in a hydraulically operated frame. Panels are 3.5 kw and 230 v each. (*Courtesy Thermatronics Corp.*)

Table 15-2 shows the characteristics of various infrared electrically heated emitters.[19]

Controls

Controls are an important part of infrared driers to assure the adequate operation of the equipment. The emitters can be controlled by periodically interrupting the current flow. An electromechanical switch can turn the power on and off at a preset cycling frequency. Maximum possible cycling frequency for the mechanical on-off controls is 10 cycles/min. This is the simplest, least expensive method of controlling the infrared radiation. The temperature might vary as much as 3°-6°C.

For a more precise temperature control, silicon controlled rectifier (SCR) is used. Two methods may be used to start a SCR device: zero voltage switching and phase angle firing. In case of zero voltage switching, the control consists of regulating the number of voltage sine curves passing through the heating element. Phase angle firing consists of removal of a part of each cy-

Table 15-2. Characteristics of Infrared Radiation Sources

	Tungsten Filament Wire		Nichrome Spiral Winding		Panel Heater	
	Glass Bulb	Quartz Lamp	Quartz Tube	Metal Sheath	Buried Nichrome Alloy	Metallic Salt
Source Temperature, °C	1650–2200	1650–2200	760–980	540–760	200–595	
Wavelength, μm						
At energy peak	1.15	1.15	2.6	3.1	4–5	
Useful range	1.15–1.5	1.15–1.5	2.6–2.8	2.8–3.6	3.2–6	
Brightness	bright white heat		cherry red	dull red	no visible radiation	
Usual size	G-30 lamp	3/8″ diameter tube	3/8″ or 5/8″ diameter tube	3/8″ or 5/8″ diameter tube	flat panels of various sizes	
Relative energy distribution, %						
Radiation	65–80	72–86	55–45	53–45	50–20	
Convection	35–20	28–14	45–55	47–55	50–80	
Resistance						
Mechanical shock	poor	good	good	excellent	varies	
Thermal shock	poor	excellent	excellent	excellent	can be good	

Fig. 15-31. Drying oven with borosilicate glass panels installed over the conveyor. (*Courtesy Thermatronics Corp.*)

cle, but allowing a full number of cycles to pass. Phase angle firing allows the measurement of voltage. It is more expensive and is mainly used for low residual heaters. Zero voltage is primarily used for higher residual heaters.

Temperature sensing devices: thermocouples, thermometers, or infrared radiation sensors can be used for temperature indication and control. In an open loop system, the temperature is controlled by manually setting the heater supply voltage. In a semi-open loop control system, a temperature sensing device is embedded into the heater. Indicating controlling pyrometer varies the power input by means of an electro-mechanical switch or by SCR. Semi-open loop control does not correct for changes in ambient temperature, coating thickness, evaporation rate and other external variables which might require an adjustment in the radiation intensity.

Closed loop control corrects this deficiency. The sensing device measures the web temperature and signals the indicating controlling pyrometer to regulate the power input into the heaters. The product surface temperature can be controlled with an accuracy of ± 1.5°C.

In order to eliminate the edge effect and to maintain a uniform temperature across the web, an overlap of the heaters is required. An overlap equal to the distance between heaters and the web is sufficient. The same results can be achieved by having separate temperature controls for the heaters near the edge of the web.

If the web stops, the radiators should not overheat and damage the product. The current flow into the heaters is stopped, if the line stops running or the web breaks. Low residual heaters are required, or the heaters should be equipped with mechanical retractors or shutters to protect the web from heat damage or igniting. Typical heating up and cooling down curves for various heating elements are shown in Fig. 15-32.

Reflectors

Infrared heaters use reflectors to direct the radiation energy toward the web. Gold porcelain reflectors are most efficient (95% reflectancy), but are expensive and used mainly for high-temperature (480°C) heating. Gold anodized aluminum is next in efficiency (85%) and price and is used for lower temperature, up to 315°C. Non-anodized aluminum reflectors are least expensive, they require periodic cleaning and should be forced air-cooled. Ceramic reflectors are 90% effective in 1.15 μm wavelength range. The ceramic surface reradiates at 3.5 μm wavelength at a surface temperature of 540°C. They require minimum maintenance.

The reflectors used in infrared driers are nonfocusing and their sole purpose is to reradiate the infrared energy that has been reflected by the coating and adjacent equipment.

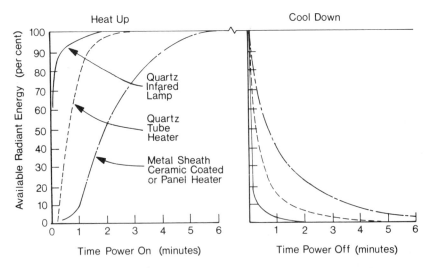

Fig. 15-32. Heating up and cooling down curves for various infrared radiation sources. (*Courtesy Fostoria Industries, Inc.*)

Air Circulation

Infrared heaters are usually used with some air circulation in order to remove the released vapor and thus increase the drying rate. In addition, the air flow cools the heaters, keeps the temperature of the coating surface lower and helps to keep the web down and away from the heaters in case of a web break. High-temperature infrared drying with medium-velocity air (45-60 m³/m²) circulation can achieve drying rates of 150 kg/(m²) (hr).

Various designs of infrared-convection driers are available. A schematic diagram of a design is shown in Fig. 15-33 and a photograph showing the modular arrangement of such units is shown in Fig. 15-34.

Gas Heated Emitters

Gas flame is another widely used source of infrared radiation. While the gas flame has only 10-15% of the available energy in the form of radiant energy, more heat can be extracted by raising the temperature of a ceramic body with hot gas.

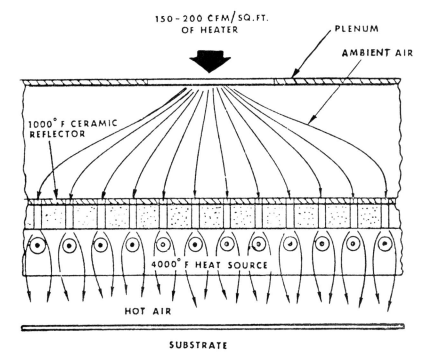

Fig. 15-33. Endview of an infrared-convection type heater. (*Courtesy Research, Inc.*)

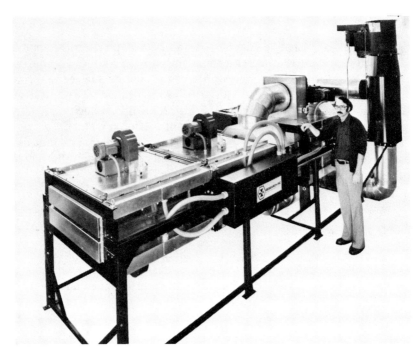

Fig. 15-34. Modular infrared drier with air circulation. (*Courtesy Research, Inc.*)

The gas-fired emitters can be subdivided into five categories.[20]

1. *Radiant tube burner* (Fig. 15-35). The air-gas mixture is fired into a tube. The tube surface becomes the radiator.
2. *Pre-mix burner,* which spreads the flame over a large, flat area.
3. *Surface combustion burners.* The fuel-air mixture is ignited on the surface of porous ceramic tile, ceramic with drilled ports or a metallic wire screen (Fig. 15-36). Initially a small flame is visible at the end of each opening. When the entire surface reaches operating temperature, the flame becomes invisible. Operating temperature range is 760-900°C. Maximum radiation output—50,000 kcal/(hr) (m²).
4. Direct impingement refractory burners which operate at radiation output of 50,000-300,000 kcal/(hr) (m²). The hot flame impinges upon refractory, which acts as an emitter with a face temperature of 980-1200°C (Fig. 15-37)
5. Catalytic burner where catalyst bed allows the gas combustion at lower temperature (315°-425°C). The radiation output is low: 2500-7500 kcal/(hr) (m²). A schematic diagram of catalytic burner is shown in Fig. 15-38.

DRYING   283

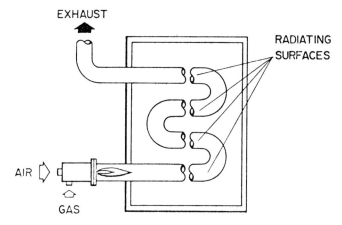

Fig. 15-35. Radiant tube burner. (*Courtesy Eclipse, Inc.*)

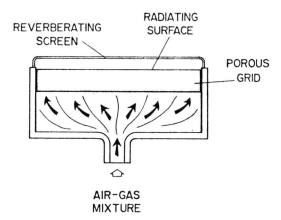

Fig. 15-36. Surface combustion burner constructed from porous ceramic body. (*Courtesy Eclipse, Inc.*)

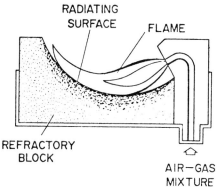

Fig. 15-37. Impingement type burner. (*Courtesy Eclipse, Inc.*)

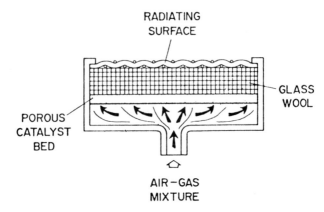

Fig. 15-38. Catalytic combustion burner. (*Courtesy Eclipse, Inc.*)

The burners are arranged in banks to provide a radiant surface of the required width and heat output. An infrared burner bank consisting of 14-cm wide perforated ceramic tile burners is shown in Fig. 15-39. The sheet metal enclosure is air cooled.

There are three basic systems for air/gas mixture delivery to the infrared burners.

1. *Proportional mixing system* where the air is delivered to the blower through a venturi tube. The gas is piped into an enclosure around the venturi and is suctioned by the flowing air. The gas flow depends on the air flow. The air-gas ratio remains constant with the change of air volume. A turn-down ratio of 10:1 is possible (Fig. 15-40). A turn-down ratio is the ratio of maximum to minimum heat output.

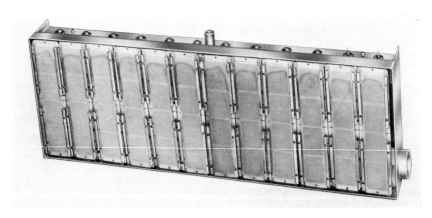

Fig. 15-39. Infrared burner bank. (*Courtesy Fostoria Industries, Inc.*)

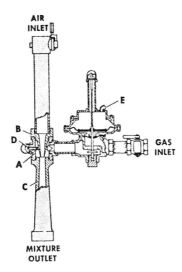

Fig. 15-40. Proportional mixer. (*Courtesy Eclipse, Inc.*)

2. *Premixing system* consisting of a centrifugal blower. A valve on the inlet regulates the air flow. The gas valve is integral with the air valve and opens and closes simultaneously maintaining a constant gas-air ratio. A downturn ratio of 40:1 is possible (Fig. 15-41).
3. *Carburetor type air-gas mixing system.* The mixture can be throttled downstream of the mixing station (Fig. 15-42). The proportional and premixing systems cannot be throttled. This system allows a very high downturn ratio.

DIELECTRIC DRIERS

Dielectric heating is based on molecular excitation of a coating in a high-frequency electrostatic field. Polar polymer molecules rotate at the same speed as the frequency of the generator. The rotational kinetic energy is converted into thermal energy resulting in the temperature rise of the polar coating or substrate. The term dielectric heating is often used interchangeably with radio-frequency, high-frequency microwave heating. Dielectric heating is a general term without reference to the frequency of the electromagnetic field, the other terms refer to a narrower range of frequencies.

The use of high-frequency electromagnetic fields to heat dielectric materials has been known for a long time and the first machines were made in the 1930's. These were operated at low frequencies (6-10 MHz), while current equipment is usually operated at a frequency of 2450 MHz.

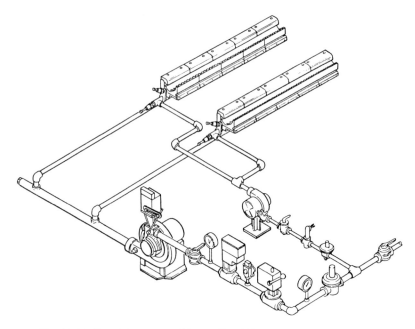

Fig. 15-41. Premixing system with centrifugal blower. (*Courtesy Eclipse, Inc.*)

The energy delivered by a radio frequency field is:

$$P = KfE^2 (\tan \delta) \epsilon \qquad (10)$$

where:

P = energy delivered to the material in watt/cm³
K = constant, dependent on units used, 55.6 x 10⁻¹⁴ for cgs units
f = frequency, Hz
E = field strength, V/cm²
$\tan \delta$ = dielectric loss tangent (dissipation factor)
ϵ = dielectric constant.

The dielectric loss tangent and the dielectric constant are the properties of the material being heated. The higher are these properties, the higher is the capacity of the material to respond to the high-frequency heating. Thus this heating method is highly selective and depends on the energy absorption characteristics of polymer, pigments, solvents and other additives used. Situations where only the coating is heated, but the substrate is not, are possible and can be utilized in handling some heat-sensitive substrates.

DRYING 287

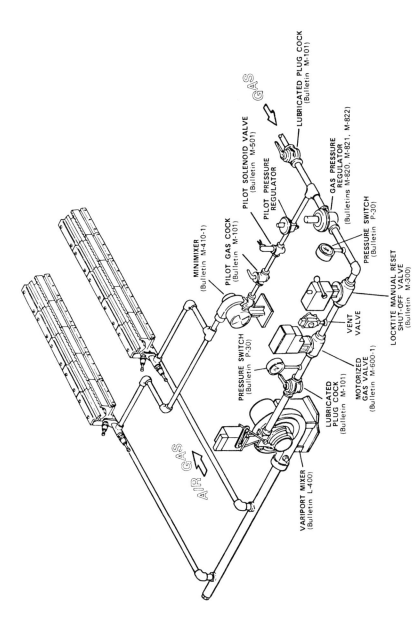

Fig. 15–42. Carburetor type air-gas mixing system. (*Courtesy Eclipse, Inc.*)

The dielectric heating is especially suitable for water based inks and coatings, because of the high-absorption of energy by water. Paper which has a moisture content of 4–7% has a low energy absorption and remains cool. Its moisture level does not change and curling problems are minimized. Table 15-3 lists the degree of response of various materials to the dielectric heating.

Dielectric heating is finding some limited applications in the web processing area. It is mainly useful to dry heavy coatings, where uniform heating by conduction would be difficult. The energy conversion efficiency in dielectric driers is good (70–75%) and comparable to convection ovens (50–65%), but the equipment costs are higher and the maintenance costs involving the tube replacement are high.

Dielectric drier consists of the following equipment:

- Power supply
- Control console
- Generator
- Concentric line coupling
- Applicator.

Power supply module consists of solid state components to generate high-voltage direct current and low voltage alternating current for the control circuit. The control console contains all instruments, switches and indicators. The generator converts high-voltage direct current to high-frequency energy which is transmitted by concentric line coupling to the applicator. The applicator consists of electrodes and the web is run between the electrodes. Several electrode designs are used and some of these are shown in Fig. 15-43.[21] Dispersed field electrode is the most practical for paper drying.

CONSTRUCTION OF DRIERS

Driers can be classified in many different ways, but the most significant classification is according to the way the web is carried through the oven. The driers may be placed in the following categories:

- Festoon driers
- Catenary driers
- Conveyor driers
- Arch driers
- Floatation driers
- Drum driers

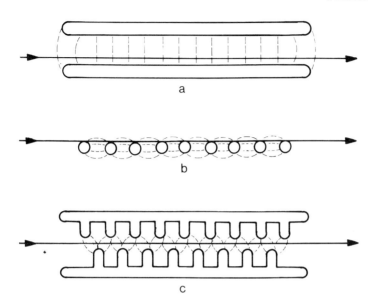

Fig. 15-43. Various electrode arrangements of dielectric driers:
a—Flat platen electrodes
b—Stray field electrodes
c—Dispersed field electrodes.

- Vertical driers
- Tenter frame driers.

Festoon Driers

This is the early type of continuous drier that developed from drying the web by hanging it up in a loft, or some other available space, even outdoors. This type of drying equipment can be occasionally found on some old machines, but generally it is only of historical interest as the first type of continuous drier. In the festoon drier the coated web moves in a loop form carried by a stick traveling on chains through the drying chamber. Low-velocity, heated air is circulated in the drying chamber, which is a large walk-in enclosure. The web residence time in the drier is long, 10-20 min, the drying rate is slow and the temperature is rarely above 120°C. Water based coatings only can be dried in this fashion. The drying rate is about 0.5 kg/(m^2)(hr). Sticks may leave permanent marks in the web. Festoon driers may be constructed as a straight pass drier with the web being rewound at the opposite end from the unwind, or the web may be carried back after making a 180° turn (cross-stick method). A schematic diagram of a festoon drier is shown in Fig. 15-44.

Table 15-3. Response of Various Materials to High-Frequency Electromagnetic Field

	EXCELLENT	GOOD	FAIR	POOR	NONE
Polymers	Alkyds	Ethyl cellulose	ABS	Gum rubber	Polyolefins
	Furane	Nitrocellulose	Acetal	Polyesters	Polystyrene
	Melamines	Methyl cellulose	Acrylics	Polysulfone	Teflon, FEP, TFE
	Phenolics	Cellulose acetate	Acrylonitrile	Silicones	TPX
	Polyamides	CAB	Kel-F	Polyvinyl acetate	
	Polyimides	CAP	Polycarbonate	Polyisobutylene	
	Urea formaldehyde	Epoxy resins	Silicones uncured		
	Polyester, uncured	Phenoxy resins	Shellac		
		PVC			
		Polyvinylidene chloride			
		Polyurethane			
		DAP compounds			
		Polysulfides			
		Casein			
		Synthetic rubbers			

Solvents	Water	Isophorone	Ethyl Cellosolve	Trichloroethylene	
		Acetone	Ethylene glycol	Toluene	
		Methanol	Ethanol	Hexane	
		2-Nitropropane	Methyl Cellosolve	Heptane	
		MIBK	Ethyl acetate		
		MEK	Diethylene glycol		
		Methyl cyclo-hexanone	Cellosolve acetate		
			n-Propyl acetate		
			n-Propyl alcohol		
			Isopropyl alcohol		
			Butyl acetate		
Substrates		Cellulose films	Polyester	PVdC coated olefins	Polyolefins
		Asbestos	PVC		
		Cotton	Kel-F		
		Paper	Glass		
		Vulcanized fibers	Nylon		
Pigments		Carbon black	Phthalo blue and green	Benzidenes	
		Titania		Rubines	
		Chromium pigments	Hansa yellows		
		Iron blue			

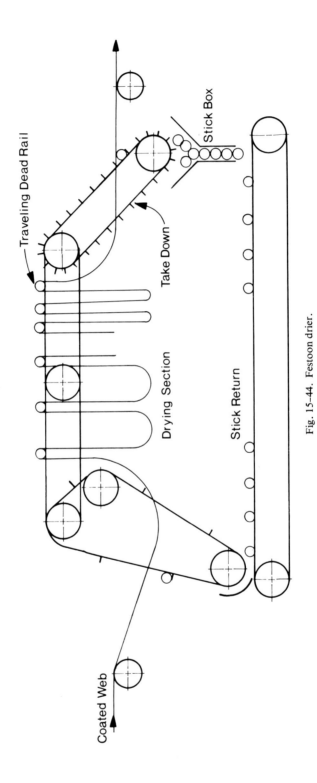

Fig. 15-44. Festoon drier.

Catenary Driers

Catenary driers, also called suspension driers, support the web only at two points: at the beginning and at the end of oven. The web assumes the catenary curve between these two points of support. Such ovens are used mainly for coil coating and they are being replaced by floatation driers even in this application. Stiff, self-supporting materials only can be used in these ovens. A schematic diagram of a catenary oven is shown in Fig. 15-45.

Vertical Driers

The advantage of a vertical drier is that the web does not have to touch any supports until it reaches the top of the tower. This arrangement is useful for saturated or two-side coated webs. The disadvantage is the need to use fairly high tensions to pull the web through. Floatation ovens are replacing the vertical ovens for handling of two-side coated webs.

A variation of a vertical tower drier is a U-type drier, where the web follows a U path. The web can be passed through the drier twice, as shown in Fig. 15-46. Such driers occupy little space and they are often used on printing equipment and laminators for flexible packaging webs, where the drying load is low.

Idler Roll Supported Driers

Idler rolls spaced at some intervals in the oven provide a support for the web. A web wrap of 2°-4° around each idler is sufficient to prevent curling. Various web paths through the oven are used. The web might just go through the oven once or it might be returned. An arrangement of idlers in an arch is often used for difficult-to-handle webs. Part of the tension force applied to the web is translated into the cross-machine directional force that prevents the edges from curling. A typical arch oven is shown in Fig. 15-47.

In high-velocity impingement ovens, the supporting idler rolls might be placed immediately below the air jet to provide a support for the web.

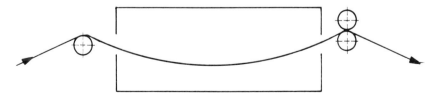

Fig. 15-45. Catenary oven.

294 WEB PROCESSING AND CONVERTING TECHNOLOGY AND EQUIPMENT

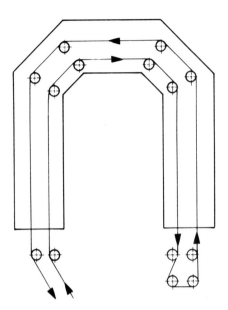

Fig. 15-46. U-type drier.

Fig. 15-47. Arch oven. (*Courtesy Faustel, Inc.*)

The idler rolls may be driven either by a positive or a tendency drive system. Positive drive systems rotate each roller at the web speed, and they are connected mechanically, or electrically, to the main drive. The surface speed of all idlers must be the same if minimum web tension is required. Tendency drive systems decrease the drag on the web and do not require a tie-in to the speed of the main drive. They remove the bearing friction and lower the roller inertia. The tendency drive might be of the dual bearing construction, or it might employ a torque clutch. In case of the dual bearing construction, the inner race of the bearing is rotated by a low torque motor. The force is delivered by a flat belt system (Fig. 15-48). The torque clutch tendency drive is a mechanical device which transmits just enough torque to barely turn the roller.

The bearings of the idler rolls are mounted outside the oven enclosure to keep it at a lower temperature. The shafts might expand despite such precautions and the bearings should be sized to accommodate this expansion.

Idler rolls are made from steel or aluminum. Aluminum rolls cannot be used for higher temperature ovens and steel is preferred. Steel is subject to corrosion and therefore the rolls might be chrome or nickel plated, coated or sleeved with Teflon. Special chrome plating is required for temperatures above 150°C.

Spacing of the idler rolls is determined by the material processed. Materials which require considerable support, such as thin films, might have the rollers spaced every 45-60 cm, foil 75-100 cm; 0.1-0.15-mm thick papers require 90-120-cm spacing, and heavy paperboard can be handled with spacing of 130-150 cm. Large-diameter rolls should be used for heavy paperboard to prevent cracking.

Conveyor Driers

Crossbar or belt conveyors are used for web support. Web is carried at no tension through the oven. The conveyor belt is supported by rollers as shown in Fig. 15-49. Heat can be applied only to the top surface in belt conveyor ovens.

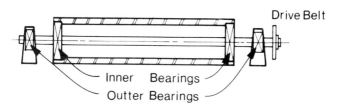

Fig. 15-48. Tendency driven dual bearing idler roll.

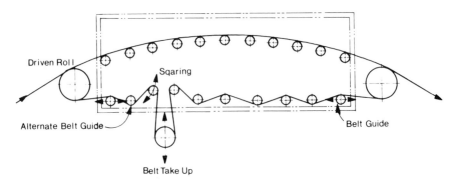

Fig. 15-49. Belt conveyor drier.

Air Floatation Driers

Air floatation driers combine drying and web supporting functions. The floatation driers have a high heat transfer rate and the web can be handled without touching the idler rollers or other supports. In order to float the web the nozzles must be used on both sides of the web, unless air foil type nozzles are employed, which can float the web from one side only.

The nozzles can be placed opposite to one another (Fig. 15-50), or they can be staggered, imposing a sinusoidal path to the web (Fig. 15-51). The curvature helps to improve the lateral stability of the web which is not very good for floatation ovens.

The airfoil nozzles can be intermixed with impingement nozzles. Vertical floatation ovens are constructed by providing air cushion support rollers (Fig. 15-52), so that the web can travel through the oven without touching idlers.

Floatation ovens have found applications in many web processing areas. Coil drying was one of the first floatation oven applications replacing catenary ovens. Floatation ovens help to process cambered, wavy edge strips.[10] Short airfoil driers are employed in the paper manufacturing industry as the first drying step after a size press (Fig. 15-53).

Drum Driers

Drums may be used to support the web while it is dried. Drum driers are usually a combination of conductive and convective heat transfer drying. Heat is transferred to the web from a heated drum and inpingement nozzles might be used to transfer heat by hot air impingement. A schematic diagram

DRYING 297

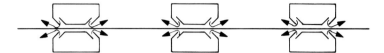

Fig. 15-50. Air floatation nozzles arranged in opposing manner.

Fig. 15-51. Air floatation nozzles arranged in a staggered pattern.

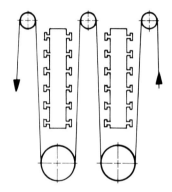

Fig. 15-52. Vertical floatation oven with air cushion rolls.

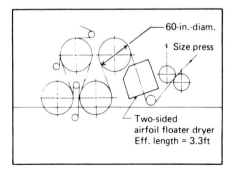

Fig. 15-53. Airfoil drier installation after a size press.[13]

298 WEB PROCESSING AND CONVERTING TECHNOLOGY AND EQUIPMENT

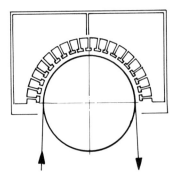

Fig. 15-54. Drum drier.

of a drum drier equipped with air nozzles is shown in Fig. 15-54. Such driers are often used in the paper manufacturing immediately after a size press as shown in Fig. 15-55.

Heated drums may be used alone without air impingement. Such driers constitute the main drying train in paper making machines, web saturators, and some laminating machines. The heat transfer is mainly by conduction.

Tenter Frame Driers

Materials which show a tendency to shrink or distort in the machine direction, because of tension or heat, may be processed in a tender frame in order to restrain the material. Coating of textiles, especially of knitted textiles, requires such equipment.

The drying of textiles in a tenter frame or in a similar drier has been studied by Beckwith and Beard.[22] A typical fabric surface temperature pro-

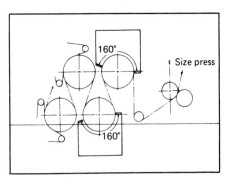

Fig. 15-55. High-velocity impingement drier where the web is supported by drums.[13]

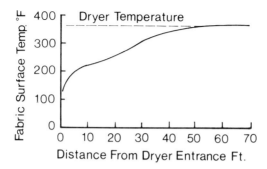

Fig. 15-56. Fabric temperature profile in a tenter frame drier.[22]

file in such textile driers is shown in Fig. 15-56 and the corresponding fabric moisture content profile in Fig. 15-57.

A mathematical model simulating the drying process of a saturated textile web has been developed. This model can be simulated on a digital computer and the effect of change of various operating variables on the effect of temperature and moisture content profiles in the fabric can be studied. The model assumes that the textile web consists of a wet layer sandwiched between two dry layers. The central wet layer has no temperature and moisture content gradients. Water is evaporated at the dry-wet layer interface and passes through the dry layer as a vapor. This simplified model has shown a good agreement with experimental data. It should be applicable for saturated textiles, non-woven fabrics and paper, but not for coated materials, where the rate of volatiles removal might depend on the diffusion rate through the coating.

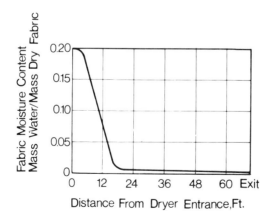

Fig. 15-57. Fabric moisture content profile in a tenter frame drier.[22]

Miscellaneous Driers

Besides the drier designs mentioned above, there are several types of driers used less frequently for web drying. Indeed, the number of combinations of various web path and air flow geometries is large. Arrangement of the web path in a spiral is sometimes used for small pilot machines in order to save the space at the expense of passing the web through a tortuous path. Various small integral drying units are available for installation on existing lines. Figure 15-58 shows a compact drying unit which can accommodate 2 m of web inside the hood and the air flow is controlled by adjusting the fan speed and dampers.

Oven Length

Ovens are usually subdivided into several sections, each section with independent air temperature and flow controls. Each section is required to perform a different function and the drying characteristics of the web must be

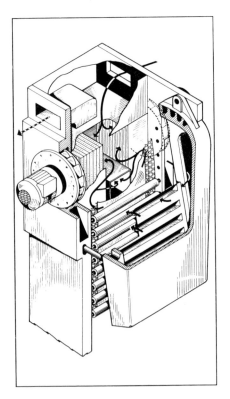

Fig. 15-58. A compact web drying hood. (*Courtesy Costruzioni Meccaniche Cesare Schiavi.*)

understood in order to arrive at a reliable estimate of the length of each section.

Heavy polymer coatings show an initial constant rate drying period and then enter into a lengthy falling rate drying period. The higher is the coating thickness, the longer is the falling rate drying period. Saturated webs, wet laid non-woven fabrics, wet textiles have much longer constant rate drying periods. Drying rate during constant rate drying period depends on the heat input, while the drying during the falling rate period is governed by diffusion rate.

In drying of polymeric coatings the constant rate drying period should be prolonged by maintaining the mass transfer rate at a low level, while drying of wet textiles or paper requires maximum heat input. These drying characteristics, as well as the amount of solvent or water to be removed, determine the oven length required at a desired running speed.

An easy access to the web at all points in the drier must be provided. Usually a drier is equipped with doors located at frequent intervals. Some driers have a clam shell design, where the top section can be lifted, exposing the web. Clam shell arrangement is especially popular with floatation ovens (Fig. 15-59). Split driers allow lifting the top section lengthwise.

AIR HANDLING

The basic air flow diagram is shown in Fig. 15-60. Hot air from a direct or indirect heater is mixed with fresh make-up air and with recirculated air and

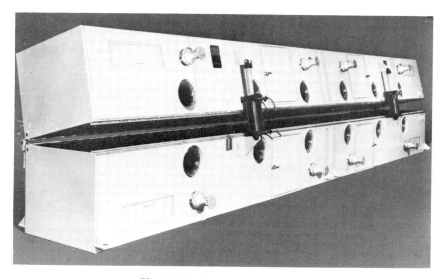

Fig. 15-59. Clam shell drying oven.

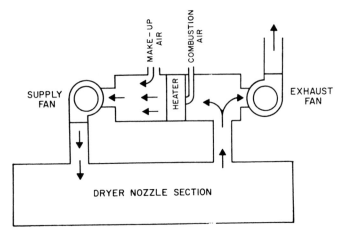

Fig. 15-60. Air flow in a web drier.[23]

is blown by supply fans through the drying nozzles onto the web surface. The spent air is exhausted by blowers. In addition to these basic features, various equipment might be employed to recover the heat from the exhaust gases, to remove the solvent vapor from the exhaust, or to preheat the make-up air.

Air Heating

Air may be heated by direct or indirect means. Direct heating is more efficient, but the products of combustion are mixed with the drying air and clean fuel is required, in order to avoid the contamination of the coating. In indirect heaters the drying air is heated through heat exchangers, and the choice of fuel is less important. Various fuels are used to heat the drying air: gas or electricity for direct heating, and gas, oil or coal for indirect heating. Steam is a frequently used heat transfer medium for indirect air heating.

Gas burners used for air heating may be line or gun type. In a line heater the air passes over the gas flame. The turn-down ratio is low, 15-25 to 1. The line heaters require a uniform distribution of air velocity, and the air velocity is limited: flame loss might occur if the air velocity is too high.

Gun type burners have a separate combustion chamber from which premixed gas and air is delivered to the burner. These hot gases might be mixed with additional air downstream of the burner. Turn-down ratios as high as 50 to 1 are possible. Gun type burners of various designs are most widely used for web drying equipment, because of their wider range of temperature control, higher heat output and better dependability.

The gas heating system consists of gas burner, piping, flame safety, explosion proof safety and electrical controls. The gas burners are sized from

125,000-2,500,000 kcal/hr of heat output. Gas can be used to heat air to temperatures in excess of 400°C.

Direct gas heaters require a certain amount of combustion air and there have been cases where the heat required to heat the combustion air was in excess of the heat required to dry the coating.[23] Under such circumstances the direct heaters are less efficient.

Oil Heaters

Oil heaters are similar to gas heaters, except that the oil must be vaporized to form a combustible oil-air mixture. Vaporization is accomplished by spraying the oil using rotating disk or cone nozzles, which employ centrifugal force to atomize the oil, or by forcing the oil through a small nozzle opening. Small oil droplets vaporize quickly to form a combustible mixture. Oil heaters are less desirable for direct air heaters, since they do not burn as cleanly as gas.

Electric Heaters

Electric heat is more expensive and therefore it is not used much to heat the drier air. The air is moved through the heating coils by a fan mounted either before or after the coils. Air can be heated to temperatures above 400°C by electrical heaters and the temperature can be controlled within ±1°C.

Steam Heating

Steam is readily available in manufacturing plants as a heating medium, and drying ovens are often equipped with steam heated heat exchangers for heating of air. Heat exchangers are constructed from finned tubes. The most generally available heat exchangers can be used for pressure up to 13.6 atmospheres (182°C). Heat exchangers for higher pressure steam are also available, but are more expensive. A good steam heating system can deliver the air at 10-15°C below the temperature of available steam.

An important advantage of steam heating, as well as other indirect heating systems, is that the external air supply is not required and higher recirculated air/fresh air ratios can be used than in the direct fired systems.

Hot Oil Heating

Hot oil heat exchangers are useful to obtain higher air temperatures without excessively high pressures that might be required by steam heating. A temperature of 260°C can be achieved at 3 atm oil pressure. A single oil heat exchanger produces about 125,000-200,000 kcal/hr.

Heat Recovery

Besides heat exchangers used for indirect heating of air supply, heat exchangers are used for recovery of heat in the exhaust gases. A simple heat exchanger for heat recovery is shown in Fig. 15-61.

Slowly rotating (10-30 rpm) wheels between the hot and cold air streams are used for heat recovery. Wheels are made from various materials: aluminum for use below 200°C, stainless steel up to 425°C and ceramic for higher temperatures. The wheels are constructed from fluted, corrugated, honeycombed materials and may be 150-600 mm thick.[24]

A heat exchanger suitable for small air flows and operating at 60% heat recovery efficiency is shown in Fig. 15-62. A refrigerant and a capillary are permanently sealed in a metal tube. Thermal energy applied to one end of the pipe causes the refrigerant to vaporize. This vapor is then condensed at the other end of the pipe by cool incoming air. This evaporation-condensation cycle goes on continuously.

Heat recovery is very important in combustion of flammable effluents. A large volume of gases must be preheated in order to conserve on the energy requirements. Various types of heat exchangers are used in this operation, including fluid bed heat exchangers which can operate up to 85% thermal efficiency.

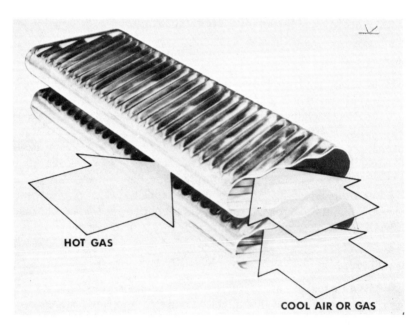

Fig. 15-61. Recuperative heat exchanger. (*Cor-Pak®*, *Courtesy C-E Air Preheater, Combustion Engineering.*)

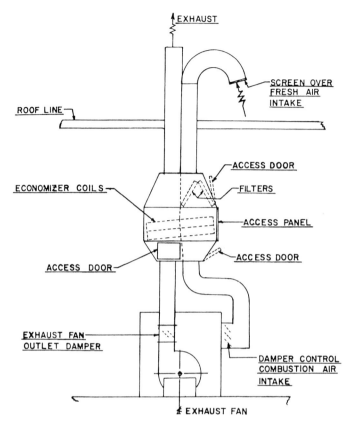

Fig. 15-62. Small-volume refrigerant containing heat recovery system.[23]

Air Moving

The power required to move the air and exhaust gases in the driers is provided by blowers and the air flow is contained by ductwork. The equations generally used for calculation of air systems and sizing of fans and blowers are applicable for driers as well.

$$\frac{P_2}{P_1} = \left(\frac{Q_2}{Q_1}\right)^2 = \left(\frac{V_2}{V_1}\right)^2 = \left(\frac{N_2}{N_1}\right)^2 \qquad (11)$$

and

$$\frac{E_2}{E_1} = \frac{P_2 Q_2}{P_1 Q_1} = \left(\frac{V_2}{V_1}\right)^3 = \left(\frac{N_2}{N_1}\right)^3 \qquad (12)$$

where:

E = fan horsepower
P = fan pressure
Q = fan flow
V = velocity of air through the nozzles or along the surface
N = fan speed.

The proper design of the drier requires optimization of the balance between the maximum possible heat transfer coefficient and minimum consumption of power. A slight improvement of heat transfer coefficient at a high increase of power might not be an economical solution. A considerable amount of energy is required to gain an increase in heat transfer coefficient as expressed in the equation below.

$$\frac{h_2}{h_1} = \left(\frac{E_2}{E_1}\right)^{0.24} \quad (13)$$

In addition the increased heat transfer coefficient does not necessarily indicate an increased drying rate. Whether the drying rate will be favorably changed by an increased heat transfer coefficient depends on the web drying characteristics.

A simple index of drier performance is the heat transfer coefficient attained with dry air at a temperature of 400°F and air energy at the nozzles of 1hp/ft² of surface.

The power required to move the air through the nozzles is calculated by the following equation.[6]

$$E_A = 0.63 \, C_D A_R \left(\frac{V}{4000}\right)^3 \frac{\varrho_t}{\varrho_{70}} \quad (14)$$

where:

E_A = kinetic air power for nozzle flow hp/ft²
C_D = orifice discharge coefficient (0.92)
A_R = ratio of orifice area to surface area
V = velocity of air flow through nozzles, or of air flow along the web surface
ϱ_t = density at temperature t
ϱ_{70} = density of dry air at 70°F.

Air is transported by fans and blowers. Fans are used for low pressures below 350 mm of water column. The main types of fans are centrifugal and axial flow fans. A centrifugal fan consists of a multiblade rotating wheel. The air intake is at the center of the wheel, air direction is changed by 90° and the air is discharged at the periphery of the wheel. The air pressure arises from two sources in a centrifugal fan: centrifugal force due to the rotation of an enclosed volume of air and the velocity imparted to the air by the blades and partly converted to pressure by the volute shaped fan casing.

The axial fans use the helical action of the rotating blades. The air does not change direction but flows along the axis of the propeller. Several types of axial fans are used.

Blowers or centrifugal compressors are used to move large volumes of air at pressure rises from 350 mm of water column to tens of atmospheres. Rotary blowers are positive displacement machines also suitable for higher pressure work.[25]

Fans are classified into several classifications according to their discharge pressure.

Class I—3.75″ water column (95 mm)
Class II—6.75″ water column (170 mm)
Class III—12.25″ water column (310 mm)
Class IV—over 12.25″ water column (over 310 mm)

Class I and II centrifugal fans are most commonly used for web driers. The air supply fan can be mounted on either side of the heater. If mounted downstream from the heater, the sealing around the heater is simplified. If the fan is mounted upstream from the heater, additional air volume is developed because of thermal expansion. The exhaust fan requires much less power than the air supply fan, usually only about 10% of that required for air supply.

The supply air to the drier and the exhaust air is contained by the ductwork. Ducts are made from black steel plate, galvanized steel, stainless steel and coated fabric. For use with flammable vapors, ducts must be made from noncombustible materials.[26] Ducts may be round, square or rectangular. Round ductwork offers the least resistance to the air flow. Elbows might have turning wanes to reduce frictional losses. The turning wanes are recommended for 45° elbows and mandatory for 90° ones, if the air volume is large.[27] Flexible duct connections are needed to accommodate duct expansion or movement. Flexible bellow type, sliding or breakaway joints are used for such purpose.

The drier is constructed from insulated panels. Tongue and groove connections are often used.

Dampers might be used to control the air flow, although this method is inefficient and the air flow is best controlled by the blower speed. Air filters are usually required on the air intake side.

Air Recirculation

It is advantageous from the energy conservation point of view to recirculate as much air as possible. The recirculation is limited by accumulation of an excessive amount of flammable vapor and creating an explosion hazard, accumulation of excessive amount of vapor and decreasing the rate of drying, or by accumulation of contaminants. In multiple zone driers some air might be recirculated by cascading: the exhaust gases from one zone might be used as supply for another.

In the solvent coating process the allowable flammable solvent vapor concentration is 25% of the lower explosive limit (LEL). This concentration may be increased to 40-50% LEL if proper instrumentation for monitoring of the flammable gas concentration is used. Combustible gas analyzer based on comparing the calorimetric value of the sample with a standard has been commonly used. The gas is passed over the gas flame in such an apparatus. Another method is to burn the gas over hot wire, or a catalyst, which is a part of a Wheatstone bridge. The heat produced by the flammable vapor causes an imbalance in the resistance bridge, which is proportional to the flammable vapor concentration. Hot wire or catalytic filament sensors cannot be used in the presence of silicones which may accumulate on the catalyst's surface, causing poisoning. The presence of other impurities, which might have a tendency to get deposited on hot surfaces, could also cause problems with such sensors.

Infrared analyzers, although considerably more expensive, are used quite extensively for this application. Such an analyzer operates on non-disperse infrared principle. Twin infrared beams compare the gas in the sample cell with the reference cell. A schematic diagram of such an analyzer is shown in Fig. 15-63. The analyzer is connected to a visual or audio alarm which is actuated when the concentration of flammable gas exceeds the preset value.

SOLVENT RECOVERY AND INCINERATION

It might be required to remove the solvent vapor from the drier discharge because of the environmental regulations, or it might be economically desirable to recover the solvent for its reuse. While there are several processes available for removal of solvent vapor, adsorption by activated carbon and incineration are the only two processes of major practical importance.

Gas Adsorption

The effluent stream is passed through an activated carbon bed and the hydrocarbon vapors are adsorbed. Adsorption efficiency is 90-99%, depending on the specific solvent vapor and on the design of the carbon bed.

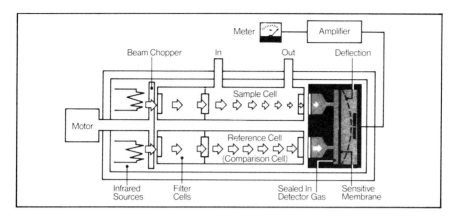

Fig. 15-63. A schematic diagram of an infrared analyzer for combustible gas. (*Courtesy Mine Safety Appliances Co.*)

When the carbon bed becomes saturated, the discharge flow is directed to a fresh bed and the saturated bed is recovered by stripping the solvent by either steam or vacuum. Steam stripping is the preferred method for most coating operations, vacuum stripping has found applications in lower volume uses. The solvent is removed along with the condensate and it is separated by decantation, if the solvent is water insoluble, or by distillation, if it is soluble. Gas adsorption technology has been reviewed by Bethea[28] and by many other authors. Figure 15-64 shows a schematic diagram of a carbon absorption system stripped by vacuum.

A gas adsorption system might be equipped with an infrared analyzer to

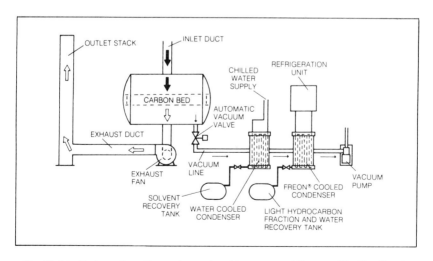

Fig. 15-64. Carbon adsorption system stripped by vacuum. (*Courtesy Met-Pro Corp.*)

monitor the solvent vapor concentration in the gas leaving the carbon bed. When an increase of solvent vapor concentration is registered, the exhaust is automatically switched over to another carbon bed and the regeneration cycle is started.

Incineration

Combustion of solvent vapor in the drier effluent is carried out simply by raising the temperature of the discharge sufficiently high to affect the oxidation of solvent vapor to carbon dioxide and water. The required temperature is about 650°-900°C and the residence time in the combustion chamber about 0.3 seconds. The combustion temperature can be decreased to 250°-450°C by employing a porous catalyst. The catalyst is platinum, palladium or rhodium deposited on a ceramic honeycomb. The contact time with the catalyst bed is about 0.3 seconds. The catalyst can be poisoned by lead, mercury and other heavy metals. Halogens and sulfur dioxide can decrease the effectiveness of the catalyst.

Incineration requires raising the temperature of a large volume of gas and it is important to employ an efficient heat recovery system in order to conserve the energy. In addition to using the hot exhaust gases to preheat the gas entering the incinerator, hot exhaust may be used to preheat water that is utilized elsewhere outside of the coating process. A photograph of an incinerator is shown in Fig. 15-65 and a schematic diagram of air flow for a two-zone drying oven equipped with an incinerator is shown in Fig. 15-66.

INERT GAS DRYING

In drying of solvent based coatings, most of the energy is spent on heating a large volume of air. Since the allowable flammable solvent concentration is 25% LEL (or up to 50% LEL if proper instrumentation is used), the recirculation is quite limited and large quantities of hot air are emitted to the atmosphere. If the oxygen concentration in the drying gases is decreased, then the solvent vapor concentration can be increased beyond the above-mentioned limits without a danger of explosion or fire.

Such a system has been designed by Midland-Ross in 1972 and the first line was installed in 1976[29,30] (see Fig. 15-67). The exhaust gases carrying solvent are burned and returned to the system instead of being discarded to the atmosphere. If burned under stoichiometric solvent-to-oxygen ratio, such flue gases contain no oxygen, and, when recirculated, provide an inert gas system. When the gas mixture is short on oxygen, flammable concentrations are not possible even at high solvent vapor concentrations. Figure 15-68 shows the hexane-oxygen-nitrogen mixtures which may form flammable gas mix-

DRYING 311

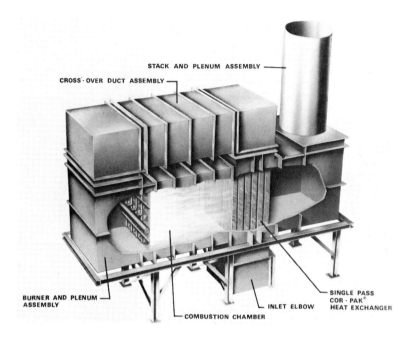

Fig. 15-65. Exhaust gas incinerator. (*Courtesy C-E Air Preheater, Combustion Engineering.*)

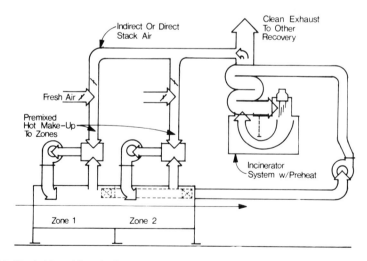

Fig. 15-66. A drier with an incinerator and heat recovery system. (*Reprinted with permission from Luedke.*[39])

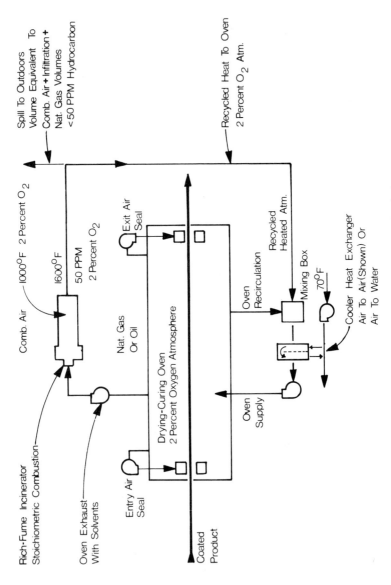

Fig. 15-67. Midland-Ross inert gas system for drying ovens.

DRYING 313

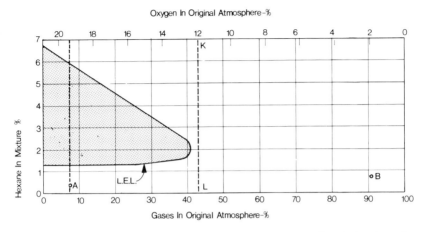

Fig. 15-68. Relation between oxygen, nitrogen and heptane concentrations in forming a flammable mixture. (*According to Janett.*[10] *Courtesy Ross-Waldron Division, Midland-Ross Corp.*)

tures (cross-hatched area). Point A represents a vapor level equal to 25% LEL, which is a normal operating level for driers. Addition of solvent vapor to the mixture represented by point A causes the concentration to rise along the dotted line entering the area of flammability. Any mixture to the right of line K-L cannot be made flammable by solvent vapor addition. The inert gas drying systems operate at about 2% oxygen concentration as represented by a point B. Such mixture does not have a sufficient amount of oxygen to become flammable on addition of solvent vapor.

A similar inert gas drying system has been designed by Pagendarm. In addition to operating in an inert gas atmosphere, solvent vapor is also separated by condensation. Such a system recovers more than 98% of the solvent. Figure 15-69 shows a schematic diagram of such a system. The exhaust gases with solvent vapor concentration of 190–360 g/nm³ are removed from the drier, 3, cooled down to 5°C in a heat exchanger, 10, and led into the condenser, 9, where they are cooled down to -40°C. the gases are returned to the drier and warmed up by passing through the secondary side of the heat exchanger, 10. Water, if removed from a web such as paper, passes through the separator, 12, and is accumulated in the separator vessel, 13. If the solvents used are not water soluble, solvent is separated by decantation. This system is viable for solvent evaporation rates above 300 kg/hr.

CURLING

Webs consisting of layers of different materials are subject to curling, if the dimensional changes take place on one side because of shrinkage or expan-

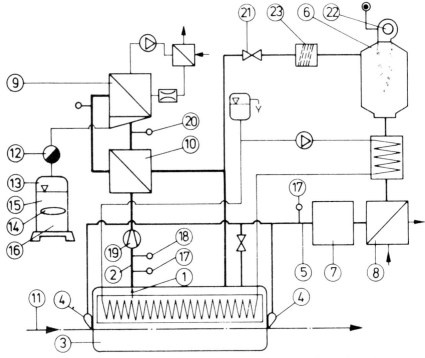

Fig. 15-69. Inert gas drying with solvent condensation. (*Courtesy Pagendarm Apparatebau GmbH.*)
1. Vapor exhaust
2. Vapor duct work
3. Drier
4. Barrier nozzles
5. Solvent-free inert gas line
6. Combustion chamber (propane 3 kg/hr)
7. Cooling
8. Heat exchanger
9. Condenser (inert gas cooled to -40°C)
10. Heat exchanger to 5°C
11. Web
12. Separator
13. Separation tank
14. Float
15. Liquid phase (solvent)
16. Water
17. Oxygen concentration monitor
18. Solvent concentration analyzer
19. Fan
20. Temperature regulator
21. Valve
22. Propane burner
23. Flame arrestor.

sion. Shrinkage of the coating may take place during drying and cause curling towards the coating. Paper is especially sensitive to curling because the fibers swell and increase in size on water absorption. Coated paper will curl if the uncoated side is wetted. Curl also can be introduced by a roll set, by deforming the material on application of excessive tension and by many other processes which either cause a dimensional change or introduce surface stresses. Curling and its prevention have been discussed by several authors. Spitz and Blickensderfer[31] have proposed that a property called curl tendency is more fundamental than curl. Haagen[32] has discussed the causes of stresses in coatings. Problems of curl in extrusion coated papers have been described by Bates and Marsella.[33] Measurement of internal stresses has been proposed by Gusmann.[34]

Absence of curling in an anisotropic material requires that the surface stresses are either absent or equal on both sides of the web. The latter condition is difficult to achieve and the stress relaxation is a much more promising and easier way to remove the curl, rather than trying to balance the stresses on both sides of the web.

Paper is a commonly used web in the converting industry and it is prone to curling because of loss of moisture in the drier. To prevent or minimize curling of coated or laminated paper webs, moisture is added at the end of drying, in order to restore the approximate equilibrium moisture content. Moisture also plasticizes cellulose and helps to relieve the stresses that might have been introduced into the web during its processing and drying.

Water is introduced by several methods. Rolls of the finished product might be stored in a high-humidity atmosphere. Moisture might be introduced by wetting the back by a roll coater, by mechanical spraying or by employing electrically charged water particles. The web might pass through a steam chamber where the steam condenses on the surface and inside the fibrous structure. The amount of condensed steam depends on the temperature of the entering web. The lower the temperature, the more steam will condense. Figure 15-70 shows a chamber of this type. Figure 15-71 shows a unit for moisture addition, where the web is passing through a condensation chamber supported by coiled rolls. The cooling rolls limit the temperature

Fig. 15-70. Steam chamber for moisture addition. (*Courtesy Pagendarm Apparatebau GmbH.*)

316 WEB PROCESSING AND CONVERTING TECHNOLOGY AND EQUIPMENT

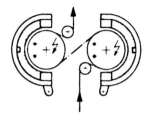

Fig. 15-71. Moisture addition by steaming the web supported on chilled rolls. (*Courtesy Pagendarm Apparatebau GmbH.*)

rise of the paper and increase the condensation rate. The introduction of moisture by such methods has been discussed by Pagendarm.[35]

Roll set is a source of curling that causes problems in sheeting, die-cutting and other converting operations. It can be especially severe on heavier substrates. Figure 15-72 shows a pressure decurling unit which eliminates the roll set. This equipment has been described by Manfredi.[36]

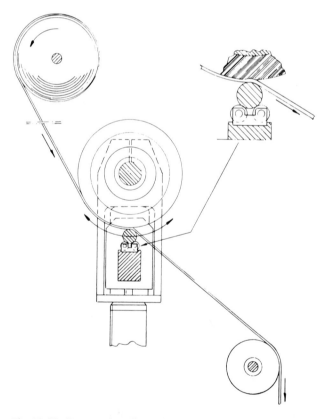

Fig. 15-72. Pressure decurling unit. (*Courtesy Moore and White Co.*)

REFERENCES

1. Hougen, O. A., McCauley, H. J. and Marshall, W. R. *Trans. Am. Inst. Chem. Engrs.* **36**:183-202 (1940).
2. Keey, R. B. *Drying: Principles and Practice.* New York: Pergamon, 1972.
3. Satas, D. *Handbook of Pressure-Sensitive Adhesive Technology.* D. Satas (Ed.). New York: Van Nostrand Reinhold, 1982, pp. 533-557.
4. Colburn, A. P. *Trans. Am. Inst. Chem. Engs.* **29**:174-210 (1933).
5. Gardner, T. A. *Tappi* **43**(9):796-800 (1960).
6. Gardner, T. A. *Pulp Paper Mag. Canada* **62**(6):T327-T332 (1961).
7. Hultman, J. D., Leekley, R. M. and Garey, C. L. *Tappi* **60**(11): 105-109 (1977).
8. Wedel, G. L. *Tappi* **63**(8):89-92 (1980).
9. Janett, L. G., Schregenberger, A. J. and Urbas, J. C. *Pulp Paper Mag. Canada* **66**(1):T20-T26 (1965).
10. Janett, L. G. *Update on Drying Systems, Coated Products, Systems and Markets.* Institute of Graphic Communications Conference, August 24-26, 1980, Andover, Mass.
11. Overly, Wm. F. and Pagel, K. J. U.S. Patent 3,629,952 (1971) (assigned to Overly, Inc.).
12. Richardson, C. A. and Lawton, D. W. *Tappi* **56**(4):86-89 (1973).
13. Lawton, D. W. *Tappi* **57**(6):105-107 (1974).
14. Villalobos, J. A. *Engineering Considerations in Through Drying.* International Water Removal Symposium, March 20, 1975, London.
15. Metcalfe, W. K. *Paper Age* **10**(3):28 (1974)
16. Daane, R. A. and Han, S. T. *Tappi* **44**(1):73-80 (1961).
17. Lee, P. F. and Hinds, J. A. *Tappi* **62**(4):45-48 (1979).
18. Cowan, W. F. *Tappi* **47**(12):808-811 (1964).
19. *Bulletin No. 50-580-80.* Fostoria Industries, Inc., Fostoria, Ohio.
20. Munce, J. *Infra-red Process Heating. SP 193 Rev. 11/74.* Eclipse Combustion Division of Eclipse, Inc., Rockford, Ill.
21. Trembley, J. F. and Loving, C. M., Jr. *Tappi,* **52**(10):1847-1850 (1969).
22. Beckwith, W. F. and Beard, J. N., Jr. *Transactions of ASME* **101**:80-84 (1979).
23. Lawton, D. W. *Energy Conservation in Drying Coating.* TAPPI Coating and Graphic Arts Conference, October 1975, Philadelphia.
24. Weiss, H. L. *Coating and Laminating Machines.* Milwaukee: Converting Technology Co., 1977.
25. Genereaux, R. P. *Chemical Engineer's Handbook, 5th Ed.* Robert H. Perry and Cecil H. Chilton (Eds.). New York: McGraw-Hill, 1973, pp. 6-21.
26. *Blower and Exhaust Systems 1973.* NFPA 91. National Fire Protection Association, Quincy, Mass., 1973.
27. Booth, G. L. *Coating Equipment and Processes.* New York: Lockwood, 1970.
28. Bethea, R. M. *Air Pollution Control Technology.* New York: Van Nostrand Reinhold, 1978.
29. Grenfell, T. N. *What's New in Oven Design? Interair Systems.* Technical Meeting on Water Based Systems, Pressure-Sensitive Tape Council, June 21-22, 1978, Chicago.
30. Hemsath, K. H., Thekdi, A. C. and Vereecke, F. J. U.S. Patent 3,909,953 (1975) (assigned to Midland-Ross Corp.).
31. Spitz, D. A. and Blickensderfer, P. S. *Tappi* **46**(11):676-689 (1963).
32. Haagen, H. *Farbe und Lack* **85**(2):94-100 (1979).
33. Bates, A. and Marsella, L. J. *Tappi* **47**(7):133A, 168A-170A (1964).
34. Gusmann, S. *Official Digest* **34**(2):895 (1962).
35. Pagendarm, R. *Wochenblatt fuer Papierfabrikation* **17**:670-673 (1979).
36. Manfredi, J. L. *Paper, Film and Foil Converter* **52**(11):78, 80 (1978).

37. Johns, R. E. *Tappi,* **61**(2):41–44 (1978).
38. Satas, D. *Some Aspects of Drying and Coating.* Proceedings Pressure Sensitive Tape Council 1983 Technical Seminar, Itasca, Illinois.
39. Wedtke, E. R., Jr. *Fume Incineration for Process Heat Recovery.* 1979 Paper Synthetics Conference Proceedings, TAPPI, 1979, pp. 305–315.

16
Ultraviolet Irradiation

Donatas Satas
Satas & Associates
Warwick, Rhode Island

Ultraviolet (UV) irradiation curable coatings have occupied an important place amongst various surface coatings used for functional and decorative purposes. UV irradiation curable coatings consist of a blend of reactive monomers or oligomers capable of free radical initiated polymerization. Photoinitiators are used with UV curable coatings and they are the source of free radicals produced on irradiation.

UV curable coatings have been developed for many different end uses. Flexible coatings for vinyl flooring and vinyl film, coatings for wood, filler sealers, basecoats and topcoats, printing inks, overprint varnish (high gloss and matte), coil coatings, solder masks, vapor barrier coatings, prime coatings for metallizing, pressure-sensitive adhesives,[1] metal coatings for name plates, electronic parts, cans, coatings for fiber optics, abrasion-resistant coatings for plastics and many other types of coatings are among the numerous applications.

Many different chemicals serve as photoinitiators. Benzoin and derivatives, aromatic carbonyls, condensed ring compounds, halogenated chemicals and amines have found application as photoinitiators.

UV radiation encompasses radiation of the wavelength 200–400 nm.* Most of the radiation sources used emit radiation in very narrow wavelength bands (see Fig. 16-1). Different photoinitiators require different wavelength and the radiation from the source and photoinitiator sensitivity should be matched. The photoinitiator should absorb ultraviolet radiation in a range which is not absorbed by the monomers or the pigment. The most commonly

---
*nm = nanometer, 10^{-9} m.

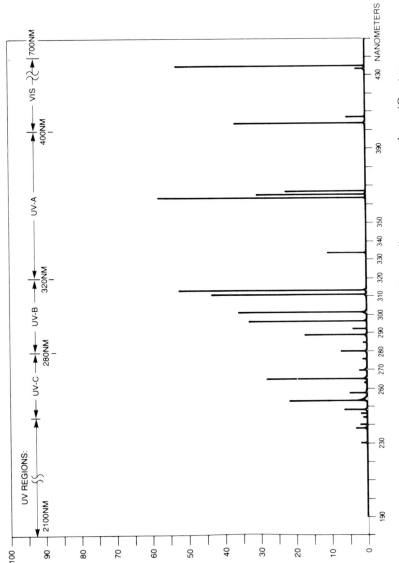

Fig. 16-1. Spectral distribution of radiation by a medium-pressure mercury lamp. (*Courtesy Canrad-Hanovia, Inc.*)

used UV source, medium-pressure mercury lamps, emit over a wide range of wavelengths and is suitable for all UV curing applications. UV light sources have been reviewed by McGinnis.[2]

The rate of curing reaction depends on the number of free radicals produced and thus upon the density of UV radiation. The relationship is not simply linear, especially in air, because of the inhibiting effect of oxygen. Once the inhibiting effect of oxygen is overcome, any additional radiation energy input rapidly increases the curing rate. The change of the conveyor speed has similar effect, since it is inversely proportional to the radiation density received by the coating.

Most of UV radiation is absorbed near the surface of the coating and thicker layers require extended irradiation time. Generally UV curing is practical for thin surface coatings only.

UV irradiation equipment consists of the following parts:

- Radiation source
- Lamp housing and reflector
- Cooling accessories
- Power supply and controls
- Shielding and safety equipment
- Conveying system.

In addition to the conveying system, coating equipment to apply the curable coating onto the web, unwind and rewind stands and other web handling equipment is required.

RADIATION SOURCES

There are several different UV radiation lamps available for curing and other applications: low- medium- and high-pressure mercury lamps, xenon gas lamps and several other types. Medium-pressure mercury lamps are by far the most important radiation sources for curing of coatings and only a short note on the other types will be sufficient.

Low-Pressure Mercury Lamps

These lamps are constructed from a quartz tube filled with gas (argon, neon) and a small amount of mercury (10^{-3} torr). The gas aids in the ionization. Two basic types of low-pressure mercury lamps are available: heated cathode and cold cathode. The radiation is mainly around the wavelength of 185 nm (resonance fluorescence) and 254 nm (resonance phosphorescence). The power rating of such lamps is 0.4–4 watts/cm. The radiation intensity of such lamps is low and the curing must take place in an inert atmosphere. The bulb temperature is low, about 40°C, and such equipment is suitable for substrates sensitive to elevated temperatures.

The main application of low-pressure mercury lamps is for sterilization and they are generally known as germicidal lamps.

Xenon Lamps

Xenon lamps can be constructed to discharge the radiation in flashes. The simplest method is to have a capacitor discharging at regular intervals. Flashing lamps provide a high-intensity, although short-duration, radiation and such lamps are useful for slow irradiation of thick layers. Xenon lamps are inefficient and the power requirements are high. The bulb surface must be cooled. This makes for a complex and expensive system.

High-Pressure Mercury Arcs

These operate at 20 atm of mercury vapor pressure and emit a broad band radiation. Some lamps might operate at several hundred atm pressure.

Medium-Pressure Mercury Lamps

These lamps are the most important for UV curing of coatings. These lamps emit in the 200–400-nm wavelength range. Figure 16-1 shows the spectral distribution of a medium-pressure mercury lamp.

The lamp consists of a straight quartz tube about 2.5 cm in diameter. Clear white quartz transmits radiation from 180 nm through the infrared wavelength. Ozone-free quartz may be used to eliminate the ozone as a radiation by-product. Such quartz is doped with special additives which cut out radiation below 220 nm.[3] The lamp is filled with inert starting gas, which is easily ionized, when exposed to an electrical potential, and a precise amount of mercury. The pressure developed in the medium-pressure mercury lamps is 1–2 atm at operating conditions. At room temperature the mercury is liquid.

There are two modifications of medium pressure mercury lamps depending on the method of exciting the mercury atoms. Electric discharge excites the mercury atoms in the electrode arc lamp. The electrodes are located at both ends of the tube and they are connected either to wires or to metal caps. When connected to an electrical power source, an arc is established between the two electrodes. As the temperature increases, mercury vaporizes and the pressure inside the lamp increases. This warm-up period might take 2–5 min. If the power is shut off, a period of 10–15 min is required before the lamp can be started: the mercury vapor must recondense before striking an arc again. Additional circuitry can be added to decrease the warm-up and restrike periods. The temperature of the inner core discharge reaches 6000°K, the outside of the lamp is at about 800°C and the electrodes might be at 300°C or lower.[4] If the outside lamp temperature drops to below 600°C, mercury starts condensing and its pressure drops, causing a decrease in radiation density. The construction of electrode discharge lamps is shown in Fig. 16-2.

The electrode discharge lamps are available in lengths of 4–200 cm and power ratings of 25–300 watts/cm, although the 75 watt/cm lamps are used most often. The power level can be varied by changing the capacitance in the secondary circuit. The lifetime of a lamp depends on the number of operating hours, the number of starts and also on the handling and maintenance. The lamps might last for 1000–5000 hr. Usually the UV radiation output is expected to decrease to 85% of the original after 1000 hr. The deterioration rate increases after 1000 hr of use.

The other type of medium pressure mercury lamps are electrodeless and they are excited by radio frequency electrical vibrations. These lamps reach their full efficiency in about 15 sec. The lifetime of electrodeless lamps is much longer. Maximum length of the bulb is 25 cm and several lamps are needed for irradiation of wider webs. The conversion efficiency of electrical energy to radiation is lower than that for electrode discharge lamps. The equipment is more expensive, because of complicated electrical circuitry and radio frequency shielding. The characteristics of various lamps are listed in Table 16-1.

Mercury lamps should be cleaned before operation using pure solvent or clean water with detergent. Gloves should be used when handling the lamps. Fingerprints, if left on the lamp, might become etched into the quartz, changing the radiation intensity at that place.

The output spectra of medium-pressure mercury lamps can be changed by addition of various metallic materials or metal halides (iron iodide). A typical 75-watt/cm input power lamp emits 4–5 watts/cm in the region of 350–400 nm, which is the most important for initiation of curing reactions. Addition of iron iodide can increase the radiation energy three times in that

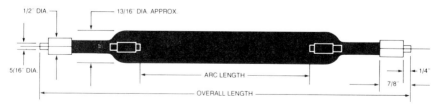

Fig. 16-2. Construction of an electrode discharge medium-pressure mercury lamp. (*Courtesy Canrad-Hanovia, Inc.*)

Table 16-1. Characteristics of Various UV Irradiation Lamps

	LOW PRESSURE MERCURY	XENON PULSING	MEDIUM PRESSURE MERCURY	
			ELECTRODE DISCHARGE	RADIO FREQUENCY ENERGIZED
Input power, watts/cm	0.4–4		40–175	215
Lamp power-watts/cm	0.4–4	0.04–4 x 10^6 at peak	75–120	120
Lamp temperature	Cool	Moderate	High	High
Arc lengths, cm	25–190	1.5–75	4–200	25
Lamp warranty, hrs	17,500	1000	1000	3000
Spectral efficiency	Excellent	Poor	Good	Very good
Radiant efficiency	Very good	Poor	Good	Fair
Overall efficiency	Fair	Poor	Good	Good
Spectral variations	None	Limited	Moderate	Extensive
Relative system costs	Low	High	Moderate	High

region.[5] This is a much better way to increase the radiation density than by going to higher power input (75–120 watts/cm). Higher power input decreases the operational lifetime of the lamps. The radiation of mercury lamps can also be pulsed similarly to xenon lamps for a concentrated high-energy pulse of a short duration.

REFLECTORS

Several types of reflectors might be used:

- Elliptical focusing
- Parabolic focusing
- Flat non-focusing.

Reflector design has been reviewed by Gamble.[7] Parabolic reflectors provide a parallel beam of radiation. Focusing elliptical reflectors are most often used with medium-pressure mercury lamps. The lamp is mounted in one of the focal points of the ellipse and the substrate to be irradiated passes at the height of the second focal point. This is illustrated in Fig. 16-3. Rubin[8] has investigated the variation of relative intensity of radiation along the vertical and horizontal axes. Figure 16-4 shows the variation of relative irradiance as a function of the vertical distance from the lamp. The maximum intensity is reached at a distance of 8 cm from the lamp. The minimum of radiation intensity is observed at 3.5 cm. The radiation intensity then rapidly increases with decreasing distance, but it is not practical to get closer to the

ULTRAVIOLET IRRADIATION 325

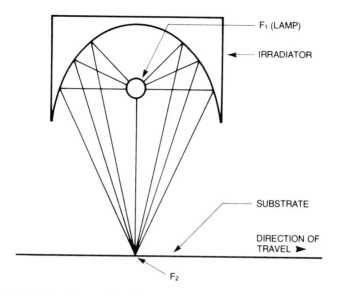

Fig. 16-3. Elliptical reflector and positioning of the lamp. (*Courtesy Canrad-Hanovia, Inc.*)

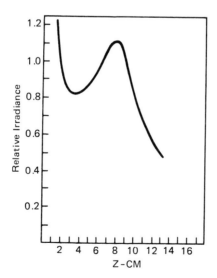

Fig. 16-4. Variation of UV irradiation with vertical distance from the lamp (z) in an elliptical reflector.

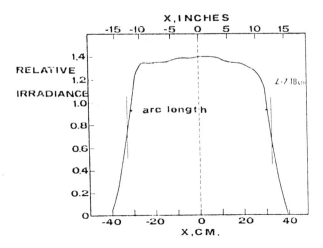

Fig. 16-5. Variation of UV irradiation across the web width.

lamp, since the reflector interferes with the web and the material could be easily damaged by the hot lamp.

Figure 16-5 shows the radiation intensity distribution along the horizontal axis across the width of the web. The radiation intensity is fairly constant across 80% of the arc length.

COOLING

Medium-pressure mercury lamps generate a large quantity of heat which has to be removed in order to avoid overheating of lamp terminals, reflector, shutter, if it is used, or the substrate that is irradiated. Water or air cooling might be used. Construction of a hollow reflector body allows the air passage and cooling of the lamp seals and sockets without overcooling the lamp surface itself. Lamp surface, if cooled below 600°C, will cause condensation of mercury. A specially designed hollow reflector body is shown in Fig. 16-6.

Water cooling of the reflector and the shutter is not sufficient, some flow of air is required to remove the generated ozone.[10]

Shutter

For irradiation of continuous web, unlike for irradiation of sheets, UV curing equipment must be capable of instant stopping in order to avoid overheating of the web that might cause distortion or even fire. Mechanical shutters might be used for this purpose.

ULTRAVIOLET IRRADIATION 327

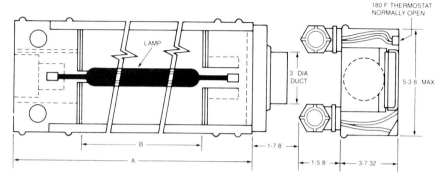

Fig. 16-6. Hollow elliptical reflector body designed for air cooling. (*Courtesy Canrad-Hanovia, Inc.*)

Power Supply and Electrical Controls

As the arc intensity in a medium-pressure mercury lamp increases, the electrical resistance decreases and without input power control the increasing current would destroy the lamp.[10] For this reason a constant wattage ballast is used with each tube. The ballast may have a capacitor switching as shown in Fig. 16-7 to allow the reduction of UV radiation output. Stabilized ballast allows to maintain the power variation as low as ± 3% when the line voltage varies ± 10%.

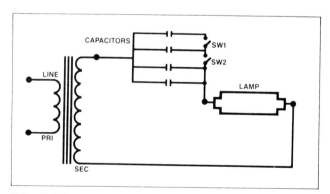

SW1 & SW2 CLOSED = 100%
SW1 OPEN, SW2 CLOSED = 75%
SW1 & SW2 OPEN = 50%

Fig. 16-7. Power supply diagram for a medium pressure mercury lamp. (*Courtesy Canrad-Hanovia, Inc.*)

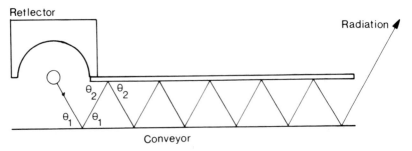

Fig. 16-8. Shielding parallel to the web permits escape of stray radiation.

Shielding and Safety Accessories

UV radiation produces ozone which must be exhausted to the outside along with a small amount of vaporized monomer. Shielding must be provided to protect the personnel from direct or first reflected UV radiation. The area of web entrance and exit requires shielding. An effective shielding is provided by an enclosure which is not parallel to the substrate surface.[11] A parallel enclosure might allow the stray radiation to escape as illustrated in Fig. 16-8, but an enclosure constructed at a 10° incline effectively blocks the escape of stray radiation as illustrated in Fig. 16-9.

The atmosphere in the vicinity of the radiation lamp is very corrosive, because of the elevated temperature and presence of ozone. Aluminum is the preferred material of construction. The internal surfaces might be painted with black elevated temperature-resistant paint. Silicone rubber sheets and silicone sealants are used to block the radiation. A photograph of a typical UV curing unit is shown in Fig. 16-10.

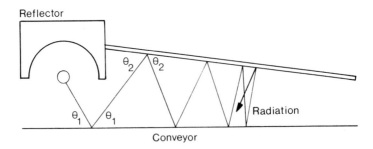

Fig. 16-9. Shielding at an angle with the web stops stray radiation.

ULTRAVIOLET IRRADIATION 329

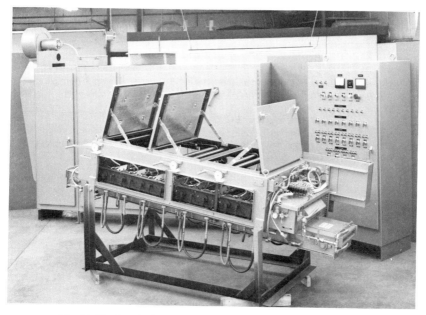

Fig. 16-10. A typical UV radiation unit. (*Courtesy RPC Industries.*)

APPLICATION OF UV-CURABLE COATINGS

UV curable coatings are usually 100% solids and solvent removal equipment is not required. Only infrequently UV curable coatings might be diluted with a solvent in order to improve its coatability. The coating might be sometimes applied, and heated in order to decrease its viscosity and increase its reactivity upon irradiation.

Coating methods generally suitable for applying lightweight uniform coatings are also used for UV curable coatings. Offset and sometimes direct gravure, wire wound rod, various roll coating methods and rotary screen have been used for coating applications over various substrates. Sheeted material may be coated by roll coaters, or for optically perfect coatings, by dip or flow coating. UV curable inks may be applied by lithography and silk-screening. Their penetration into flexo and gravure printing has not been extensive.[9]

REFERENCES

1. Dowbenko, R. *Handbook of Pressure-Sensitive Adhesive Technology*. D. Satas (Ed.). New York: Van Nostrand Reinhold Co., 1982, pp. 586-604.
2. McGinnis, V. D. *UV Curing: Science and Technology*. S. P. Pappas (Ed.). Norwalk: Technology Marketing Corporation, 1980, pp. 96-132.
3. Lienhard, O. E. *SME Technical Paper* **FC75-335** (1975).
4. Coppinger. *Radiation Curing* 3(2):4 (1976).
5. Panico, L. R. *SME Technical Paper* **FC 75-326** (1975).
6. Searls, R. *SME Technical Paper* **FC78-544** (1978).
7. Gamble, A. A. *J.O.C.C.A.* **59**:240-244 (1976).
8. Rubin, H. J. *J. Paint. Tech.* **46**(588): 74-81 (1974).
9. Bean, A. J. and Bassemir, R. W. *UV Curing: Science and Technology*. S. P. Pappas (Ed.). Norwalk: Technology Marketing Corporation, 1980, pp. 185-228.
10. Roffey C. G. *Photopolymerization of Surface Coatings*. Chichester: John Wiley & Sons, 1982.
11. Kirk, N. *Radiation Curing* 1(3):24-29 (1974).

17
Electron Beam Irradiation

William G. Baird, Jr.
Cryovac Division
W. R. Grace & Co.
Duncan, South Carolina

While the use of accelerated electrons to produce a picture on a television screen or readout on a computer terminal is well known, an extension of this technology to much higher voltages to carry out various chemical reactions on a commercial scale has been less widely publicized.

Electron beam processing is actually a very large industry and even confining the scope of this chapter to web processing still results in a formidable list of successful commercial applications. These include manufacture of shrink films and bags, vulcanization of rubber, electrical insulating materials, wire and cable, pipeline wraps, foamed polymers, cloth finishing, web sterilization, adhesive, print and coat curing, to name a few.

This chapter will concern itself only with the electron beam processing equipment, controls, safety and auxiliaries. Theoretical aspects of radiation chemistry are beyond the scope of this chapter. Readers who wish to become more familiar with this field of science should refer to the many books and papers on the subject. A partial list of these is to be found at the end of the chapter.

While x-rays or radioactive isotopes such as Co_{60}, Ce_{137} could be used for processing web materials, their very great penetrating power makes them inefficient for this application. Thus the electron accelerators that will be discussed here are those in the 150–800-kV range ruling out the very high-voltage 1–5-MEV machines and equipment such as linear accelerators (4–20 MEV), etc., which would require excessive festooning of the web to be of use.

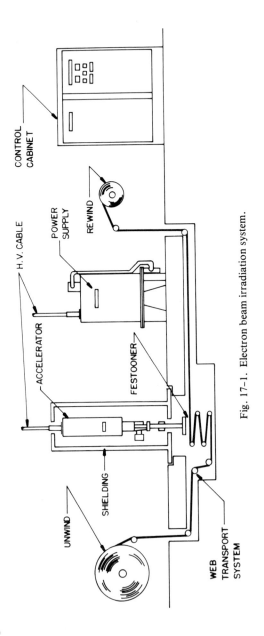

Fig. 17-1. Electron beam irradiation system.

ELECTRON BEAM GENERATOR SYSTEM COMPONENTS

A complete electron beam generator system consists of a dc power supply, accelerator tube, X and Y scan axis circuits, voltage and current control networks, vacuum equipment, safety devices and shielding. In addition, auxiliaries such as window cooling systems, web festooners and inert gas sources as needed.

The Power Supply

All of the currently available accelerators have a source of stabilized high-voltage dc. For the lower end of the voltage range (150-300 kV), power supplies similar to those used in medical x-ray and radiography suffice. They consist of a high-voltage transformer, suitable tube or solid state rectifiers and associated control, measuring and switching equipment.

For higher voltage 300 KV and up, other, less orthodox dc power supplies

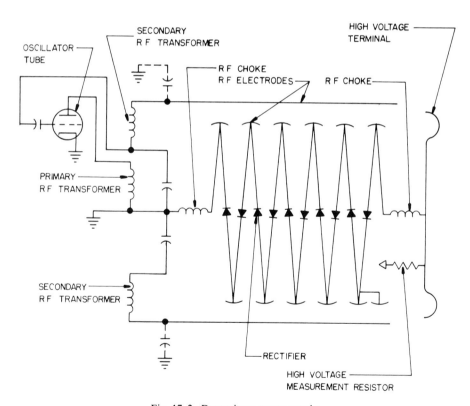

Fig. 17-2. Dynamitron power supply.

334 WEB PROCESSING AND CONVERTING TECHNOLOGY AND EQUIPMENT

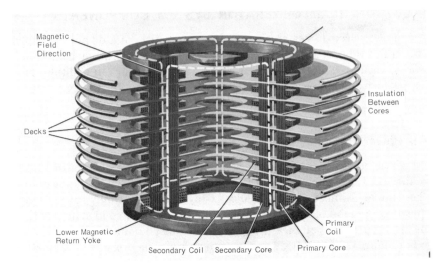

Fig. 17-3. Insulating core transformer.

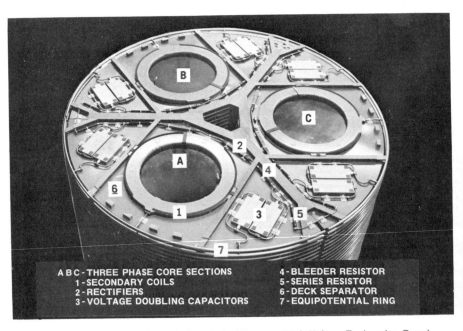

Fig. 17-4. Insulating core transformer deck. (*Courtesy High Voltage Engineering Corp.*)

ELECTRON BEAM IRRADIATION 335

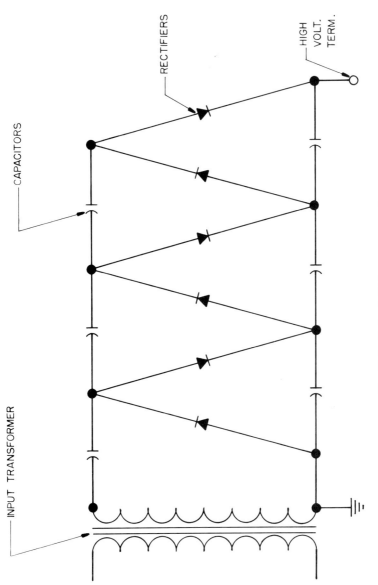

Fig. 17-5. Cockroft Walton power supply.

may be used. These include Van de Graf generators, Dynamitron, Insulating Core Transformers, Cockcroft Walton and others. The latter three types of supplies are currently used in most of the equipment sold in the 500–800-kV range.

The Dynamitron power supply basically consists of radio frequency oscillator operating in the frequency range which is capacitively coupled to a series of solid state cascaded rectifier sections to produce the desired high voltage. A control console containing a microprocessor is used to control the operation.

The Insulating Core Transformer (ICT) is a three-phase, inductively coupled transformer which operates at 50 or 60 cycles directly off the power main. Three-phase utility power is fed to a voltage regulator which adjusts the ICT input voltage over the 0–500-volt range. The regulator output is fed directly to the primary windings of the ICT. When the primary windings are energized, a magnetic flux is established in the three secondary legs. A copper secondary winding and solid state rectifier network is associated with each segment of each secondary leg. The output of these rectifier decks are added in series to produce the required high voltage dc.

The Cockroft-Walton is a single- or multi-phase, series fed, capacitively coupled rectifier cascade. Utility supplied power is fed to a voltage step/up frequency converter which typically increases frequency to between 2 and 4 kHz. This high frequency ac is fed to a powerstat which directly feeds a three-phase, multi-stage, capacitively coupled, dc rectifier cascade. The output of the individual sections of the cascade are added in series to produce the required high-voltage dc.

Accelerator Tube

There are two distinct types of accelerator tubes, one based on familiar cathode ray tube techniques employing a point source of electrons, one or more accelerator anodes together with associated focus coils, X and Y scan coils and a thin foil window. This type of tube may be situated at a remote site from the power supply (the high voltage being supplied by cables) or may be coaxially mounted in the center of the power supply.

The linear cathode accelerator has a continuous line filament electron source which is surrounded by a focusing/modulator structure. It is provided with a suitable window through which the high-voltage electrons are emitted. With this type of arrangement, no X and Y scan is needed, thus greatly simplifying the circuitry of the system.

The electrons in order to do useful work must exit the tube through a window. This is usually 1.5–7-mil aluminum or titanium foil. In doing so, losses in the window result in heating, which limits the current which can be ob-

ELECTRON BEAM IRRADIATION 337

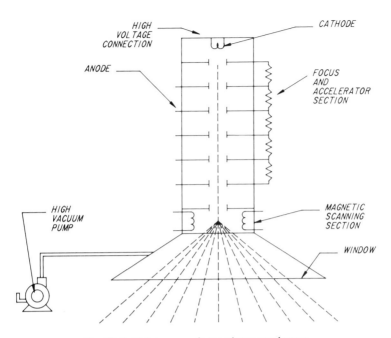

Fig. 17-6. Point source electron beam accelerator.

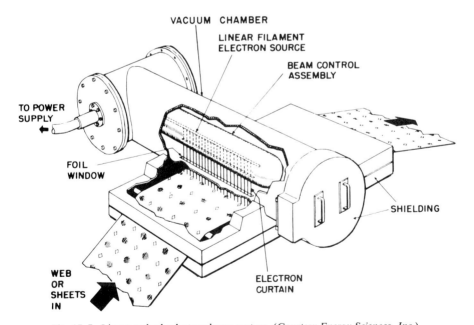

Fig. 17-7. Linear cathode electron beam system. (*Courtesy Energy Sciences, Inc.*)

338 WEB PROCESSING AND CONVERTING TECHNOLOGY AND EQUIPMENT

Fig. 17-8. Forty-eight-inch-wide 250-kv accelerator window. (*Courtesy High Voltage Engineering Corp.*)

tained. Suitable X and Y scan axis techniques are used to spread the heat load. In most cases, liquid or forced air is used to provide adequate cooling to prevent window failure and to allow higher beam current output.

Depending on the voltage and thickness of the web, the technique of festooning is often used. In this multiple pass system, the electrons passing through the top web then impact the second pass web (opposite side) thus after several passes most of the energy is absorbed. Using this system, the dose is usually more uniform than a single pass.

Vacuum System

In order for the electron beam accelerator to be operational, it must be evacuated to a high level. All of the accelerator tubes operate in a high-vacuum range of 10^{-6}–10^{-8} Torr and hence exacting detailed maintenance and operating standards must be followed to prevent leaks outgassing, etc. While mechanical pumps are used for roughing, the final low-vacuum requires ion diffusion or turbo molecular pumps to attain operating levels.

Fig. 17-9. Two-roll web festooner.

Safety and Shielding

The operation of an electron beam generator system employs high dc operating voltages from which personnel must be protected. This is easily accomplished using standard safety procedures.

The radiation resulting from both, high-energy electrons and secondary x-rays or Bremsstrahlung (as a result of the impact of electrons with metal) in the system, pose a very specific hazard. The primary protection from harmful radiation is by providing shielding of sufficient thickness to contain essentially all of the radiation. This will range from a few centimeters of lead for lower voltage machines in the 150–500-k V range to 3–4 m of concrete for very high 1–5-MEV units. Units most suitable for web processing can be fitted with integral compact shielding which negates the necessity for vaults or below ground pits. Federal and state regulations require suppressing radiation to 100 mrem/week in restricted areas. For proper operation and maintenance access to the accelerator must be provided. This necessitates a

Fig. 17-10. Linear cathode accelerator (72" x 300 kv) with integral shielding lowered showing window and safety interlocks. (*Courtesy Energy Sciences, Inc.*)

foolproof mechanical and electrical interlock system so that accidental start-up with the shielding open is not possible. This system should be redundant and of a fail-safe type.

If inert gas blanketing of the web is not used, some of the oxygen present will be converted by the high-energy beam to ozone which is extremely toxic. Thus ventilation to remove ozone becomes a necessity. Fortunately, the stability of ozone is quite poor so that a suitable exhaust blower and vent stack of moderate height above the roof is sufficient.

Maintenance

Electron beam processing systems require unique maintenance capabilities. Without proper maintenance personnel and techniques, the unscheduled downtime for an electron beam processing system can reach 30–40%. Most vendors of electron beam systems will supply field service for their equipment. As with most field service agreements, this type of arrangement

ELECTRON BEAM IRRADIATION 341

Fig. 17-11. Ozone exhaust arrangement for two 500-kv accelerators.

necessarily requires some amount of downtime associated with the response of the service personnel and their supply of repair parts. By far, the most efficient method of maintaining an electron beam processing system is with an in-house staff drawing from a well-stocked repair parts inventory.

The personnel required for an in-house staff must be able to comprehend the far-reaching concepts of radiation processing equipment. These requirements include a very wide range of technologies; i.e., physics, dosimetry, health physics (radiation safety), beam optics, electrostatics, high voltage, low and very high vacuum, magnetics, dc and ac power, rf power, analog and digital logic, microprocessors and programmable controllers. Training of personnel is offered by the electron processing equipment manufacturers.

Critical repair parts for most processing systems are not normally obtainable through local electrical/electronic or mechanical suppliers. This makes a good supply of repair parts a necessary part of any business using electron beams. The amount of spares is dependent on the locality of the operation and its access to the equipment vendor's supply of parts. For continuous production, this author recommends enough parts for two major failures of the processing system. The nature of repairs on electron beam processing equipment can be very costly. The value of the spare parts for a processing system very often exceeds the value of that system.

ELECTRON BEAM SYSTEM PARAMETERS

Having described the electron beam generator system, it is well to consider the parameters which affect the choice of a system necessary to do the job efficiently. The important factors are:

- Acceleration voltage, which determines penetration
- Accelerator current, which together with voltage determines power
- Working width
- System efficiency factors.

From these and from consideration of the physical and chemical characteristics of the web to be treated (i.e., density, thickness, width), a dosage, rate of dosage and web speed can be derived.

The accelerator power W is a product of the acceleration voltage V times the acceleration current A

$$W = VA \tag{1}$$

The working width is determined by the scan and window width, or cathode and window length in the case of a linear cathode machine.

For the electron beam to do its job it must contact and penetrate the moving web. For surface and chain reactions sufficient penetration is easily accomplished with relatively low voltages, 150–250 kV. However, for uniform crosslinking reactions throughout the web, penetration and absorption parameters must be taken into consideration if proper dosages are to be delivered.

Radiation dosage is defined as:

$$\text{Dosage} = \frac{\text{Energy}}{\text{Mass}}. \tag{2}$$

The common unit for irradiation dosage is the rad, which corresponds to absorption of 100 ergs of radiation per gram. An SI unit coming into general use is the gray (gy), and 1 gray is equal 100 rads. Dose rate is usually expressed as Mrad/sec, throughput as Mrad·kg/hr.

The penetration of an electron varies nearly in proportion to its energy and inversely to the density of the absorbing medium. A very useful way to consider these interrelating relationships for a given beam energy is to express the range in terms of the weight of material to be irradiated per unit of its area. This is expressed in g/cm^2, which is the product of the thickness (cm) and the density (g/cm^3). A common practice is to express thickness as

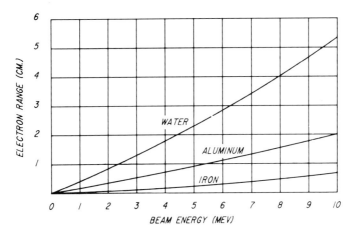

Fig. 17-12. Electron absorption curves in various materials.

equivalent thickness at unit specific gravity which is obtained by multiplying the thickness by the specific gravity of the material being irradiated.

While the above calculations relate to the total beam energy which is all delivered in the ranges specified, the dose is not uniform. The dose varies in depth as shown in Fig. 17-13. Note that the dosage (measured as ionization) is maximum at a point about one-third through the total range. This is the effect of secondary electrons which are given off as a result of the collision of the primary electrons with the base material. At a depth of two-thirds, the relative dose is equal to that at the surface.

To equalize dose, a process called "cross-fire" is used where the material

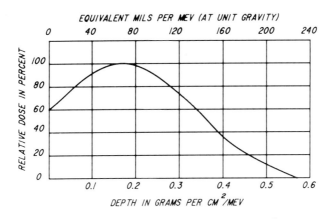

Fig. 17-13. Ionization intensity versus penetration irradiation from one side.

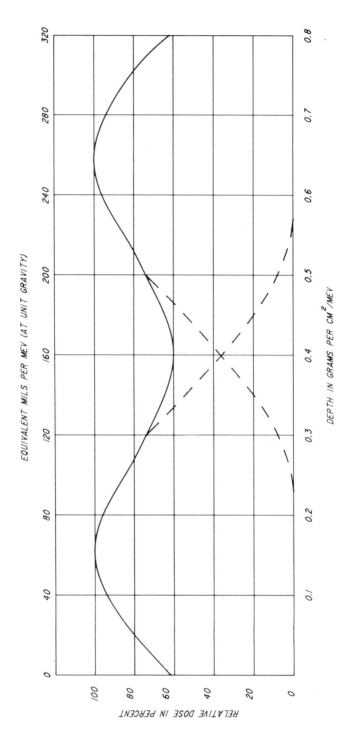

Fig. 17-14. Typical cross-fired depth profile at various efficiencies.

is bombarded from each side. This can be done in line with two accelerators or by passing the material through the process a second time inverted. Crossfire can also effectively double the effective voltage of a machine allowing thicker materials to be used. Of course, these effects are the most pronounced for thick materials. For thin webs, the problem is the efficient utilization of the beam.

Another important parameter for scanned machines is side to side uniformity. This is taken care of by careful programming of the scan circuitry. Of course, the linear cathode type machine minimizes this problem. In addition to the cross-web scan, most machines utilize a web direction scan to minimize heating of the window.

ELECTRON BEAM POWER OUTPUT AND PRODUCTION CAPACITY

Web products, unlike gas, liquid, powder or discrete units, cannot be changed in configuration. Width, thickness and density are fixed by other operating factors. Therefore, a decision must be made regarding voltage and power of equipment as well as whether to cross fire, festoon or run straight through.

Actual accelerators are rated at top maximum voltage and current thus setting their power capability. In actual practice the voltage is left constant and the current or product speed varied to attain correct dosage. Efficiency of beam utilization depends on losses in the system (window heating, air ionization, over scan, etc.) and must be estimated for each application. A good rule of thumb to use states that at 100% efficiency 1 kW of electron

EFFICIENCY %	DOSE IN MEGARADS								
	.01	.1	1	2	10	15	20	50	100
100	80000	8000	800	400	80	53.3	40	16	8
90	72000	7200	720	360	72	48	36	14.3	7.2
80	64000	6400	640	320	64	42.6	32	12.8	6.4
70	56000	5600	560	280	56	37.3	28	11.2	5.6
60	48000	4800	480	240	48	32	24	9.7	4.8
50	40000	4000	400	200	40	26.6	20	8.0	4.0
40	32000	3200	320	160	32	21.4	16	6.4	3.2
30	24000	2400	240	120	24	16	12	4.8	2.4
20	16000	1600	160	80	16	10.7	8	3.2	1.6
10	8000	800	80	40	8	5.3	4	1.5	0.8

POUNDS PER HOUR PER KW OUTPUT

Fig. 17-15. Dose versus pounds throughput at various efficiencies.

WIDTH OF SHEET INCHES	THICKNESS IN MILS (0.001")				
	2	10	50	100	125
5	3000	600	120	60	48
10	1500	300	60	30	24
15	1000	200	40	20	16
20	750	150	30	15	12
25	600	120	24	12	9.6
30	500	100	20	10	8

PROCESSING RATE
FEET PER MIN. PER KW AT ONE MEGARAD

Fig. 17-16. Processing rate for various thickness and width.

beam output can irradiate approximately 800 lb of product at unit density per hour to a dosage of 1 Mrad.

Another important parameter for web products is linear rate. It is possible to express production capacity in terms of an area or linear rate rather than in pounds. This will hold true as long as the relationship of the average pro-

Fig. 17-17. Linear beam accelerator mounted on a coated line. (*Courtesy Energy Sciences, Inc.*)

ELECTRON BEAM IRRADIATION 347

Fig. 17-18. Curing line using 300-kv ICT unit with self-contained shielding. (*Courtesy High Voltage Engineering Corp.*)

duct thickness to voltage is known. A rule of thumb is one millampere of beam current will deliver an average dose of 1 Mrad to approximately 2400 square inches of product per minute regardless of actual voltage provided the thickness is two-thirds of the range for that voltage. At this thickness, the average dose is 85% of the maximum dose.

For example, for thin films such as low-density polyethylene (density 0.92), one kilowatt will process 30,000 ft^2/min at 1 Mrad if the film is 1" wide and 0.001" thick with an overall beam utilization of 70%. Figure 17-16 gives figures for other linear speeds, widths and thicknesses.

Many surface treating applications such as curing of adhesives, inks, coatings, etc., depend on free radical catalization of the polymerization reaction. Thus relatively low-voltage and currents are sufficient. For these applications, penetration considerations give way to reaction rates as an important parameter. There is no good rule of thumb to use in such cases and tests must be run to determine these. Most all equipment manufacturers, however, have facilities to enable a potential user to determine the correct system parameters.

It should be pointed out here that for the aforementioned type of reaction oxygen is often a poison or at least competes for electron energy which is

348 WEB PROCESSING AND CONVERTING TECHNOLOGY AND EQUIPMENT

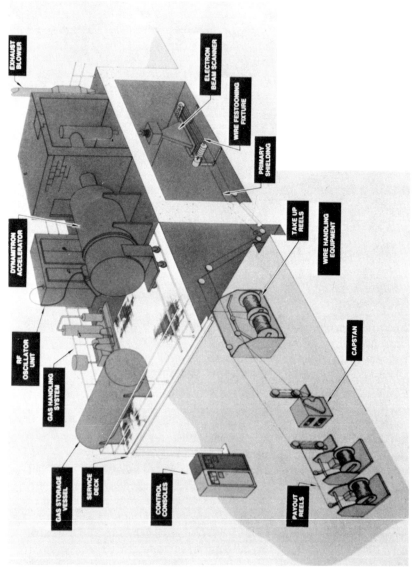

Fig. 17-19. Typical wire irradiation facility using dynamitron accelerator. (*Courtesy Radiation Dynamics Corp.*)

ELECTRON BEAM IRRADIATION 349

Fig. 17-20. Accelerator and controls (500 kv x 50 MA) on shrink film line. (*Courtesy Cryovac Division, W. R. Grace & Co.*)

converted to unwanted ozone. For such applications, the web is usually blanketed with an inert gas such as nitrogen or carbon dioxide.

An important factor in considering maximum dose rate in all cases is that there is a build-up of heat in the product which must be taken into account. For a material of unit density (water):

$$100 \, \text{Krad} = \frac{0.239 \, \text{cal}}{\text{g}}, \tag{3}$$

which means an increase of $0.239°C/100$ krad absorbed. In addition, any exothermic chemical reaction taking place will add to the heat build-up.

Often forced air or inert gas is used to cool the accelerator window and in doing so takes care of the product cooling problem.

Electron beam processing is a very practical and widely used commercial process for carrying out chemical reactions such as crosslinking, curing, sterilization, etc., on web materials. In most cases, it is usually more energy-efficient than normal processing methods, such as (a) high-temperature chemical crosslinking, (b) hot air drying in the case of laminating or printing, (c) high-temperature or chemical sterilization. The use of radiation curing for laminating adhesives and printing inks eliminates or minimizes the use of costly solvents. This in turn reduces or eliminates the need for emission control systems to meet environmental standards set by government agencies.

In the case of curing or crosslinking, it generally provides a clear-cut reaction with no residual promoters or crosslinking agents that might cause degradation of the product or problems with governmental agency regulations on toxic residues in products for food and drug applications.

It is a reliable cost-effective system and should be given serious consideration whenever a web material needs treatment of the type for which electron beam processing is suitable.

REFERENCES

1. Charlesby, A. *Atomic Radiation and Polymers.* New York: Pergamon, 1960,
2. Chapiro, A. *Radiation Chemistry of Polymeric Systems.* New York: Interscience, 1962.
3. Hoffman, A. *Applied Industrial Radiation Chemistry of Monomers and Polymers.* Fifth International Congress of Radiation Research, Seattle, 1974.
4. Harmer, D. and Ballantine, D. *Chemical Engineering* **78**(9): 98, (10):91 (1971).
5. Silverman, J. *Radiation Physics and Chemistry* **14**(1),(2),(3),(4),(5),(6) (1979).
6. Silverman, J. *Radiation Physics and Chemistry* **18**(1),(2),(3),(4),(5),(6) (1981).
7. Holl, P. *Industrie-Lackierbetrieb* **48**(9):313 and (10): 362 (1980).
8. *Proceedings of International Conference Radiation Processing for Plastics and Rubber,* Brighton, England, 1981, Plastics and Rubber Institute, London.
9. *Handbook of Electron Beam Processing, 2nd Ed.* Burlington, Mass: High Voltage Engineering Corp.
10. Nablo, S. V., Denholm, A. S. and Checkland, J. A. *A 250-kv Selfshielded Electron Processor for Extrusion Coated Wire Applications.* Proceedings 15th Electrical/Electronics Insulation Conference, October 19-22, 1981, Chicago, Ill. (IEEE Publication No. 81, CH IFIF-8).
11. Nablo, S. V., Quintal, B. S. and Checkland, J. *Selfshielded Longitudinal Electron Processors for Fibers, Filaments, Tubing and Wire.* Proceedings IAEA International Conference on Ind. App. of Radiation Technology, September 28–October 2, 1981, Grenoble, France.
12. Tripp, E. P. *Paper, Film and Foil Converter* **55**(11):55–61 (1981).
13. Nablo, S. C. and Tripp, E. P. *Advances in Electron Curing for High Speed Converting.* Proceedings TAPPI, 1978 Paper Synthetics Conference, September 25-27, 1978, Orlando, Fla.

14. Nablo, S. V. Progress toward practical electron beam sterilization. *Sterilization of Medical Products,* Vol. II. E. Gaughran and R. F. Morrissey (Eds.). Montreal: Multiscience Publications, 1981.
15. *Accelerator Requirements for Electron Beam Processing,* Technical Information Series TIS 78-11, Radiation Dynamics Corp.

18
Film Orientation

Frank Jacobi

Brueckner-Maschinenbau*
Siegsdorf, West Germany

Both process and equipment for the orientation of films have had a considerably shorter history than, for instance, paper manufacturing, converting or textile finishing technologies. Yet, current equipment and design used for film orientation has been derived from these predecessors.

In describing modern machinery, we differentiate between commercial equipment and proprietary designs, which are mostly in-house built constructions of film producers and of which little detail has been published. The film orienting equipment is subdivided into the following types:

- Sequential film orienting machinery
- Simultaneous tubular equipment
- Simultaneous flat sheet orientation
- Other methods.

SEQUENTIAL FILM ORIENTING EQUIPMENT

The term "sequential" means that the orientation of film is carried out in one orthogonal direction at a time; i.e., first in the machine direction on a hot roll stretcher or similar equipment and then, in the transverse direction on a downstream installed tenterframe. Sequential orientation of films is the predominant process, in terms of the numbers of equipment on stream and also, their production volume. The productivity, line speed and output are

---

*Currently Frank Jacobi Ingineurbuero, Stephanskirchen, West Germany.

higher than those of other processes. Equipment of this nature has been used for the manufacturing of large volumes of products.

Machine Direction Orienting Equipment

Most polymers orient well at a certain temperature which is usually above room temperature.

Preheating systems are designed to raise the film temperature to the required level prior to orientation. One way to heat the film is to use driven, heated rolls in diameters of 200-800 mm. The most advanced designs feature double shell rolls with spiral leads for thermal fluid which is pumped through the rolls at turbulent flow conditions for enhanced heat transfer. Another method is to preheat the film by hot air convection, or infrared radiation, while it is traveling through the preheater. Some equipment designs use a combination of hot roll contact and convection or radiation heating.

Most equipment of current design has means of compensating for the thermal growth of the film by employing positively driven rolls with stepped diameters, or stepped drive ratio, or uses tendency driven rolls.

Some hot polymers are corrosive, and therefore roll surfaces are coated, chrome plated or made of corrosion resistant materials.

Machine Direction Draw Sections

Orientation in the machine direction occurs because of a speed differential between slow and fast draw rolls. Slow rolls retain the draw forces applied by the fast rolls. A schematic diagram of such an operation is shown in Fig. 18-1.

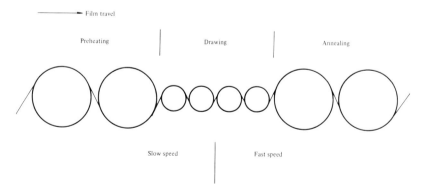

Fig. 18-1. Machine direction draw section.

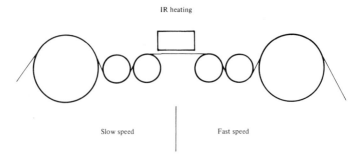

Fig. 18-2. A single-draw stage with infrared heating section.

Single-draw stage consists of two driven slow rolls and two driven fast rolls, followed by an annealing roll assembly. The temperature, initiation of the draw and the draw ratio are the variables necessary to achieve the required orientation. The temperature of draw rolls is controlled for optimal friction coefficient and small neck down: draw distance is adjustable. The design permits extension to two or multiple-draw stages by adding more pairs of draw rolls. Such equipment is used for biaxially oriented polypropylene film.

Figure 18-2 shows a single-draw stage with two driven and heated slow rolls and two fast rolls. Orientation is enhanced by an infrared heating stage. Such a design allows to increase the web temperature while the film is traveling in the draw stage itself. Sticking problems to rolls are avoided and necking down is increased. The tendency to form surface defects is reduced since the film in contact with the rolls is always below the melting temperature. Equipment of this or similar design is widely used in the polyester film making industry.

Machine direction draw starts anywhere between slow driven and fast driven rolls as soon as the draw roll temperature has been reached by means of heat transfer from the idler rolls. Roll diameters are relatively small (40-100 mm) to reduce the free travel distance between tangential points of rolls for neck-in control.

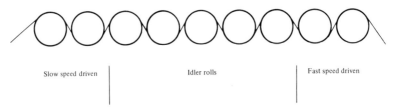

Fig. 18-3. Machine direction draw with multiple temperature controlled idler rolls.

Annealing and Cooling Function

Oriented film products usually require annealing after machine direction orientation or even shrinkage under temperature controlled conditions. The purpose is to improve the thermal stability of the finished product. Annealing roll assemblies are very similar to preheating rolls, except they are usually equipped with variable speed controls between rolls to enable controlled shrinkage during annealing. Particularly for polyester films, such rolls are cooled to avoid surface defects between process stages.

Cooling rolls are of simpler construction than preheating or drawing rolls. They are normally hollow and cooling fluid is fed and removed through rotary unions. Internal protective coating against corrosion is sometimes used.

Operating Conditions

Rotating roll equipment is hazardous and a source of possible injury, particularly during thread-up. Except for very old designs, modern equipment in operation features mechanical threading devices, such as driven belts, chain systems with clamps to hold the film, and pneumatic web conveying systems. Yet, manual threading is used quite often with roll equipment.

Some machine designs are vertically arranged, such as those used for thin PVC films, where two roll bearing assembly frames open up and the web can vertically drop into the roll assembly. Operating speeds and material throughputs vary depending on the equipment used. Reported operating speeds of biaxially oriented polypropylene film systems are over 200 m/min on the machine direction orienter (Fig. 18-1). Some polyester film lines are known to run at even higher speeds, leading to film throughputs in excess of 3000 kg/hr.

Machine direction draw ratio and draw force are important criteria for equipment design. Commercial size units in service have draw force capabilities up to 5000 kg, particularly for thick polyester film. Draw ratios are determined by the orientability of the used polymers:

Polymer	Approximate Machine Direction Draw Ratios
Polypropylene	4–6
Polyethyleneterephthalate	3–4.6
Polyvinylchloride	1.5–2.5
Polystyrene	3–3.5
Polyacrylate	4–6

Draw ratios indicated also apply to the transverse direction.

Most commercial film orienting equipment is installed to operate continuously around the clock, seven days a week. Large plants require an extremely long period of time to reach equilibrium of process conditions after a complete or partial stop.

Machine direction orientation equipment is exposed to high forces, temperatures and speeds; wraps of film around hot rolls occur and rolls are subject to damage. A significant criterion of current designs is the ability to keep downtimes low by incorporating design features which permit easy maintenance.

Tentering Equipment

The second orienting stage in the sequential film orientation is transverse direction drawing, or the tentering process. The tenter consists of chain assemblies and attached grippers which hold the film on both sides. Then the chain tracks start to diverge from each other and the film stretches in the cross direction. Similarly to the machine direction orientation, tentering requires initial control of the film temperature.

Tenter Heating Systems. Most tenter heating systems use circulating air as a heating medium. Pressurized by circulating radial or axial fans, air passes through heat exchangers or mixes with hot gas in directly fired ovens and impinges on the passing film through nozzle systems. Nozzle design—slot or hole—determines the uniformity of air flow and therefore of heat transfer and ultimately, the film temperature.

Counterflow air systems are preferred. Process heat sources, besides natural gas in direct fired ovens, are saturated steam, electrical heat exchangers or thermal fluid circulation through radiators. Temperatures range from 80°–100°C to more than 200°C in polyester film crystallizing zones.

The oven must be properly insulated in order to avoid heat losses and to maintain a controlled temperature level. Tenter inlet and exit ends are sources of heat loss as slots for web and tenter travel remain open. Many polymers and additives emit fumes when heated. Therefore, tenter ovens have to have exhaust means.

A very viable way of heating films in tentering is by infrared radiation, especially for films which absorb infrared radiation well, such as polyester films. Infrared heating is clean and efficient. However, if the web should break and the film come into contact with the hot radiator surface, this could cause contamination. Hence, radiators should be quickly retractable to avoid such contamination. It has been reported that infrared heaters en-

FILM ORIENTATION 357

Fig. 18-4. Tenter oven with heat insulating panels and air circulation fans. (*Courtesy Brueckner Maschinenbau.*)

hance the film stretching operation in tenter ovens, adding some heat right at the start of the actual draw stage.

Tenters. A tenter consists of chain, tracks and clips with the necessary chain drive system. Clips are connected to the chains on either side of the tenter and the chains travel in tracks which are adjustable in the cross direction. Looking at a top elevation of a tenter, the installation is reminiscent of the image of a funnel where the narrow end is the feed section of the tenter chain. Tenter systems are differentiated according to the means of supporting the chain in the tracks.

Sliding tenters use the sliding friction of the chain in the track, utilizing wear materials which have low friction coefficients and long wear time. Sliding tenters are usually lubricated by oil. Sliders are known for their durability, simplicity, ease of maintenance and speed potential. Most high output equipment uses this technology.

Roller bearing tenters originally were used in areas where a high load at relatively low speeds has prevailed. Equipment features lubricated or sealed bearings. Guide rails or, preferably, runways of the roller bearing surface are in sections, or continuously bent in order to reduce shock load on the bearings. Advantages of roller bearing systems, particularly the sealed type, are relative cleanliness and a low friction coefficient. Disadvantages are higher maintenance requirements as compared to tenters. Some designs have combined sliding and roller bearing features.

Tenter Tracks. Tenter tracks must be adjustable over certain width ranges. They are parallel in the preheating section, diverge in the draw section, and toe-in in the annealing/crystallizing/cooling sections.

Track joints between tenter sections are adjustable within certain angle ranges to permit adjustment to the necessary tenter configuration. For width adjustment, either motorized or manually operated threaded spindles are widely used. Chain track supports at tenter joints have to permit machine direction movement required to compensate for thermal expansion as well as changes of the track configuration.

Tenter Clips. The industry uses various types of clips for varying duties such as for heavy duty polyester or light-gauge polypropylene capacitor films. Clip designs are very different for each purpose. The most common clip designs have one or more of the following features:

- Single gripper or multiple gripper
- Swiveling gripper (in MD)
- Edge pinners to reduce edge curling
- Springs to hold the clip closed
- Dual function springs for forced opening and closing of the clip.

The most common feature of film clips is that once film is in the clip, and the film is pulled by diverging tracks, the vertical component of the gripper holding force will increase.

Operating Conditions. Threading in most known tenters is manual which means that threading occurs at tenter speeds well below actual operating speed. Tenter operating speeds for thin-gauge films are reported to be in excess of 300 m/min and most other equipment operates at speeds close to 200 m/min. Draw ratio, draw angle, and draw forces depend largely on the processed polymer. The application of sequential orienting equipment is practically unlimited and most known polymers can be processed in this way.

SIMULTANEOUS ORIENTING EQUIPMENT

The term "simultaneous orientation" is most often used in connection with flat sheet orienting. Yet, tubular orientation is also a simultaneous process.

TUBULAR FILM ORIENTING EQUIPMENT

The tubular film orientation starts with the extrusion of a thick wall pipe, cooled down in a water quench system with temperature controlled non-

FILM ORIENTATION 359

Fig. 18-5. Tenter clips at turn-around sprocket. (*Courtesy Brueckner Maschinenbau.*)

turbulent flow. In addition, some producers use impingement with chilled air on the inside of the pipe.

Orientation starts in a preheating zone where infrared or hot air impingement heats the film to process temperature. Nip rolls are used to separate the casting section from the orienting section. Modern designs have rotating heaters or air die assemblies for temperature equilibration.

In most equipment, orientation starts from an elevated platform in the downward direction. The machine direction draw is defined by the speed differential between nip roll assemblies located before and after the draw stage. Insertion of a compressed air pipe into the draw stage allows an increase of pressure inside the bubble sufficient to stretch the film in the cross direction. Before reaching the secondary set of nip rolls, cooled air is blown on the film to stabilize the orientation. Orientation control is relatively easy and is accomplished by maintaining constant pressure inside the air bubble and machine direction draw ratio between the drives of the nip roll assemblies.

Some bubble orientation film lines for polypropylene films are reported to require tower heights of more than 40 m. Packaging films require annealing-heat setting after orientation in order to reduce inherent shrink back. For this purpose, commercial lines use either driven and heated annealing rolls or tenterframes.

Operating Conditions

Threading of bubble film lines is usually done with pull-in rope in the orienting section, similar to the standard blown film line. There is no report of any automatic or mechanical threading system in double bubble film orienting equipment.

Operating speeds at the discharge end for a modern 240 cm proprietary tubular film orienting system are similar to those of polypropylene tenter orienting systems. Draw ratios are about 50:1 (for polypropylene).

Bubble orienting equipment has found commercial application for polypropylene only, except for some small nylon film orientation uses.

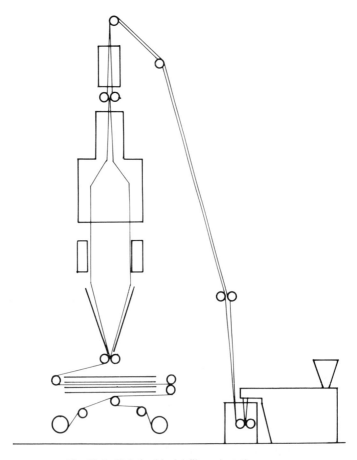

Fig. 18-6. Tubular biaxial film orientation process.

Simultaneous Flat Sheet Orienting Equipment

Materials made on flat sheet simultaneous orientation equipment are predominantly nylon films, some polypropylene, polyester, and high-density polyethylene. The widest machine known to be in service has a 4 m outlet width.

Spindle Unit. This system, similarly to the sequential tenterframe, uses circulating clips with grippers in diverging tenter tracks. Transport of individual clips occurs through rotating spindles with an increasing thread distance in the draw section, whereby pins of each clip are guided within the thread, and this way the machine direction distance between the clips is increased and the film is drawn in the machine direction. Although in limited use commercially, the system is technically interesting, since one step of the flat sheet sequential orientation is eliminated. Specific loads on spindles and grippers however, considerably limit speed and unit throughput. Clips and grippers are of a different design than those of ordinary tenterframes, because transverse and machine direction loads have to be taken at the same time by each clip.

Linear operating speeds of simultaneous orienting equipment are lower in comparison to sequential orienting equipment due to mechanical complexity. A schematic diagram of such a unit is shown in Fig. 18-7.

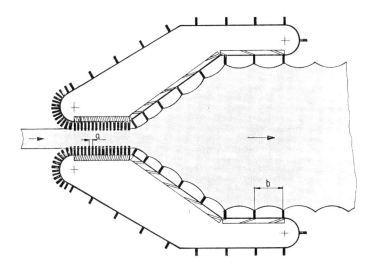

Fig. 18-7. Simultaneous orienting spindle unit.

362 WEB PROCESSING AND CONVERTING TECHNOLOGY AND EQUIPMENT

Panthograph Design. The panthograph orientation equipment is similar to that of the spindle unit, except that the machine direction distance between grippers changes by means of a scissor type chain assembly where the scissors are close together in a preheat section and far apart after the orienting zone. Mechanical problems with this type of equipment are greater than with the spindle unit due to the self-locking nature of this equipment.

OTHER METHODS

Disk Orienting Equipment

This equipment is derived from textile finishing systems introduced to the industry during the 1930's. Disk orienting equipment comprises two disks which rotate in the machine direction. The disk axis is horizontal. The disks can be adjusted to an angle providing a narrow end where the film is grabbed by pins on the disk circumference. The film travels along the radius, being stretched in the cross-direction and after 180° travel, is released from the pins oriented in the transverse direction. No commercial use of this equipment has been reported.

Grooved Roll System

Grooved rolls are generally known as spreaders in the film industry. Experimental lines were designed to laterally orient film using grooved rolls. Rolls

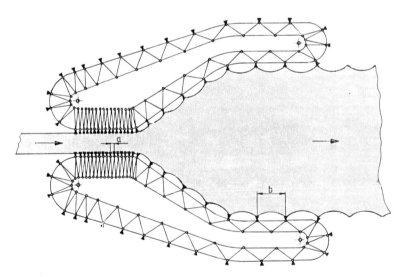

Fig. 18-8. Panthographic orienting unit.

are spirally grooved and the spirals, if rotated, stretch the film from the center to both left and right sides of the roll.

"Umbrella" Type

Another experimental design has used the principle of an umbrella for tubular orientation whereby the draw ratio in the cross direction was changed by opening or closing the umbrella.

Flat Table

This equipment has been used for transverse direction (TD) orientation of calendered PVC films between two pairs of driven nip rolls. A spherically rounded and heated body is installed between the nip roll assemblies and the incoming film pulls over this body, providing TD orientation to some degree. A speed differential between the nip roll pairs creates machine direction orientation.

19
Winding, Slitting and Splicing

Thomas S. Greiner
Jagenberg
New York, N.Y.

A continuous web emerging from a production machine must be wound to a roll for transportation and further processing. Such a roll must fulfill one or more of the following requirements:

- It must be of the required footage or diameter
- It must be of the desired width
- It must be wound on a mandrel or core of specified diameter
- It must be within acceptable tolerances of density
- It must be endless.

Due to the variety of compositions of webs, machine designs, uses and processing requirements, a multitude of equipment has been developed to fill the needs of particular applications. We shall identify equipment and techniques for various groups of applications in the light of the foregoing basic functions.

PRIMARY WINDERS

When examining apparatus for winding webs we can distinguish between primary and secondary (and sometimes tertiary) winders. A primary winder serves simply to wind up a web as it emerges from a piece of production equipment. If the manufacturing or converting process is continuous, such a winder must be able to transfer the web on the fly to a second mandrel or spool without interrupting production.

Reel Spools

The web, which is processed on primary winders, is most frequently wound up on a spool (Fig. 19-1). A spool is a mandrel that can be used for winding and unwinding continuous webs. No core is required or generally used. A spool can have an outside diameter of six or eight inches and up to 24 inches and more, depending on machine width and speed. Its journals carry its own antifriction bearings in housings which enable the user to deposit the spools in various supports on rewind and unwind equipment. This is essential when production of the web is to be continuous.

Spools are generally reversible and the ends of their journals are fitted with coupling devices. These are usually square male members (protected by a bell guard), or internally geared female members. They are clutched to a drive in case of center driven windups or to a brake when they are used in unwinds.

Other types of mandrels, such as shafts, can be used, but generally only on light duty applications since they carry the full weight of the winding roll, which is usually wound to a fairly large diameter.

The Reel

While the word "reel" frequently refers to the mill roll or jumbo roll, the paper maker uses it for the primary winder at the end of a paper machine or a coater. The function of the reel is to wind up a web to a given length or diameter and to effect flying changes so that the process is not interrupted when one roll is complete.

Many designs are in use, but the one most frequently encountered is the horizontal rail or Pope-type reel (Fig. 19-2). The spool winding the web is held by air cylinders in a slide against a driven steel drum. The winding mill roll is thus surface driven by the reel drum. While the mill roll grows, a new spool is placed into a pair of swing-arms, called primary arms. When the winding roll is fully grown, the primary arms are swiveled to bring the new spool into contact with the reel drum which is wrapped by the web. Paper is usually transferred to the new spool by an air jet, sometimes with the aid of

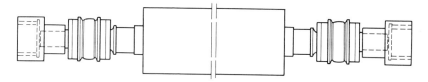

Fig. 19-1. Reel Spool.

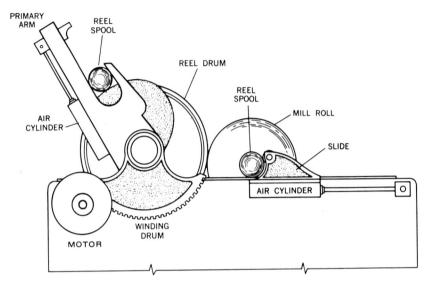

Fig. 19-2. Surface driven horizontal reel.

water. The full roll is then moved along the rails to the end position where it is lifted out of the slide bearings by crane. The slide returns to the position next to the reel drum. The primary arms with the new roll are rotated to the horizontal position where the spool is transferred to the horizontal rail and held by the slide. The primary arms are now relieved of their load and are rotated back to receive a new spool. An electric powered drive is generally used to rotate or index the primary arms.

Other types of reels or continuous windups are used. They have two or more windup positions and the mandrels are center driven through belts, slip clutches or eddy current couplings. The web is wound up alternately on either position.

The Turret Winder

Some coaters and other web processing machines operating at reasonably modest speeds and winding fairly small rolls use turret type winders (Fig. 19-3). These are winders with two positions located at the ends of opposing arms which can be rotated or indexed. Each position is center driven by a dc motor or an ac motor with an eddy current clutch. These winders are designed for use with core shafts or spools.

As the roll winds up on one position, the spool or dressed shaft is loaded on the other. At the time of transfer, the arms are rotated till the new mandrel contacts the web. The core or spool has been fitted with an adhesive tape

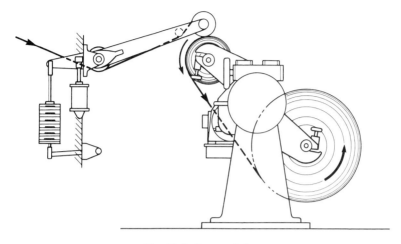

Fig. 19-3. Turret winder.

to pick up the web. Heavier webs require severing by hand or by means of an air-loaded knife built into the winder.

Unwinds for Primary Winders

If the primary winder is not placed directly following the machine which produces the web, there has to be an unwinding device for the base stock. Many different types of unwind stands are in use, ranging from simple rollstands to fully automatic reel changers.

The first consideration must be given to the question of whether the process is continuous or discontinuous. In the latter case (a calender, for example), a very basic type of unwind, such as shown in Fig. 19-4, may be sufficient. It consists of a pair of side frames with bearings to accommodate

Fig. 19-4. Shaftless unwind stand.

either a reel spool, a shaft or a pair of cones (those for shaftless unwinding). There may have to be a mechanism to adjust the center distance of the frames to each other, if rolls of different widths are to be processed. As a rule, one of the bearings is adjustable in machine direction to take up slack in a web. All unwind stands are equipped with brakes to maintain a certain amount of tension in the web. If the process is to be continuous, we must employ a type of unwind, which is designed to change rolls on the fly. Such a flying splice unwind is similar in design to the continuous winders discussed previously. A flying splice unwind used frequently is built on the level rail principle, similar to the horizontal reel. A pair of rails will accommodate the unwinding spool clutched to an electrical or mechanical brake. A second set of bearings receives the new roll of base stock. A pattern of adhesive for splicing is painted or taped on the outside of the new roll.

The new roll must be accelerated until its peripheral speed matches that of the web. For this purpose the mandrel of the new roll is clutched to an electric motor (which can also serve as a regenerative brake in the unwinding mode), and a reflective tape is applied to the side of the roll. This will time the rpm of the roll until a speed match is achieved. The speed matched roll is then brought to the vicinity of the expiring one and a so-called marriage roll brings the web into contact with the new roll. As soon as the adhesive bonds the web to the new roll, an automatic knife cuts the tail of the expiring roll. The new roll is then transferred to the bearings of the spent one.

The same principle is followed by the unwind version of the turret winder, as shown in Fig. 19-5. The new roll and the expiring web are brought together by indexing the turret. The turret type unwind is used mostly on rolls of either moderate width or moderate diameter, or when the stock is wound on cores rather than spools. The level reel type is preferred for heavy-duty, high-speed machines.

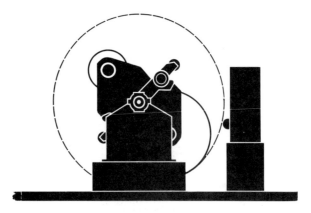

Fig. 19-5. Turret unwind.

WINDING, SLITTING AND SPLICING 369

SECONDARY WINDERS

Though terminology is by no means uniform, this term is used to describe machines for rewinding a web that has already been wound up at the end of a processing machine. This rewind operation becomes necessary to achieve the final quality and dimensional specifications enumerated at the beginning of this chapter.

The Re-Reeler

The re-reeler is probably the most basic form of secondary winder. Its purpose is to accept a web wound up on a spool and rewind it onto another spool. In this process it will:

- Trim the edges of the web
- Build the rewinding roll to the desired diameter
- Achieve reasonably uniform wound-in tension
- Provide means for splicing to make the web endless.

Figure 19-6 is a schematic of such a re-reeler.

Core Shafts and Expanding Chucks

The parent roll or mill roll of a secondary winder may be wound on a spool or on a core. We have previously dealt with spools. Rolls wound on cores can be unwound with the aid of a core shaft. A core shaft for unwinding is usually of the expanding type. There are many kinds of expanding core shafts.

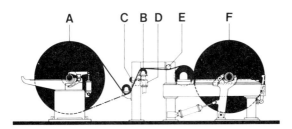

Fig. 19-6. Re-reeler:
         A—Unwind section with reel spool ejector
         B—Guide rolls [the one following the slitters (C) is used for tension sensing]
         C—Trim slitters
         D—Trim removal
         E—Spreader roll
         F—Rewind section (in this case, like a horizontal reel without the primary arms).

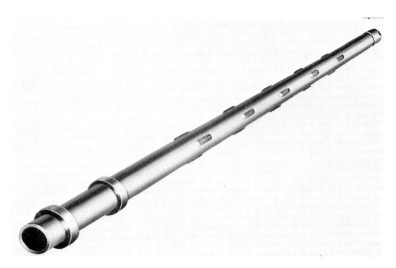

Fig. 19-7. Lug-type expanding shaft.

Fig. 19-8. Expanding chuck.

WINDING, SLITTING AND SPLICING 371

Some incorporate an air tube which, when inflated, expands segments or leaves of the shell of the shaft. They are known as leaf type expanding shafts.

A shaft design better suited for unwinding is the lug type expanding shaft as shown in Fig. 19-7. The lugs extend for the purpose of transmitting torque (brake torque in this instance) and grip the inner surface of the core. There are mechanical and pneumatic versions of the lug type expanding shaft. Buttons instead of lugs are used sometimes as the expanding element.

Shafted unwinds have frequently made way for shaftless ones. If the cores are sturdy enough in relation to the weight of the roll, it is sufficient to insert expanding core chucks (see Fig. 19-8) and transmit torque through them. The advantages of dispensing with shafts are obvious. Only one expanding chuck is needed if there is only one brake. A plain chuck suffices for the side without a brake, since no torque is transmitted on that side. Shaftless unwinding requires, of course, a stand designed to adjust the distance of the two frames. Both frames must travel to insert the chucks into the core.

Unwinds for Secondary Winders

Those unwinds which accommodate rolls on spools are the same as the ones on primary winders. Figure 19-9 shows a simple fixed stand. The handwheel moves the takeup bearing to adjust for a slack edge. This unwind can be shaftless provided the rolls are loaded by crane with a sling or roll grab.

Figures 19-10 and 19-11 illustrate two types of shaftless self-loading stands. These unwinds can pick up rolls from the floor and left them to operating position after the chucks are inserted. The stand shown on Fig. 19-10 uses hydraulic cylinders to lift the roll. A hydraulic motor is used frequently on this type of stand for the purpose of travel in cross-machine direction.

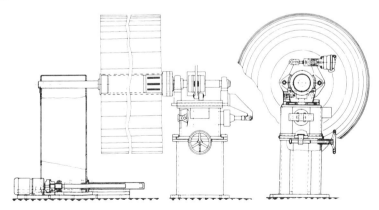

Fig. 19-9. Crane loading unwind stand.

372 WEB PROCESSING AND CONVERTING TECHNOLOGY AND EQUIPMENT

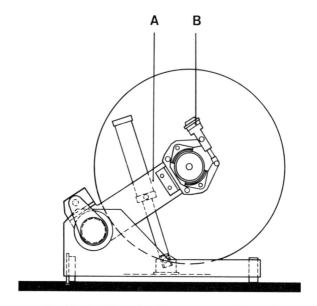

Fig. 19-10. Self-loading lift arm type unwind stand.

The stand shown on Fig. 19-11 has delta shaped frames and hoists the roll on their inclined planes. The lifting motor operates a self-locking lead screw which moves a chuck assembly mounted on a slide. A second motor is used for traversing the frame. The torque of that motor can be transmitted through another lead screw, but a rack and pinion transmission is preferable. Electric motors rather than hydraulic ones are used on this unwind.

The turret type unwinds can be used here with or without flying splice. Some of these turret unwinds are built for shaftless operation. Please observe that all shaftless stands require certain minimum roll diameters for floor pickup.

Fig. 19-11. Self-loading ramp type unwind stand.

Unwind Brakes

Some type of brake is required on any unwind to maintain web tension. Unwinding brakes are built for continuous duty as opposed to stop or holding brakes. Brakes used on unwind stands can be of the mechanical or electrical type. Electrical brakes are motors which function as generators when applying tension to an unwinding web. They are not only sensitive and virtually maintenance free, but have the added advantage of feeding back power to the supply of the user. Apart from size, their principal drawback is cost. The initial expense of electric or regenerative brakes is considerably higher than that of mechanical brakes. They are used primarily on large applications where their excellent response to tension sensing signals and the savings in power can be brought to bear. These regenerative brakes will do double duty as payout drives in the threading phase of large winders and will speed acceleration of the parent roll.

Of the various types of mechanical brakes used, the pneumatically operated ones predominate by a wide margin. Pneumatic brakes can be drum brakes or disk brakes; they can be air cooled or water cooled.

Drum brakes are used on most larger applications and on many small ones as well. They are quite easy to maintain and are frequently built to work with several pairs of brake jaws operated by air cylinders. These cylinders are often of the frictionless diaphragm or rubber element types. Figure 19-12 is a drawing of a fairly typical three-cylinder water cooled drum brake. A rotary union connects the water supply to a hollow shaft mounted on the head of

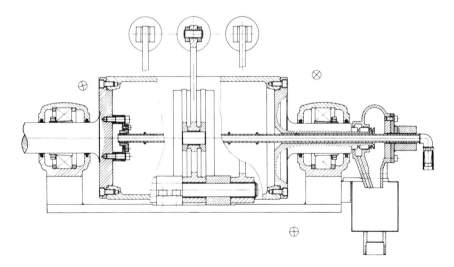

Fig. 19-12. Water cooled pneumatic drum brake.

the drum. The water returns between the outer and inner walls of the shaft and flows into a drain.

Figure 19-13 shows a four-cylinder drum brake connected to a spool-type unwind. An air operated clutch connects the two elements. The brake assembly can be side shifted or oscillated together with the frames of the unwind. The motor for this lateral movement is seen in the foreground. Paper unwinds are often designed to oscillate with a stroke of about $\pm 2''$ so that high- or low-caliper areas in the web can be spread more evenly across the width of the rewinding roll. This means, of course, that wider than usual edge trims must be taken.

Disk brakes can be used on some applications instead of drum brakes. They are chosen primarily for their sensitivity and convenient size. A disk brake found frequently on unwind stands is illustrated by Fig. 19-14. This type of brake employs an air tube as expanding element to create friction against a disk. Such disk brakes may be water cooled or air cooled. Caliper disk brakes are used sometimes on light duty applications. All disk brakes can develop a tendency to squeal when the disk surfaces are not quite level.

Web tension as created by the unwind brake must be controlled. Most winders are fitted with some type of tension control which acts on the unwinding brake. Most winder controls require constant tension, meaning that the actual tension of the web remains at the set value throughout the winding

Fig. 19-13. Reel spool oscillating unwind with water cooled drum brake.

WINDING, SLITTING AND SPLICING 375

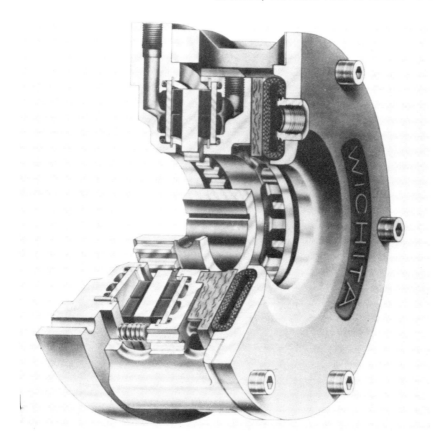

Fig. 19-14. Air tube disk brake.

process. Since brake torque is a function of friction power times the roll diameter, it is evident that brake power must be reduced as roll diameter builds down if constant tension is to be maintained.

An all-pneumatic tension control system for a winder can be seen in the diagram of Fig. 19-15. The unwind brake is an air operated drum brake (G). Its cylinders are set for a certain air pressure by regulator valve (B) on the control desk. When the air pressure in these cylinders equals the set pressure there will be an equilibrium of pressure in diaphragm valve (D). If increasing web tension dislodges tension sensing roll (F), its movement will disturb this equilibrium and cause air to bleed through control valve (E). When the web goes too slack, the brake will be tightened by additional air pressure gated in through control valve (E) till air pressure on both sides of the diaphragm is again equal.

376 WEB PROCESSING AND CONVERTING TECHNOLOGY AND EQUIPMENT

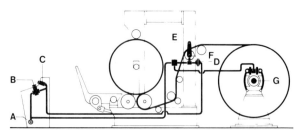

Fig. 19-15. Schematic of pneumatic tension control.

Fig. 19-16. Pair of draw shear slitters.

SLITTING SYSTEMS

Slitting the web is an essential operation in practically all secondary rewinding. Sometimes it is confined to trim slitting, but in most cases the rewinders must divide the parent roll into two or more narrower rolls.

The Shear Cut

There are several types of slitters in use, but the one most frequently employed is shear cut slitting. The idea of the shear cut is to imitate as closely as possible the action of the blades of a pair of scissors or shears. Every shear cut slitter has two types of knives, a top (or front) slitter and a bottom (or back) slitter. Figure 19-16 shows a typical pair of shear cut slitter knives. The bottom knife is cup-shaped to provide a ledge or support for the web. It is usually power driven by an individual electric motor or a common shaft for all slitters on the machine. The top slitter blade is usually beveled and overlaps the bottom slitter by approximately 1.5-3.0 mm. The top slitter penetrates the web so that a slight amount of dislodging takes place at that point. The top slitter is friction driven from the bottom slitter.

The sheet run through most shear slitter systems is so designed that the web passes tangentially between the slitter knives. This system is commonly called draw shear and is illustrated by the schematic of Fig. 19-17. In order for the web to pass through the slitters in a straight line, it must be supported upstream and downstream by a pair of idler rolls. If the web is not surface-sensitive, a slitter table can be used instead of the idler rolls. It is important to

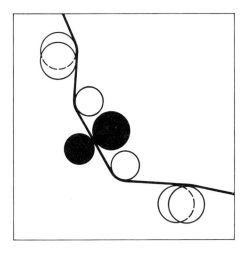

Fig. 19-17. Sheet run through draw shear slitters.

avoid sheet flutter as web is drawn through the slitter section. Consequently, the closer the supports are to the slitter knives the better. That is no problem with stationary surfaces as exemplified by the slitter table. When idler rolls are used, however, they must be quite small in diameter to keep the open draw as short as possible. For this reason we find that sectionally sleeved idler rolls are used on high-speed machines. These rolls are generally grooved to provide an escape for the air trapped under the sheet and thus avoid "skating" of the web. Figure 19-18 provides a view of the slitter section of a large winder. The back slitter is driven by a variable speed motor. The slitters are moved by rack and pinion and are connected by an air cylinder while they travel.

It is an essential requirement of draw shear slitting that the two slitter knives and the web meet at exactly the same point. If the web contacts the bottom (female) knife first, there will be no cut. If the top (male) knife is the first to touch the web, it will penetrate it without the action of the bottom knife. The result will be a ragged cut because of the displacement of the material without the shear action of the bottom knife.

On properly adjusted shear cut knives, the top slitter will be mounted at a

Fig. 19-18. Motor driven draw shear slitters.

slight (about 1/2°) angle against the bottom knife. The angle must be closed in the upstream direction to the web so that the cut takes place at the first point at which the two knives overlap.

When draw shear slitters are power driven, it is necessary for the bottom slitter to have a peripheral speed slightly larger than that of the web. This will prevent slack areas from forming in the slitter section, leading to jams and web breaks. Slitter motors are dc or variable-frequency ac motors, slaved to the drive of the winder. Some draw shear slitters are designed to be driven by the web, in which case there is no overspeed. This requires a special slitter design and is not suitable for all types of webs.

An alternative to the draw shear slitter is the wraparound shear. In this case, the bottom slitter element is not an individual circular knife, but a sleeve made of a case hardened steel forging with a number of cutting grooves ground into its surface. Each of these grooves can be used as a bottom shear cut slitter. A pair of knives of a wraparound shear slitter is illustrated by Fig. 19-19.

The sleeves, called anvil sleeves or slitter bushings, are mounted on a driven expanding or clamping shaft. If the cuts are to be made in certain areas only, plain, ungrooved sleeves can be used where not cutting takes place. Since the sleeved bottom slitter roll is wrapped by the web, there is no need for supporting idlers. The sheet run is illustrated by the diagram of Fig. 19-20. In a wraparound shear the bottom slitters are driven at web speed or very close to it. The wraparound slitter rolls are usually so designed that small gaps are left between the individual sleeves on the shaft so that a sleeve can be positioned to bring one of its grooves to the exact spot of the intended cut.

The Score Cut

The older score cut, still used on some applications, uses the same sheet run as the wraparound shear. In the score cut the bottom element is simply a hardened sleeved roll, called a platen roll. The top slitter has a rounded edge and is pressed against the platen roll by pneumatic or hydraulic action. The effect is that of a knife cutting against a hard surface.

Setting a score cut slitter is simplicity itself since only the top slitter need be set. The cut quality, however, is considerably poorer than that of the shear cut and leads to dusting conditions on many types of webs.

Slitter knives need periodic regrinding. On shear cut slitters, the bottom knives are generally harder than the top ones and thus need grinding less frequently. On score cut slitters, the platen roll sleeves tend to become scored if cuts are made in the same location for some time.

Fig. 19-19. Pair of wraparound shear slitters.

Other Types of Slitters

Some simpler types of slitters are used on some non-paper applications, primarily plastic films and foils. The most basic is the razor cut, which uses a stationary blade against a stretched web. The razor blade has no opposite member.

The burst cut is yet another method used occasionally to slit films. It resembles in appearance the wraparound shear except that the grooved sleeves are not ground with cutting edges as required for the bottom slitter of

Fig. 19-20. Sheet run through wraparound shear slitters.

a shear cut. The top slitter blade is simply inserted into a groove in the opposing roll sleeve and causes the web to burst at the point of insertion while the land adjacent to the groove supports the web. Neither the burst cut nor the razor cut can be used on high-speed paper or paperboard winders.

SPREADERS

A sheet spreader of some type is unavoidable in any winding operation. Some kind of spreading device is used most frequently on the open draw between the point of slitting and the point of contact of the web with a drum or other support at the place of windup. Many winder designs incorporate a second spreader just ahead of the slitter section so as to eliminate wrinkles or baggy areas in the web before it enters the slitters.

Spreaders can be of the fixed bar type when the web is not surface sensitive or prone to dusting. The simplest of such spreaders is a bowed pipe, which can be rotated into and out of the web as needed. If some areas across the width of the machine are to be spread more than others, a sectionally adjustable spreader is incorporated in the winder design. This type of spreader is found frequently following the slitters.

If the web is surface-sensitive or tends to dust, a rotating type of spreader is used. The most popular of these is the rubber-sleeved spreader roll, first introduced by Mount Hope Machinery Company and still generally referred to as a Mount Hope, although it is produced by several other companies as well. These spreaders may have fixed bows or, if needed, adjustable or variable bows. Some of them have sectional steel sleeves rather than rubber

382 WEB PROCESSING AND CONVERTING TECHNOLOGY AND EQUIPMENT

Fig. 19-21. Sectionally adjustable rotating spreader.

covers. Spreaders with adjustable sectional rolls, as shown in Fig. 19-21, combine the advantages of fixed bars and rotating rolls.

SURFACE WINDERS

A web may be wound up to a roll by having its core or shaft contact one or more driven drums. Such winders are called surface winders in contrast to center winders, in which the rewinding roll is suspended from a driven shaft. By far the majority of secondary winders are of the surface driven type. There are some surface winders with center wind assists and some center winders with auxiliary contact with a driven drum, but those are not as common and will be described later.

The Two-Drum Winder

The most popular of all secondary winders is the two-drum winder. This type of winder features two power driven drums which drive as well as support the rewinding roll. A rider roll, which may be driven or not, generally rides the apex of the rewinding roll.

Figure 19-22 shows a schematic of a two-drum winder as used on many paper, paperboard and non-woven fabric applications. The winder as shown

WINDING, SLITTING AND SPLICING 383

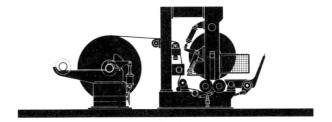

Fig. 19-22. Two-drum winder.

features an unwind for reel spools with a pneumatic spool ejector. This ejector saves time when changing parent rolls. At the completion of the wind, the empty spool is ejected from the bearings onto the outrigger arms so as not to interfere with the loading of the new reel. The spool can be picked up by crane when the winder is operating. Thus, no time is lost removing the empty spool before the new reel can be loaded.

The web first contacts a lead-in idler roll and then wraps the tension sensing roll. Tension sensing may be all pneumatic or by means of load cells mounted under the bearings of the tension sensing roll.

The next element in line is a spreader, which may be fixed or rotating. The lead-in roll and the spreader also serve for the web to maintain a constant wrap angle on the spreader roll so that its tension-sensing function is not distorted by angle changes.

The slitters following this particular winder are of the wraparound shear type. Draw shear slitters could be installed at this point, in which case there would be idler rolls preceding and following the slitters. Another spreader is installed after the slitters to help maintain cut separation. Most winder designers attempt to locate the slitter section close to the rewind elements to lessen the chance of interweaving of slit webs.

Having passed the second spreader, the web wraps the rear drum of the winder and emerges between the two drums, where it is wound up on cores with or without a core shaft. The rider roll seen on top of the rewinding roll is of special importance at the beginning of each set of rolls. It serves to produce a tight start around the cores. It is usually driven by separate motors or by belts from the front drum.

The core chucks must be retracted, at the completion of the wind, if the winder is shaftless. The roll ejector is a pneumatic model on this schematic, with air cylinders mounted between the uprights and the pivot bracketed to the front upright. On other winders, the ejector rotates around the rear drum. The ejected set of rolls is pushed onto a raised cradle, which is the lowered to floor level. Hydraulic lift tables can be seen instead of cradles on older and some smaller winders. The cradle is now generally preferred, because of its compactness and the avoidance of cutting the floor for a table. Then again, the cradle is very useful as a sturdy roll guard, especially for shaftless winding.

The configuration of this winder provides for an overhead sheet run. This allows the operator free access to the slitter and spreader elements during the run of the winder. It is less favorable, however, for webs which wrinkle easily, because of the large wrap angle on the tension sensing roll. The lead-in roll is also wrapped to a considerable extent when the parent roll diameter diminishes, and even more so when the web is unwound from the bottom. The latter is necessary when the sides of the sheet are to be reversed; i.e., when the inside of the parent roll is to become the outside of the finished rolls. The strong wrap of the idler rolls can be avoided by relocating slitter and spreader elements to provide for a fairly flat sheet run into the winder. This, however, is done at the expense of accessibility of these elements during the winding operation.

Shaftless Winding

The late 1960's started a change in the art of two-drum winding by the advent of shaftless winders. Prior to that time, most, if not all, such winders used core shafts to anchor the rewinding rolls in the machine. This was not too much of a problem on small winders since the 3″ ID core, the most popular, did not require a very heavy shaft. Wide machines, however, could not be handled without mechanical shaft loading and pulling equipment, which consumed space and made roll handling awkward.

Another problem arose on fairly wide winders. As demands for larger and larger diameters surfaced, roll quality suffered more and more from uneven caliper profiles of the web. These caused individual slit rolls of a single set to vary in diameter. Toward the end of the winding operation, a low caliper roll was lifted from the winder drums by adjacent high caliper rolls, which, in turn, were compacted by the increased nip pressure against the drums.

Most present two-drum winders are of the shaftless type; i.e., the cores are placed in the saddle between the drums at the beginning of the wind while chucks are inserted into the end cores. Torque-limiting motors or hydraulic cylinders maintain enough lateral pressure for a snug contact between the adjacent cores and the rider roll prevents the cores from rising up at the beginning of the wind. All rolls, high caliper or low caliper, will settle in firm contact with the drums, especially when the web wraps one drum as illustrated in Fig. 19-23. Since there is no shaft, each individual roll is free to assume a different diameter, depending on the caliper of the web. This contributes considerably to uniform density of the finished roll.

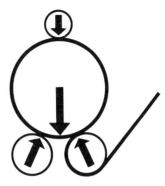

Fig. 19-23. Threading between drums for stability.

Roll Density

The problem of uniform roll density cannot be solved, however, simply by shaftless rewinding, especially when the material itself has little bulk. Since the two-drum winder not only drives the rewinding set on its surface, but also supports it on its two drums, the nip pressure between the drums and the rewinding set of rolls keeps growing as their diameter increases. This will cause rolls to become denser near the periphery, especially when they are wound to large diameters.

A number of measures are taken to counter this problem. The first is rider roll pressure. While it is useful to exert a considerable amount of pressure with the rider roll at the beginning of the wind, this pressure must be relieved as the rewinding set of rolls grows. Progressive rider roll relief, pneumatic or hydraulic, is now standard on most two-drum winders. Rider roll drives are helpful too in that they can be adjusted to provide progressive draw in relation to drums.

By far the most useful means of influencing roll density of two-drum winders has proven to be programmed torque control between the drums. In order to see how we can influence wound-in tension and consequently roll density by varying drum torques, we must look at possible web threading patterns. Figure 19-24 illustrates two such patterns. In Diagram I the web enters over the top of the rear drum and proceeds to the front drum. Differential drum torque would have only little influence on such a winder because the draw from *a* to *b* is short and the drums are hardly wrapped. The same would be true if the web were to come over the front drum and proceed to the rear drum from there.

In Diagram II the web comes up between the drums, wraps the rear drum about 180°, then wraps the core and touches the rider roll before coming into

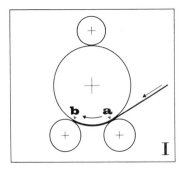 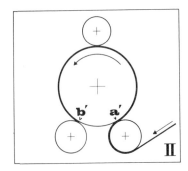

Fig. 19-24. Two threading patterns.

contact with the front drum. The draw between the drum from *a'* to *b'* is about 300°. This enables the winder to influence winding hardness by varying the torque between the two drums.

This is accomplished, as a rule, by driving each winder drum with a separate dc motor. (Some smaller winders have only one drive motor and a mechanical speed variator to the other drum. The effectiveness of these systems depends on the design of the mechanical variator.) Since a tight start is needed at the beginning of the wind, the front drum motor is programmed to exert more torque than the rear drum motor at that time. As the roll diameter grows, the program shifts load from the front to the rear drum. At about 30" diameter, the two drives share the load about evenly. Beyond that, it is the rear drum that assumes more and more of the load to wind the web more loosely and counteract growing roll density.

Reducing Downtime

The two-drum winder is popular because of its simplicity and its sturdy design. Speeds of 2500 m/min and greater are reached on these machines. Since, however, winding is inherently a batch operation, it is not always easy to keep up with a production machine even if that were to operate only at half the speed of the winder. On such critical applications it is important to keep downtime between sets at a minimum.

A number of features have been developed to accomplish this. In situations which require frequent changes of slitter settings, the slitters can be moved by electric motors and settings can be programmed by computers. The operator can enter the next slitter setup while the machine is processing the previous order. The computer can be used even to calculate the most economical assortment of roll widths considering the tonnage required for each width as well as the trim of the production machine.

WINDING, SLITTING AND SPLICING

Roll set changing itself can be automated by sequencing and pre-insertion of cores, so that downtime for a set change (from the same parent roll) can be kept as short as 35-40 sec. Such systems are in use primarily on large winders.

Winders with Individual Rewind Stations

The measures described earlier, taken on two-drum winders to produce controlled roll density, are effective to a point. There are applications, however, for which the two-drum winder is a less than satisfactory tool for producing evenly wound rolls. These applications concern primarily webs of high density when they must be wound to large diameters, say to more than 100 cm. Coated publication type papers (LWC), especially for rotogravure printing, would be a good example. When the rolls of such material reach fairly large diameters, torque control between the winder drums becomes ineffective and the rolls will turn out excessively dense, with too much wound-in tension. This will lead to overstretching of webs and bursts which destroy the usefulness of the roll.

Since it is not possible to rewind such webs suspended on center shafts across the machine, the solution is a winder which rewinds each finished slit roll individually by supporting it on its core. Most of these winders are single-drum winders. Figure 19-25 is a schematic of such a machine. The single drum is driven by a dc motor. The rewind stations are staggered, each accommodating a single rewinding roll. The rolls are surface driven by the winder drum. They are pressed against the drum by pneumatic or hydraulic cylinders. Thus the nip pressure between drum and rewinding roll can be programmed independently from the weight of the rewinding roll.

The sheet run of such individual station winders is somewhat more complicated than that of a two-drum winder. In the example illustrated by Fig. 19-25, the web, coming from a conventional spool unwind, passes through a pre-trimmer at the entry to the winder. The pre-trimmer is used when the trim to be taken on the web is wider than the removal at the main slitters can

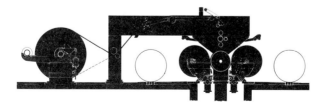

Fig. 19-25. Single-drum winder.

388 WEB PROCESSING AND CONVERTING TECHNOLOGY AND EQUIPMENT

handle. Since the web is threaded overhead, there are tape drives as well as air tables to mechanize this function. The first of the two spreaders doubles as a tension sensing roll. There is a vertical run through the draw shear slitters, after which the ribbons alternate to wind up on the rewind stations on either side of the drum. Rider rolls (twin rolls on this machine) aid in the start of each rewinding roll and then swing up on pivoting arms. The finished rolls are lowered to the floor by tilting the stations. The rolls are removed from the winder on slat conveyors in the floor.

Figure 19-26 illustrates an individual station with the core inserted between the chucks. Electric motors are used to position the stations and to insert and retract the chucks.

The rolls produced on such winders are not only of more uniform and preselectable density, but they are free to assume quite different diameters. It is possible even to splice one individual roll or to terminate it and start again on the same station without disturbing the others.

Winders with individual stations are, however, not without drawbacks. Slitters are far less accessible and two-side frames must be set for each individual finished roll. Core chucking as well as threading and taping the webs to the cores consumes more time than on regular two-drum winders. It stands to reason, therefore, that high production winders of this type include a large degree of automation. Slitters as well as stations can be set by servo

Fig. 19-26. Individual rewind station.

motors controlled by a computer program. While this increases purchase cost, it can be justified by much higher productivity. Even so, these winders frequently require more personnel than two-drum winders.

Some winders with individual stations are not true surface winders. On some of them the rewinding rolls on the stations are driven centrally by hydraulic motors. On other winders they are surface driven and use center wind assists.

Duplex Winders

Duplex winders, if surface driven, follow a similar principle of winding up staggered rolls in two positions against a drum. Instead of individual stations, however, we find common rewind shafts on either side. They share with the individual station winders the advantage of positive separation of rewinding webs. This precludes interweaving, which can be vital on narrow slitting, which is the main application of duplex winders.

While the use of duplex winders eliminates interweaving problems, they are generally much slower and require much more setup time than two-drum winders. Their use is therefore confined to converting operations, in which the roll separation feature is vital.

Tertiary Winders

While tertiary winders may be of the same design as secondary winders, they process webs which have been wound up twice before. Consequently, they perform a specialized function which was impractical to carry out at an earlier stage of the production process. Salvage winders should be mentioned at this point, though those may be secondary or tertiary machines. Their job is to rewind rolls which were previously wound in an unsatisfactory way.

CENTER WINDERS

Some tertiary winders are true center shaft driven winders. In contrast to the surface winders, they include no drums which drive the rewinding rolls on their periphery. A common core shaft of suitable size is driven to wind up the set of rolls. Since the rewinding roll (or rolls) does not contact a driving surface and is not supported, there are no serious roll density problems. Yet, these winders have a host of other problems.

The rewinding roll is suspended on a shaft, which is supported by bearings at its ends. The shaft must therefore be strong enough to carry the full-diameter finished roll without more than minimal deflection. If the shaft deflects, it will cause the rewinding roll to sag, which results in interweaving of the adjacent strips. Rolls will be stuck together. The thickness of the shaft is limited, however, by the desired final core ID and by the weight the operators can lift to load and extract the shaft.

The minimum OD of the shaft thus depends on the width of the machine and the diameter to be wound. There is yet another parameter: the width of the total web in relation to the center distance of the rewind shaft bearings. The wider the web in relation to the capacity of the winder, the less is the danger of deflection. The most unfavorable condition occurs when rewinding a narrow web to a large diameter. In that case, the entire load will be carried by the center section of the shaft causing maximum deflection. Rewind shafts used by these winders are generally expanding shafts, mostly of the leaf type.

Driving a center winder is more involved than driving a surface winder. If the winder is to run at a uniform web speed, the center shaft must be driven at high rpm at the beginning and at gradually reduced rpm as the rewinding roll builds. In other words, the motor must run at high speed and low torque at the start and gradually build up torque and reduce speed. This means a constant horsepower rather than a constant torque drive.

SPLICING

Splicing is frequently an integral part of a rewinding operation. Flying splice unwinds have been discussed in the early part of this chapter.

Splicing in a rewinding operation may become necessary for any one of the following reasons:

1. There is a web break in the parent roll.
2. The parent roll expires before the specified diameter of the rewound roll has been reached.
3. The parent roll has a defective part, which must be cut out.
4. The web breaks during the rewinding operation.

There are generally two types of splices: butt splices and lap splices.

Butt Splices

Butt splices are used on thick webs when the double layer of a lap splice causes problems in subsequent processing. In a butt splice the edges of the material are joined by means of tape without overlapping the material itself. This cannot be done on the fly. Most butt splicers work manually with the aid of a splicing table, though mechanical splicers exist.

For the purpose of butt splicing the web must be held stationary for a short time. If such splicing is to be done on the fly, a so-called accumulator is installed ahead of the splicer. The accumulator can store a certain length of web, usually in the form of a festoon. This gives the splicer enough time to join the tails of the web without stopping the machine. It stands to reason that this splicing method is suitable for fairly low-speed applications only.

Lap Splices

The majority of splices are lap splices, meaning that there is an overlap between the ends of the web which are spliced. The flying splicers on turret and similar unwinds all produce lap splices. The two laps are bonded by means of an adhesive applied to one of the tails. It must be understood that an extra thickness of web passes the equipment following a flying splice unwind. If this can damage the equipment—a coater, for instance—there is need for a splice sensor, which causes the critical unit to disengage while the splice passes.

Splicing is quite a common part of a rewinding operation. There are many methods of splicing webs, either by means of applying adhesive to one tail, or by attaching the tail with adhesive tape. The winders are stopped for this purpose.

While splices on a secondary winder can be made conveniently at the parent roll, it is often of importance that such splices be made under tension to avoid puckering or wrinkling of the web in the area of the splice. On a two-drum winder which threads the web between the drums there is a convenient and effective method of making such splices under tension.

When a web break occurs, the rider roll is to be lifted and the winder rethreaded. A strip of single-sided adhesive tape with a release-coated back is placed on the full length of the naked front drum. A strip of double-sided adhesive tape is applied over the first tape, as shown in Fig. 19-27. A cut is made across the rewinding roll to produce an upper and a lower tail, as per Fig. 19-28. The protective cover of the two-sided tape is peeled back and the lower tail is fastened neatly to the adhesive strip on the front drum. The overhanging part of the lower tail (P_2) is folded back and torn off at the edge of the tape.

The winder is started now at crawl speed. The nip pressure at the front drum will cause the double-sided adhesive tape to peel off the release-coated back of the single-sided tape and adhere to the newly made tail (see Fig. 19-29).

As soon as the exposed tape at the edge of the new tail contacts the rear drum, it attaches the tail to the new web, which wraps the rear drum. The splice is made under tension and nip pressure, so there will be no wrinkles or puckers in the area of the splice.

392 WEB PROCESSING AND CONVERTING TECHNOLOGY AND EQUIPMENT

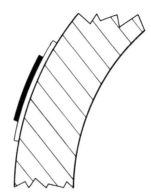

Fig. 19-27. Splicing tape on front drum.

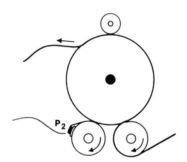

Fig. 19-28. Winder prepared for splicing.

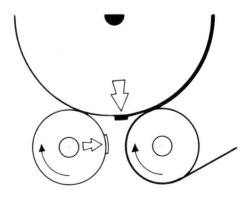

Fig. 19-29. Splicing tape contact transferred from front to rear drum.

Fig. 19-30. Completed splice.

As soon as the spliced tail reaches the front of the winder, the machine is stopped. The web is folded back and torn off at the edge of the splice, as indicated in Fig. 19-30. The splice is now complete, but it can be reinforced by applying a strip of single-sided adhesive tape over the outside of the splice.

REFERENCES

1. Rienau, J. M. A primer on slitting and rewinding basics- part I. *Paper, Film and Foil Converter,* **53** (4), 61-66 (1979).
2. Rienau, J. M. A primer on slitting and rewinding basis-part II. *Paper, Film and Foil Converter* **53** (5), 172-178 (1979).
3. Greiner, T. S. Winders with individually programmed stations and slitters. *Tappi* **63** (10), 39-42 (1980).
4. Greiner, T. S. The automation of roll set changes on fast winders. *Tappi* **64** (10), 59-64 (1981).
5. Greiner, T. S. Winders and winding. *Pulp & Paper Canada* **83** (4), 61-65 (1982).

20
Web Handling

Donatas Satas
Satas & Associates
Warwick, Rhode Island

Web when processed must be flat at the point of coating or lamination, it must track well, remain centered on the machine, and form straight-edged rolls without wrinkles.

It is required that the web is at some tension when processed and the level of required tension might vary throughout the processing line. The web might have a tendency to form wrinkles and special rolls might be required to keep the web spread in the lateral direction.

Neither the web nor the machine are perfect, and some steering of the web is usually required. While coating and laminating are continuous operations, web winding is not, and various methods are employed to overcome the discontinuous winding. Web accumulators are used for slower operations.

All of these techniques of tracking the web, maintaining its tension at a proper level and keeping the web wrinkle-free constitute web handling operations, which are discussed in this chapter. In addition, coating thickness measurement techniques and some miscellaneous slitting and winding methods are also covered.

TENSION

The web when processed must be at some tension in order to maintain the lateral stability, to remove the wrinkles, to present a smooth, flat surface at points of coating, laminating and other operations, and to wind the web into stable rolls.

Various equipment is used to impart and maintain the web tension in the web train from the unwind to the rewind stations. Often the tension is varied along the web train in order to meet the processing requirements. High tension might cause permanent deformation of the web, wrinkles and curling. It is advisable to maintain the tension at 25% of the value which causes distortion.[1] It is expected that the tension can be controlled within ±10% on normal web handling equipment.

Unwind and rewind tensions depend on the torque exerted on the roll. The relationship between the torque and tension is expressed in the equation below.

$$T = F \cdot r \quad (1)$$

where:

T = torque
F = tension force
r = roll radius

The tension is usually expressed as the force per unit width in order to eliminate the effect of the web width. If the torque remains constant, the tension will increase with decreasing roll radius. The curve for the constant torque winding is a rectangular parabola (see Fig. 20-1c). If the roll is unwound at a constant torque, the tension towards the end of the roll might increase to a sufficiently high value to rupture or damage the web. This is avoided by adjusting the unwind tension as the roll diameter decreases. Unwinding at constant tension might be required for some operations (Fig. 20-1a). Rewinding is usually carried out with the tension decreasing with increasing roll diameter, but decreasing slower than along the parabolic curve (Fig. 20-1b). This is accomplished by tension controls which regulate the brakes or clutches, or the current of the electrical motor.

Tensioning Devices

Tension is produced when driven roll is pulling against a resistance, such as unwind braking. The most primitive way to produce unwind tension is a drag belt in contact with the unwind roll surface (Fig. 20-2). The drag belt imparts the tension directly to the web by frictional forces acting between the belt and the web surface.

The drag belt may be used on the brake drum producing a constant torque (Fig. 20-3). Modern equipment uses air or magnetic brakes instead of such simple devices as drag belts. Since such brakes are constant torque devices,

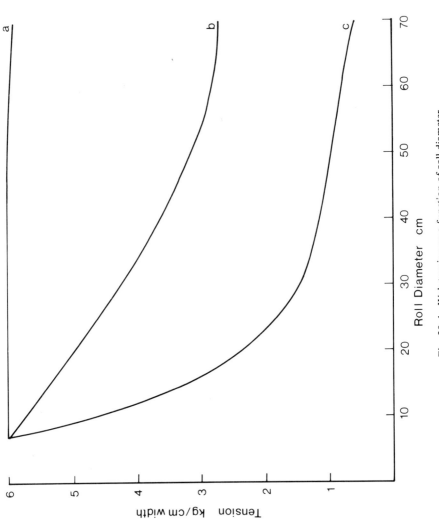

Fig. 20-1. Web tension as a function of roll diameter.

WEB HANDLING 397

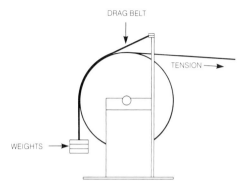

Fig. 20-2. Roll surface drag belt. (*Courtesy John Dusenbery Co., Inc.*[2])

an adjustment of the torque as a function of roll diameter is required in order to obtain the required unwind tension. The torque can be adjusted by a device consisting of a follower arm connected to a control valve (Fig. 20-4). The control valve regulates the air pressure into the air brake according to a programmed pattern. Analogous arrangement is made with magnetic brakes. Similar arrangement may be used for the rewind as well. Figure 20-1 shows various tension control possibilities of rewind or unwind rolls.

In addition to the winder, tension is produced by any driven roll. The web tension required for proper processing of the web might be quite different from the tension required to wind a good roll. Therefore pull rolls are often located just before the winder and the winding tension becomes completely isolated from the tension in the preceding sections. The pull rolls might be used for some other function, such as cooling, surface calendering, embossing. A set of pull rolls is shown in Fig. 20-5. The force F_1 is larger than F_2. In addition to the pulling roll arrangement shown in Fig. 20-5, other pull rolls are shown in Fig. 20-6.

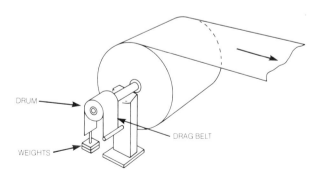

Fig. 20-3. Drag belt over brake drum. (*Courtesy of John Dusenbery Co., Inc.*[2])

398 WEB PROCESSING AND CONVERTING TECHNOLOGY AND EQUIPMENT

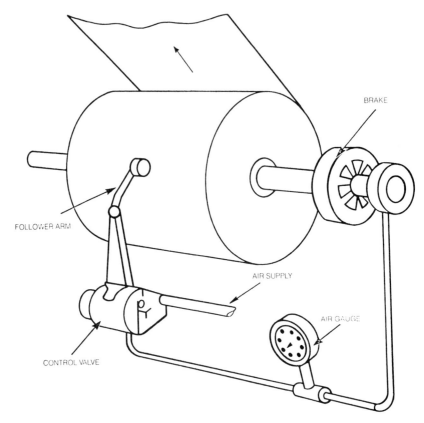

Fig. 20-4. Roll follower arm tension control. (*Courtesy John Dusenbery Co., Inc.*[2])

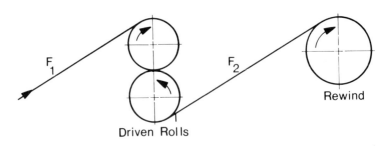

Fig. 20-5. Pull rolls before rewind.

WEB HANDLING 399

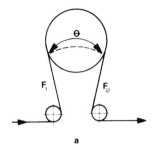 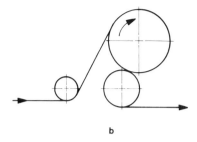

Fig. 20-6. Pull rolls:
a—Without a nip roll
b—With a nip roll.

The relation between the web tensions and coefficient of friction and wrap angle are shown in the equation below.

$$F_1/F_2 = e^{\mu\theta} \qquad (2)$$

where:

F_1 = web tension on the incoming side
F_2 = web tension on the outgoing side
μ = coefficient of friction
θ = wrap angle

If the tension F_1 as shown in Fig. 20-6a is increased and F_2 remains constant, the wrap angle θ increases and the web will show a tendency to wrap around the roll unless the location of the idler is changed. If F_1 is decreased than the wrap angle θ decreases. This might lead to the loss of traction around the pull roll. In order to avoid such problems, a nip roller as shown in Fig. 20-6b is used. The nip roller is placed on the bottom of the driven roll permitting maximum wrap.

If tensioning of the web is required before some operation, an arrangement similar to the one shown in Fig. 20-5 may be used. The web is wrapped in "S" fashion around the rolls which are equipped with brakes. This allows an increase of tension downstream from this tensioning device.

Tension Sensing Devices

In order to control the tension it is necessary to measure it and to use the transmitted signal to actuate a tensioning device.

One of the most often used tension sensing devices is the dancer roll. A dancer roll is a counter-balanced roll in contact with the web. It can move depending on the magnitude of tension. A dancer roll can be used for web speeds up to 700 m/min and for web tensions above 2.5 kg. Multiple dancer roll systems are required for lower tensions.[2] The dancer roll can move in either direction from its null position and to give a signal to the tensioning device to either accelerate or decelerate.

Many different dancer roll designs are possible. The dancer roll might move vertically or horizontally; it might be pivoted and move in an arc. A single dancer roll, or an assembly of several, might be used. The dancer roll is counter-balanced to minimize the effect of the roll dead weight. Counter-balancing is accomplished by counter-weighing, or by employing pneumatic cylinders.

The advantage of dancer rolls over other tension sensing devices is that they can move and thereby attenuate small disturbances in web tension. The dancer roll assembly can store some material and eliminate short duration deceleration or acceleration.

Figure 20-7 shows a vertically moving dancer roll. It may be counter-balanced by a weight on a chain which travels on sprockets. Horizontally moving dancer roll assembly is constructed in a similar fashion. Figure 20-8 shows a pivoted dancer roll used in conjunction with an unwind. Vertical and pivoted dancer rolls are used most often.

Multiple dancer rolls are used for materials which have to be processed at a low tension. A single-roll dancer assembly requires a force equal to 50% of its inertia to move it from null position. Increasing the number of dancer rolls and web legs pulling, decreases the tension required to move the dancer

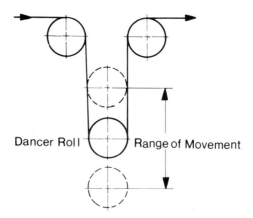

Fig. 20-7. Vertical dancer roll.

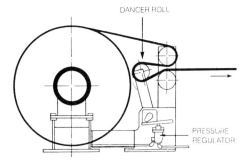

Fig. 20-8. Pivoted dancer roll used for unwind. (*Courtesy John Dusenbery Co., Inc.*[2])

rolls. Therefore, multiroll dancer assembly can handle materials at lower tension and respond to smaller tension changes.

Load Cells and Force Transducers

These load sensing devices are mounted under an idler roll on one, or more often, on both sides. The roll with load sensing devices on both sides is more accurate, the measured tension value does not require centering of the web and it is not affected by non-uniform web tension across the width.

Pneumatic or hydraulic load sensors require a roll movement of 0.25-1.25 mm. Electronic load cells require a movement of only 0.05-0.4 mm from zero to full load position. Piezoelectric or magneto-elastic load cells are used for electronic sensors.

The tension idler rolls can be mounted in various ways as shown in Fig. 20-9. A web tension roller with transducers is shown in Figs. 20-10 and 20-11.

WEB GUIDING

Due to imperfections within a web and various outside influences, most webs have a tendency to move laterally when pulled through the equipment. Loose edges, varying tension, telescoped and uneven unwind roll, misalignment of the rolls are amongst the reasons causing the wandering of the web. Neither the web, nor the equipment are perfect and therefore corrective devices are used to keep the web laterally aligned.

The web guiding systems may be located at the following places in the processing train:

- Unwind
- Intermediate
- Rewind

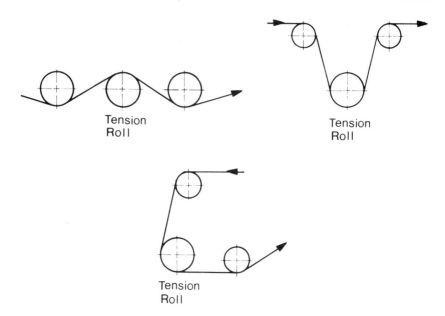

Fig. 20-9. Various ways to mount tension rolls.

Web guiding at the unwind is required if the supply rolls are not evenly wound, or to insure proper positioning at the start of a process. Without web guiding provisions at the unwind, rewinding of imperfect rolls on separate equipment would be required.

Intermediate guiding is used to place the web accurately before some operation, such as coating or laminating, or generally to correct the lateral shift of the web caused by the cumulative effect of various factors.

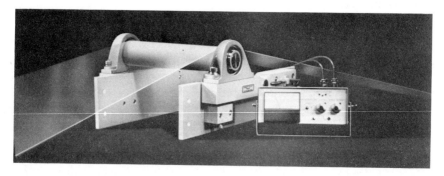

Fig. 20-10. One-roller web transducer. (*Courtesy Tensitron, Inc.*)

WEB HANDLING 403

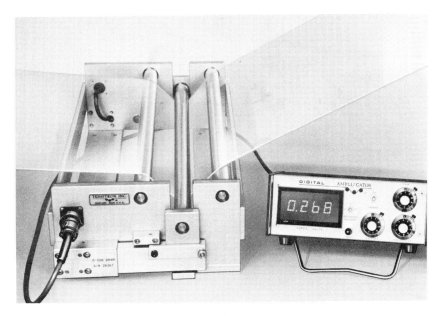

Fig. 20-11. Three-roller web transducer. (*Courtesy Tensitron, Inc.*)

Web guiding at rewind is required to correct miscellaneous wandering of the web, especially if intermediate guiding is not used and where it is important to produce an evenly wound roll. A fixed rewind stand with intermediate guider feeding will also provide an evenly wound roll. Almost all of the web processing lines require guiding at least in one location, often in two and three locations, especially when the process has several operations. Only the machines with a short web span, such as some slitters and rewinders, might be used without any means of lateral web guiding.

The web guiding systems consist of three main elements:

- Sensing device
- Controls
- Actuating method

Figure 20-12 shows the basic elements for an edge guide.

The sensing device detects the lateral position of the web and delivers an error signal to the controller which converts that signal to a higher power level force, used to move the edge guide. The edge guide causes the web to move and correct its previous lateral position.

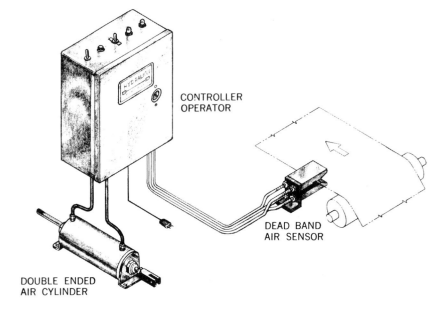

Fig. 20-12. Web guiding system. (*Courtesy Hydralign, Inc.*)

Sensing Devices

The sensing devices are usually used to detect the edge of the web, but also may be used to detect a printed line or an interior edge as in case of laminated materials. Sensing one edge is sufficient for most of the materials, but sensing of both edges, centerline guiding might be used for keeping the material centered on the equipment.

The sensing devices are classified according to the method used to detect the web's position.

Mechanical Edge Detector

This consists of a hinged plate, or palm, which follows the edge of the web (Fig. 20-13). When the edge travels too far in either direction, a pneumatic or electric signal is delivered to the control and an appropriate correction is made. These detectors are rugged, low-maintenance devices, but suitable only for fairly stiff materials. Limp webs might just curl against the plate.

Pneumatic Sensors

These employ air to sense the position of the web's edge. Vacuum sensors were used originally. They required direct contact with the web and therefore

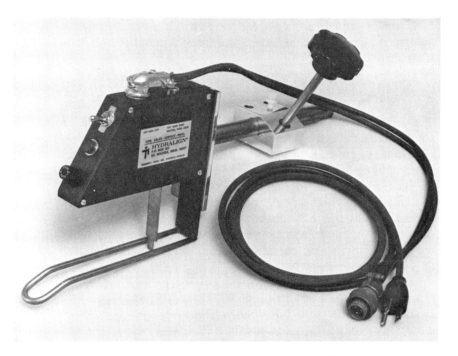

Fig. 20-13. Mechanical edge detector. (*Courtesy Hydralign, Inc.*)

were not suitable for all webs. Because of this and also because of excessive maintenance they were replaced by positive pressure pneumatic sensors (Fig. 20-14) which are commonly used devices for sensing web's edge position. The air stream is emitted from the bottom opening and reaches the sensing orifice in the top finger of the sensor. Interposed web changes the air flow and generates an appropriate signal to the controller.

Several variations of the pneumatic sensors are available. Each finger of the detector might be supplied with low pressure air. The air orifices are located opposite one another and the impinging air streams restrict the flow. Interposing web disrupts the balance and causes the appropriate response.

Two air jets might be employed. The distance between the jets is the dead-

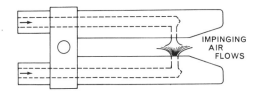

Fig. 20-14. Pneumatic edge sensor.

406 WEB PROCESSING AND CONVERTING TECHNOLOGY AND EQUIPMENT

band and no guiding takes place until the web moves out of the deadband. The standard deadband might be 1-2.5 cm wide and the air jets operate at a pressure of 70 kPa.

Photoelectric Sensors

These are also often used for the edge detection. Two basic types are in use: transmitted light and reflected light.

In the transmitted light sensors, the light is emitted by a light source and detected by a photocell. The web placed between disrupts the light transmission generating an appropriate error signal. For an on-off control two light sources and two photocells are used and the distance between the two determines the width of the deadband. This distance may be adjustable on some sensors. For extreme accuracy a single photoelectric cell and light source is used for sensors which operate with proportional (servo) control. Figure 20-15 shows a transmitted light sensor with an adjustable deadband width.

Sensing of the edge of a transparent film cannot be accomplished with transmitted light photoelectric sensors. Reflective light sensors are used for such films. Use of various filters can be helpful for semi-transparent webs. A diagonal photoelectric sensor is used for open weave textiles and other materials which are more opaque if looked at at an angle. Figure 20-16 shows such a sensor.

Fig. 20-15. Photoelectric sensor with two photocells and adjustable deadband. (*Courtesy Hydralign, Inc.*)

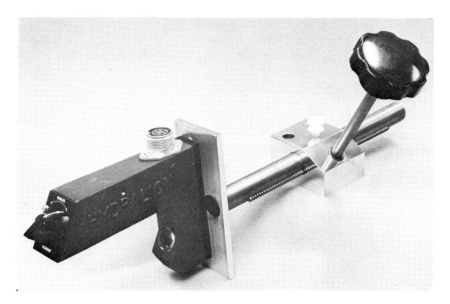

Fig. 20-16. Diagonal photoelectric sensor. (*Courtesy Hydralign, Inc.*)

Reflected light sensors may be used with glossy surface materials. They are also suitable for sensing a printed black line, or an interior edge of a laminate, if the light reflectivity is different between the two components.

Ultrasonic Edge Detector

This consists of a transducer which converts electrical energy into acoustic energy. This sensor employs a very low energy level and is especially suitable for light materials (below 0.1 mm thick).[3]

Controller

The controller is a device which receives the signal from a sensor, interprets it and transfers the low-energy signal to a higher energy force, which is used to cause a corrective action by the web guide. The controller is characterized by either the mode of control or by the way the energy is received and transmitted.

On-off and proportional modes are most often used for edge control. On-off control detects when the web edge leaves the deadband and signals for a corrective action. The corrective action lasts until the edge is again returned to within the deadband. An excessive correction can cause continuous oscillation. In some special cases, a pre-designed continuous oscillation has been utilized to prevent winding a heavy edge upon itself. Continuous oscillation distributes the heavy edge over a wider band, thus avoiding forming a roll with a heavy edge and slack center.

The width of the deadband used in on-off control is adjustable and usually is between 0.25 and 12 mm.[4,5]

In the proportional control mode, the rate of correction depends on the magnitude of deviation, unlike in the on-off controller, which does not sense the magnitude of error and calls for a correction of the same rate until the edge reaches the deadband. The proportional controller also might have a deadband in some cases. In order to measure the magnitude of error, the sensor's photoelectric cell is electronically split, creating a proportional left-right signal with a small balance point. The accuracy of proportional controllers is better. For the control accuracy of ±1.5 mm, on-off control is sufficient. For more accurate web guiding, proportional control must be used. Mechanical sensors are primarily used for on-off control but are also suitable for proportional mode. The pneumatic sensors react to a deviation of 0.125 mm; photoelectric proportional sensors have an accuracy of 0.05 mm.[5]

The controllers are also categorized according to the method of the signal reception and the method by which the controller signals the edge guide. Thus we recognize all pneumatic, or all electrical, controllers, pneumohydraulic, electrohydraulic and other combinations. The signal from the sensor to the controller is usually either pneumatic or electrical and the higher force actuation to the guide is often hydraulic, although pneumatically and electrically actuated guides are also used.[6] The popularity of the hydraulic guide actuators is due to the small size and high power of hydraulic cylinders and to the fast response because of the incompressibility of hydraulic fluid.

In selecting a web guiding system, attention should be paid to the resonance which must be avoided in order to have a stable control. It is generally thought that most of the web processing machinery is incompatible with controllers with resonant frequencies higher than 6-8 hertz.[7]

Guides

The web guides consist of hardware required to change the lateral position of the web. The guiding equipment is actuated by a controller applying power to the hydraulic or pneumatic cylinder, electrical or hydraulic motor or other devices used to convert power into mechanical movement. Linear movement is required by most of the web guides and therefore cylinders are the most frequently used devices for web guides. The guides are subdivided according to their location in the web train.

Unwind Guiding

Imperfectly wound or telescoped rolls require guiding and this is usually done by laterally shifting the unwind stand. The stand might be mounted on low friction bushings or grooved wheels running on "V" tracks. The lateral movement of 5–25 cm is generally used. The first idler is attached to the laterally moving carriage and the sensor is stationary-mounted immediately after the first idler (Fig. 20–17).

Unwind stand might be the only guided position required for short-span operations such as slitting, inspection or rewinding.

Rewind Guiding

The arrangement is physically similar to the unwind guiding. The rewind roll is mounted on a movable carriage which follows the web in order to produce an evenly wound roll. The idler just before the rewind roll is stationary and the sensor is located on the entry side of the idler and mounted on the movable carriage. There should be sufficient friction between the stationary idler and the web to prevent slippage.

Rewind control is needed to wind even rolls. Intermediate guiding, if placed a short distance before the rewind, can eliminate the need for movable rewind stand.

Intermediate Guiding

The intermediate guiding system is used to position the web properly before some operation, such as coating or laminating, or at the end of a long free

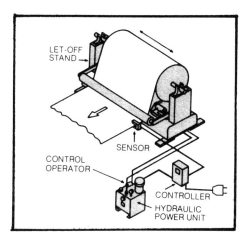

Fig. 20-17. Unwind stand guiding. (*Courtesy Hydralign, Inc.*)

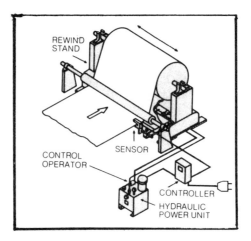

Fig. 20-18. Rewind stand guiding. (*Courtesy Hydralign, Inc.*)

span, such as after a drying oven, to correct the wandering of the web. There are two basic types of intermediate guiding systems:

- Steering type
- Positive displacement type.

Steering type guiding is based on the principle that the web will move laterally upstream from a guiding roller until the web becomes positioned perpendicularly to that roller. Consequently, cocking the roll will cause the web to move laterally upstream from that cocked roll. Long free entry upstream spans are normally required for this system to operate: the minimum free span is two to three times the web width. This steering type system is especially useful after a long free span; i.e., at the end of the oven. The sensor for the web steering system is mounted immediately after the steering roll. Web might wrinkle if the free entry span is not sufficiently long.

Single-roll or two-roll steering assemblies are used. They are usually operated by either hydraulic or a pneumatic cylinder. A single-roll system requires a web wrap of 90° and thus a 90° change of the web direction. Figure 20-19 shows various installations of a single steering roll. The shaded roll represents the guide roll, solid line-idler roll and preferred web path, interrupted line-alternate idler roll and alternate web path. The distance a should be two to three times the web width, distance b one-half the web width and c approximately 20 cm.

Two-roll web guiding does require a change of web direction, as shown in Fig. 20-20. The distances a, b, and c are the same as in Fig. 20-19. Figure 20-21 shows the roll inclining mechanism.

Inclining the guide roll in relation to the web causes bending of the web up-

WEB HANDLING 411

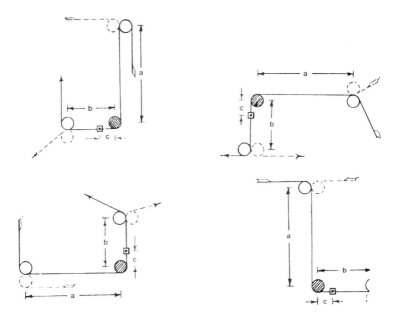

Fig. 20-19. Typical installations of single-roll steering systems. (*Courtesy Hydralign, Inc.*)

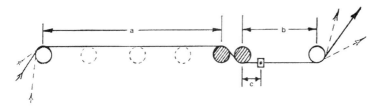

Fig. 20-20. Two-roll web steering. (*Courtesy Hydralign, Inc.*)

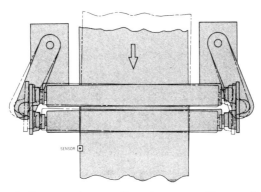

Fig. 20-21. Roll alignment of a two-roll steering system. (*Courtesy Hydralign, Inc.*)

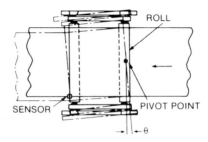

Fig. 20-22. Two-roll displacement guide. (*Courtesy Hydralign, Inc.*)

stream from the roller and redistribution of stresses. The stresses are increased on the short side of the web and decreased on the long side. The greatest stress is just downstream from the roller preceding the guide roll and the lowest is at the guide roll itself. The stress should not be high enough to cause the tearing or a permanent deformation of the web. Obviously, the longer the entry span, the lower the stress at any given point along the web. The larger the angle of inclination of the guide roll, the larger the stress.

The guiding roll should best pivot around a point located at the entry span prior to the guide roll. Ideally, the pivot point should be about three-fourths of the distance between the guide roll and the first upstream idler (entry roll).[7] The center of pivoting closer to the guide roll, and especially if the guide roll is pivoted around one of its journals, yields a less stable system and increases the stress concentration.

A positive displacement guide (offset pivot guide) makes the lateral web position correction within the web contained between the guide rolls. Only short entry and exit spans are required, unlike in the case of a steering guide. Figure 20-22 shows the top view of a two-roll displacement guide. Figure 20-23 shows a schematic diagram with the approximate dimensions. The en-

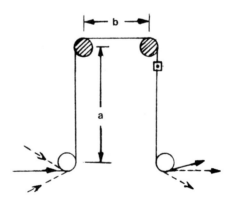

Fig. 20-23. Entry and roller spaces of a displacement guide. (*Courtesy Hydralign, Inc.*)

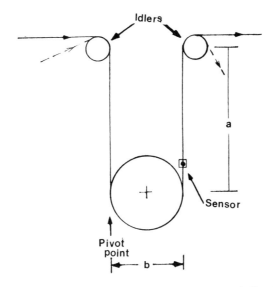

Fig. 20-24. Single-roll displacement guide. (*Courtesy Hydralign, Inc.*)

try span a is normally equal to the web width, the roller span b is usually two-thirds of the web width and the distance between the exiting roller and the sensor is as close as practical (20 cm). The positive displacement guide can be mounted in several different ways. For narrow web width or stiff materials, a single-roll displacement guide may be used. The roll diameter of the displacement guide should be two-thirds of the web width equivalent to the distance between the guide rollers in a two-roll guide. Figure 20-24 shows a single-roll guide.

While usually the web is guided to keep it properly aligned on the equipment, in some cases certain parts of the equipment might be guided to meet the web. The rails of a tenterframe are sometimes guided to meet the web, and the trimming knives can be kept centered over a web by guiding the shaft with the blades.

SQUARING ROLLS

Webs sometimes have baggy edges or centers which cause formation of wrinkles and other problems in processing such webs. Idlers equipped to change their camber are used to redistribute uneven tension because of a baggy edge. The baggy edge is made to travel a longer distance. Figure 20-25 shows a squaring roll used for such a purpose. The end A is moved vertically to compensate for a baggy edge on that side of the web. The roll must be moved in the plane of perpendicular bisector of the wrap angle. Cocking the roll at a right angle to the perpendicular bisector will cause the roll to act as a

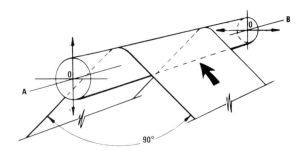

Fig. 20-25. Squaring roll. (*Courtesy Yorkshire Industries, Inc.*)

steering roll and introduce a lateral movement of the web. After the roll has been adjusted at the end *A* some horizontal adjustment may be made at the end *B*. This will compensate for the tendency of the web to move laterally. Rolls with bearings which allow such correction automatically are available. A wrap of 90° is required for the squaring roll to be effective.

Crowned rolls may be used to remove the effect of a baggy center. The roll has a larger diameter in the center and the web must travel a longer distance at that location equalizing the longitudinal stresses. Crowned rolls are not convenient to use because of a fixed crown which cannot be adjusted to accommodate various degrees of bagginess.

SPREADER ROLLS

Occasionally wrinkles might be formed as the web is processed and spreader rolls are used to move the web laterally from the center out in order to eliminate the wrinkles. Such spreader rolls are especially needed just before the web enters a nip or a processing station (a coater or a laminator), where surface waviness might be ironed to permanent wrinkles causing a defective product.

Several types of devices are used to remove wrinkles:

- Bent bars
- Bowed rolls
- Slat expander rolls
- Grooved rolls.

Bent bar causes lateral spreading when a web is dragged over it. The bow of such bars may be adjustable. Their use is limited to the materials which are not damaged by being dragged over the bar.

Bowed rolls represent an improvement over the bent bar: the roll rotates and the web is carried without rubbing against a surface. The construction of the bowed roll is shown in Fig. 20-26. The bow of the roll is set so that the

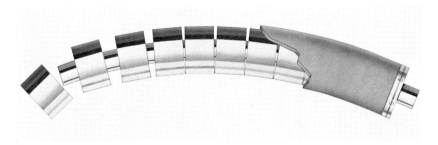

Fig. 20-26. Construction of a bowed roll. (*Courtesy Mount Hope Machinery.*)

web approaches the roll on the concave side and leaves on the convex side. Because of the curvature only the center of the roll rotates in a direction parallel to the machine direction. The points on the roll away from the center rotate at an angle to the machine direction, thus causing an outward spread of the web.

The spreader roll should be followed immediately by another roll: the exit span should be short, while the entry span should be long.

The direction of the bow is usually parallel to the web (Fig. 20-27a). The web is spread evenly from the center to the outside edges. If the center of the web is baggy, the bow is set at an angle, as shown in Fig. 20-27b. This way the center travels a longer distance than the edges and the stresses are distributed more evenly across the web. If the edges are slack, the bow of the roll is set at a downward angle, as shown in Fig. 20-27c, causing the edges to travel a slightly greater distance.

Fixed bow and variable bow expander rolls are available. The axle of the fixed bow roll is curved to a predetermined radius, while the variable bow roll curvature can be varied to suit the web requirement (Fig. 20-28). An excessively large bow can cause permanent deformation of a web. Usually a bow ratio of 6-8% is sufficient. The bow ratio is the chord of the arc of the roll face divided by the roll length. Some applications of expander rolls are shown in Figs. 20-29, 20-30 and 20-31.

Slat expander rolls consist of series of segmented wood or aluminum bars located around the circumference of the roll. The bars move in the cross-machine direction spreading the web. The bars move outward from 0° through the 180° point on the roll's circumference and inward from 180° to 360° as they pass over internal cams. The web is in contact with the bars in the area where the bars are moving outward.

The web should have a 90°-180° wrap around the slat expander for an effective spreading. The slat expanders are mainly used in the textile industry for relatively slow-speed operations. They tend to be noisy and may cause wrinkles in thin films.

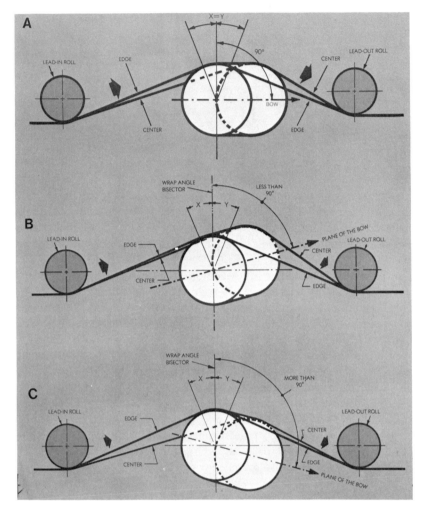

Fig. 20-27. Wrapping of the bowed rolls. (*Courtesy Mount Hope Machinery.*)
a—Direction of the bow is parallel.
b—Bow at an angle above horizontal plane.
c—Bow at an angle below horizontal plane.

Grooved Rolls

Grooved aluminum or rubber covered rolls are used for spreading lightweight webs, especially films, which might be easily deformed on a bowed roll. Grooved rolls also help to remove the air trapped between the roll surface and the web. Several types of groove patterns are available. Herringbone pattern is the most popular. It has slanted grooves starting in the cen-

WEB HANDLING 417

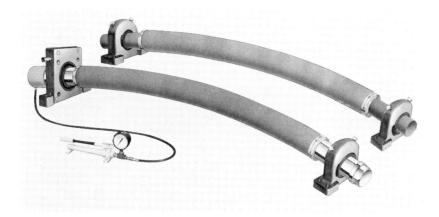

Fig. 20-28. Fixed bow expander (top) and variable bow expander (bottom) rolls. (*Courtesy Mount Hope Machinery.*)

ter of the roll. The web is spread outward from the center as it passes over the roll. A brake applied at the end of the shaft increases the spreading action.

Helical grooved rolls have opposing center to edge spiral patterns (Fig. 20-33). Rolls with dual helical grooves are more effective than with a single helical groove. Rolls with circular grooves cut at the right angle to the roll axis are also used.

A wrap of 60°-90° is used. The grooved rolls serve as idler rolls just before nips, laminating and coating stations.

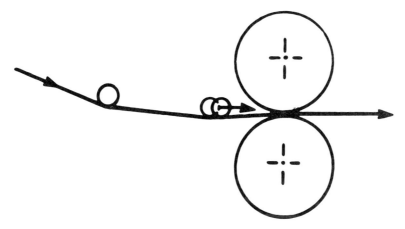

Fig. 20-29. Expander roll before a roll coater to eliminate wrinkles. (*Courtesy Mount Hope Machinery.*)

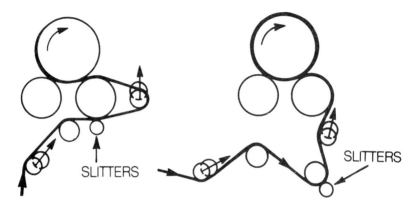

Fig. 20-30. Expander roll before slitting to eliminate wrinkles and after slitting to prevent interleaving of slit webs on a surface winder. (*Courtesy Mount Hope Machinery.*)

SLITTING AND SLICING

Some webs, such as various tapes, are used in narrow widths and their manufacturing presents special problems and require specialized equipment. The slitting process in general is discussed in Chapter 19 and the material below deals only with some special aspects of narrow-width slitting and winding.

Razor Slitting

In addition to shear and score slitting, razor slitting is often used for less demanding applications. Razor slitting is simple to set up and the equipment is quite inexpensive. At higher speeds (above 500 m/min), razor blade slitting

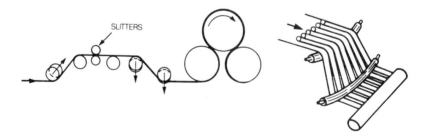

Fig. 20-31. Use of two expander rolls after slitting. The first roll spreads the slit strips apart and the second positions the strips parallel but with sufficient gaps to prevent interleaving. (*Courtesy Mount Hope Machinery.*)

WEB HANDLING 419

Fig. 20-32. Slat expander. (*Courtesy Louis P. Batson Company.*)

might not be acceptable: heat may be generated and the edges of thermoplastic materials may be burnished.

The blades may be mounted to cut unsupported web, or they may be placed in a grooved roll. The use of such rolls helps to get a cleaner, smoother and more accurate cut. The shifting of the material under the blade

Fig. 20-33. Helical pattern grooved roll. (*Courtesy American Roller Company.*)

is minimized by the use of a grooved roll. Mounting of the blades is made easier by the use of a comb. The comb provides many narrowly spaced slots for insertion of standard razor blades. Figure 20-34 shows a detail of such a comb.

Slicers

Instead of slitting while the material is being rewound, rolls can be sliced without rewinding to produce smaller rolls. A rotary blade slices through the wound material and through the core, leaving a clean, smooth edge. Slitting without rewinding was originated in the textile industry.

Successful use of the slicing method requires that the rolls are wound well. There is no opportunity to inspect the material and to take out bad portions of the coated or laminated product.

There are several methods of slicing with a rotary knife. Displacement cutting, the most commonly used method, employs a circular knife which is bevelled on one side. When cutting, the material is displaced until the narrow roll is separated from the master roll. The need for the material to be displaced requires a low coefficient of friction between individual plies and reasonably loose wind, and the roll width that is cut must be narrow. Soft elastic materials, such as plasticized vinyl film, or pressure-sensitive adhesive tape, are easily displaced because of their elasticity.

Removal cutting entails a removal of a narrow strip of material between the cuts. Double-edge circular knife cuts into the roll making two parallel incisions. The knife is followed by a narrow tool which removes the material between the incisions. This technique allows the making of wide rolls and to cut rigid materials.

Compression cutting requires material which can be compressed, such as

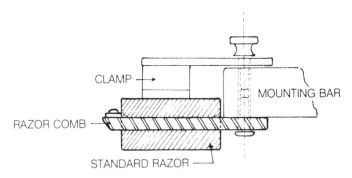

Fig. 20-34. Detail of a razor comb. (*Courtesy Arrow Converting Equipment, Inc.*)

foamed or solid elastomer tape, pressure-sensitive tape, or alike. Material is compressed instead of being displaced. Bayonet type knife is used for compression cutting.

Slicing is an accurate method to produce slit rolls. The cut widths may be produced to the consistency of ± 0.005 mm. The edges are perfectly aligned, since the material is not rewound.

Pressure-sensitive adhesive tapes are often produced by slicing. This method is especially suitable for narrow-width extensible backing tapes, such as plasticized vinyl film tapes, which are difficult to wind in the narrow-width without stretching the backing. Very narrow and accurate width tapes used in the graphic arts are also produced by slicing. A slicer capable to cut two rolls simultaneously and equipped with a turret for non-stop slitting is shown in Fig. 20-35.

Many paper products, such as toilet tissue and towel stock, are sliced at high production rates. A large (610-mm diameter) rotating blade cuts through the prepared logs. Such machine is shown in Fig. 20-36. Automatic slicers of this type can produce 70-300 cuts/min.

Core Cutting

Cores are an important part of successful slitting and winding, especially if differential winding is used. While precut cores may be purchased, many manufacturers choose to cut cores at their own facility. A close-up of a core cutter is shown in Fig. 20-37.

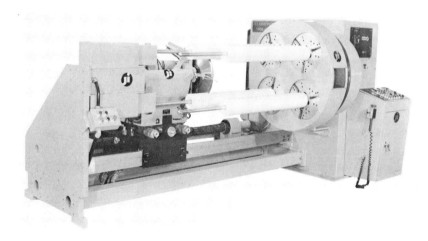

Fig. 20-35. Turret roll cutter. (*Courtesy Judelshon Industries.*)

422 WEB PROCESSING AND CONVERTING TECHNOLOGY AND EQUIPMENT

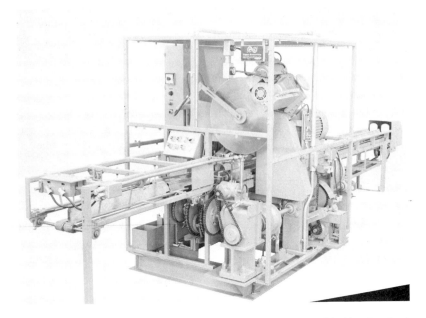

Fig. 20-36. Paper roll slicing machine. (*Courtesy Paper Converting Machine Co., Inc.*)

Winding

Center winding is the most frequently used process. In center winding, the force is transmitted from the winding shaft through the core and previously wound layers. The winding shaft is driven by a slip clutch or a dc motor. Duplex center winder is the most commonly used machine, especially for

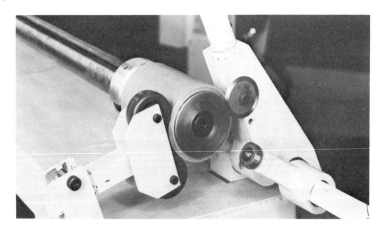

Fig. 20-37. Core cutter. (*Courtesy Arrow Converting Equipment, Inc.*)

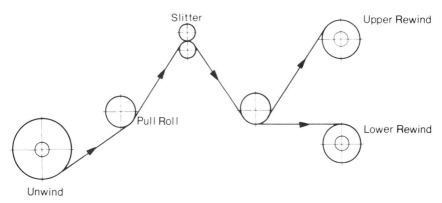

Fig. 20-38. A schematic diagram of a duplex slitter-rewinder.

products like pressure-sensitive tapes. In the duplex winder, the slit material rolls are alternated between the upper and lower rewinds. The rolls do not touch one another at the edges and do not adhere together because of tacky edges (as in some pressure-sensitive tapes), or interwoven material. A schematic diagram of a duplex slitting arrangement is shown in Fig. 20-38. A photograph of such a machine is shown in Fig. 20-39. For the ease of re-

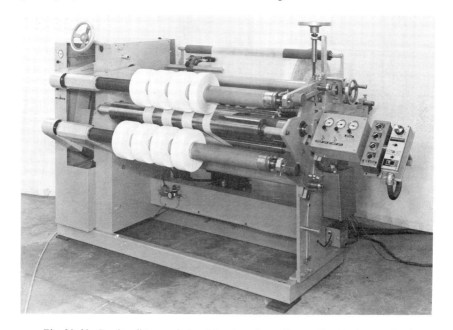

Fig. 20-39. Duplex slitter rewinder. (*Courtesy Arrow Converting Equipment, Inc.*)

moval of finished rolls, cantilever construction of the rewind rolls is often used, especially for the small-size slitters. The basics of slitting and rewinding have been reviewed by Rienau.[8]

Pressure-sensitive tapes and many other narrowly slit materials are wound utilizing differential rewinding. In this system, each individual roll is isolated from the remaining rolls. Each core can slip independently on the shaft thus allowing a uniform adjustment of the web tension. This is accomplished on a duplex machine by alternately placing cores and spacers on the mandrel. The spacers are keyed and can slide axially on the mandrel, while cores can slip around the mandrel. The level of friction between individual cores and spacers is determined by compressed air induced force along the length of the mandrel. Thus the tension on the slit strip being wound depends on the air pressure used to compress the core and spacer edges. This arrangement allows to wind materials of nonuniform thickness at the same tension. Coated and laminated materials might vary in thickness across the width and this arrangement prevents winding thinner sections loosely and stretching of heavier sections.

Various accessories are used with differential winding. Core adapters might be used to control the friction between the mandrel spacer and the core.[8] This allows the control of tension, keeping it at a low level that might be required when slitting lightweight, low-modulus films or other webs. Core adapters are also used to wind on cores of larger diameter than the mandrel.

Core drivers may be used where the winding tension must be higher than that generated by frictional forces between the core and spacer edges (usually over 50 kg). Core drivers provide a high-friction surface against which the core is slipping.

Top-riding rolls might be used to smooth out the upturned edges or to remove the wrinkles. This is especially needed for slitting of coated foil and similar materials. Top-riding rolls also help to force out the air layer that might be carried by the web when wound at high speeds. The top-riding roll should be positioned at the point where the slit web contacts the rewinding roll. The web can also be wrapped around the top-riding roll for maximum effect (see Fig. 20-40).

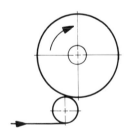

Fig. 20-40. Top-riding roll for rewind.

Good winding is obtained by employing the combined surface and center winding. Contact pressure by the rolls used for surface winding helps to eliminate the air entrapment. In addition, the radial force is acting on the roll because of the web tension. In order to overcome the variation of thickness, individual arms are provided for each of the rolls. This allows the rolls to maintain a contact with the drum regardless of the size of the adjacent roll.

Spooling

Narrow tapes and ribbons are difficult to wind and large-diameter rolls are not stable and may easily telescope and collapse. There are many applications which require large rolls for automatic high-rate feeding. Spooling is generally used to provide a stable large roll of narrow-width material. The traverse mechanism winds the tape at an adjustable pitch over a width of 25 cm or less. The rolls are as large as 40 cm in diameter and might contain 30,000 m of slit material of 3-mm minimum width. A spooling system is shown in Fig. 20-41.

ACCUMULATORS

Slower machines often have only a single rewind and unwind. This requires that the machine is stopped when the supply roll is finished, or when a sufficiently large rewind roll is made.

A short stoppage can be avoided by employing accumulators which store the material for the use during the change of the rolls.

Fig. 20-41. Spooling system with shear slitting head. (*Courtesy Arrow Converting Equipment, Inc.*)

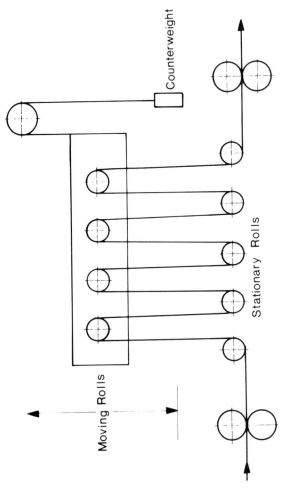

Fig. 20-42. Schematic diagram of a festoon accumulator.

Festoon Accumulator

Festooning is the most often used method to store a limited length of web for slow machines. The operation principle is shown in Fig. 20-42. The bottom row of idler rolls is stationary and the top roll moves in the vertical direction. When it becomes necessary to load up the accumulator, the in-feed nip is accelerated moving more web into the accumulator than what is taken out by the outgoing nip. During the filling of the accumulator the top row of idler rolls rises to accommodate the excess web length. When the accumulator is filled up, the in-feed nip is stopped, a new supply roll is placed in position and the splice is made, while the web is fed from the accumulator to the processing line. Similarly for the accumulator at the rewind end, the outcoming nip is stopped, while the rewind roll is removed and a new core is placed. The accumulator takes up the web from the process while the new rewind is prepared.

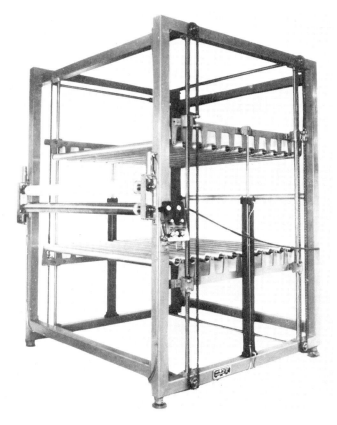

Fig. 20-43. Festoon accumulator. (*Courtesy Arc Machine Co., Inc.*)

J-Box (Scray) Accumulator

This storage method has been developed by the textile industry. The textile webs can be folded, unlike paper and films, which must be kept in tension in order not to be damaged. Figure 20-44 shows a schematic diagram of a J-box. When the supply roll is running out, the in-feed nip is accelerated and the J-box is loaded with folded fabric. This accumulated material is then used while the supply roll is changed and the splice is made. Since the web is packed without tension and rather loosely in the J-box, an edge guiding mechanism is needed to feed the fabric straight out of the accumulator.

THICKNESS GAUGING

When a coating is applied over a web, it is important to control its thickness in order to produce a uniform product. The coating thickness is adjusted either manually or automatically. In either case, the thickness must be first measured.

Manual gauging is still the most frequently used method and it is quite sufficient for most coating applications. It can be performed on the run, or samples can be taken and the coating thickness checked while the machine is stopped. A simple thickness gauge is sufficient to check the coating thickness of a dried coating. The disadvantage is that the coating is checked not immediately after it emerges from the coating head, but after drying. Wet gauges of several designs are available to measure the thickness of a wet coating, although not as accurately as that of dry. If the web is being saturated, then the weight pick-up must be determined.

Continuous thickness gauging is employed to measure the coating profile across the web and to determine the variations in the machine direction because of changing coating conditions. The adjustment might be performed manually or the loop might be closed and the thickness controlled automatically.

Several types of sensors mounted on a frame are used. The sensors provide the information related to the thickness of the web and also the location at which the measurement was taken.

In *nuclear gauges,* two measurement techniques are used: non-contact transmission and backscatter method.

In non-contact transmission the radiation source is mounted above the web on a frame. The detector head is mounted opposite the source and contain an ionization chamber and solid state electronic circuitry. Both traverse simultaneously across the web at a speed of 25 cm/sec. The amount of the radiation transmitted depends on the mass of the interposing web. The ionizing chamber converts the detected beta radiation into electric current. The following isotopes are used for the nuclear gauges.[9]

WEB HANDLING 429

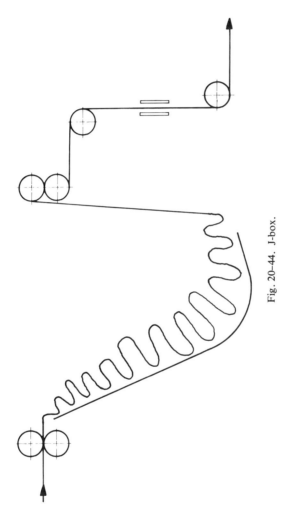

Fig. 20-44. J-box.

Promethium 147 for 0.0025-0.15-mm thick webs
Krypton 85 for 0.025-1.25-mm thick webs
Strontium 90 for 0.2-7.5-mm thick webs
Americium 241 for webs above 5-mm thick

Americium 241 is a weak gamma ray emitter and is used for backscatter sensors.

The selection of the proper isotope for the given web is important. If the radiation density is too high, it might be difficult to determine accurately the absorbed radiation. The beta radiation system measures the mass between the source and the ionizing chamber, including the mass of air. The air density might vary depending on the temperature and thus affect the reading. Provisions are made to compensate for such deviations.

In the backscatter method, the radiation energy is directed to the web and the photons of energy that are returned to a scintillation counter are measured. The signal is proportional to the mass of the web. Backscatter sensors are slower, but they are suitable to measure web thickness up to 50 mm, they are not affected by ambient conditions and the gap between the sensor and the web, they are also less expensive than the transmission gauges. The accuracy of both transmission and backscatter gauges is $\pm 1\%$.

Magnetic Sensors

These sensors measure a change in eddy current associated with the magnetic field that is set up between the sensor, which is mounted above the web and the magnetic reference mounted below the web. The sensor might be riding on an air cushion without touching the web's surface, or it might be in contact with the web. In either case, a change in the thickness changes the distance between the sensor and the reference and thus the eddy currents associated with the magnetic field.

Magnetic sensors measure the web thickness directly and are not sensitive to the coating density. (accuracy: $\pm 0.5\%$).

Infrared Sensors

These can be either transmission or reflection type. The reflection type sensor utilizes a near infrared photometric analyzer sensitive to fixed wavelengths of near-infrared radiation. The energy absorbed by the web depends on its thickness. The accuracy is $\pm 1\%$.

WEB HANDLING 431

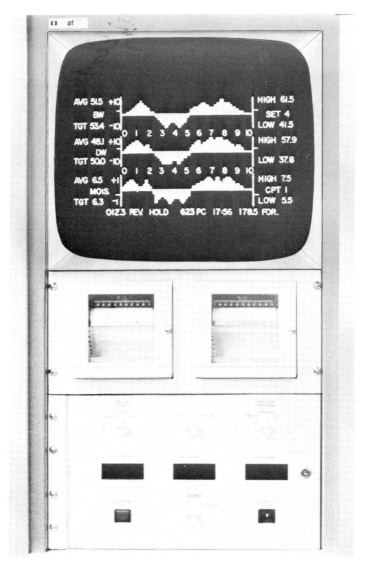

Fig. 20-45. Thickness profile display. (*Courtesy Ohmart Corp.*)

Fig. 20-46. Thickness measuring gauge on a traversing frame. (*Courtesy Ohmart Corp.*)

Ultrasonic Sensors

These are based on the measurement of an ultrasonic pulse which travels through the material, bounces off the back surface and returns to the head. Accuracy of the measurement is 0.0025 mm.

Control

Continuous web thickness measurement provides thickness information which may be continuously displayed on a video screen as shown in Fig. 20-45. If we are interested in measuring the coating thickness by employing beta radiation gauges, the measurement must be obtained in two locations: before and after the coating. The coating weight is then obtained by the difference. This compensates for the variation of the substrate's weight. A typical gauge mounted on a traversing frame is shown in Fig. 20-46.

The coating thickness adjustments may be made manually, or the loop may be completed and the microcomputer may be converting the results of

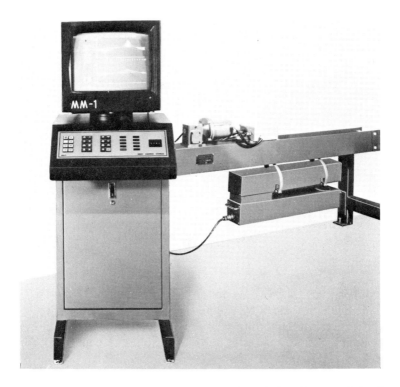

Fig. 20-47. Thickness gauge, microcomputer and a video display. (*Courtesy Indev Control Systems, Inc.*)

the measurements to the required signal for automatic adjustment of the coating head.

REFERENCES

1. Booth, G. L. *Coating Equipment and Processes.* New York: Lockwood, 1970.
2. *Tension Control Systems.* Randolph, N.J.: John Dusenbery, 1981.
3. Gerdes, R. *Paper, Film & Foil Converter* **55**(8):36 (1981).
4. Weiss, H. L. *Coating and Laminating Machines.* Converting Technology Co., Milwaukee, Wis. 1977.
5. Private communication. Lawrence R. Sundberg, Hydralign, Inc., Walpole, Mass.
6. Buisker, R. A. *Paper, Film & Foil Converter* **54**(6):54 (1980).
7. Feiertag, B. A. *Selecting a Web Guiding System.* Converting Machinery/Materials 2nd Conference, Philadelphia, October 17, 1979.
8. Rienau, J. H. *Paper, Film & Foil Converter* **53**(5):172 (1979).
9. Galli, E. *Plastics Machinery and Equipment,* May 1982, pp. 23, 25–28; July 1982, pp. 17, 19–21.

21
Sheeting

Joseph E. Radomski
Lenox Machine Co., Inc.
Lenox, Massachusetts

Paper and paperboard are converted from a continuous web into sheets of a specified length and width by equipment known as a rotary sheeter (Fig. 21-1). The sheeter usually consists of an unwinding section, for one or more rolls, a slitting section, a metering and cutting section, a delivery section and a collection section (Fig. 21-2).

In some cases, the sheeter is mounted directly after a web manufacturing or converting process, such as a cylinder board machine, a coater or a laminator, in which case the unwind section is replaced by web-carrying rolls and a method of synchronizing sheeter speed to web speed. Sheets are made by passing a web at measured speed between a pair of knives, at least one of which is rotating, also at measured speed. Sheet length is determined by the path circumference of the rotating knife and the relationship of knife rotation speed to web speed. Thus, if a 36" knife path circumference is utilized and web speed equals the peripheral speed of the knife, sheet length would be 36". If knife speed is halved, sheet length would be doubled; if knife speed is doubled, sheet length would be halved.

Sheet width is determined by web width and the placement of slitters across the web, which may slit the moving web into multiple, individual webs before cutting into sheets.

A sheeter is called a simplex sheeter if it contains only one set of knives, and the entire web width is cut by these knives. A duplex sheeter (Fig. 21-3) consists of two knife sections in tandem, following the slitting section, with a dual delivery station and dual collection stations. Thus the slit webs can be passed through the first knife or the second knife, creating two sheet lengths simultaneously. The use of a duplex sheeter, once common, is now restricted

SHEETING 435

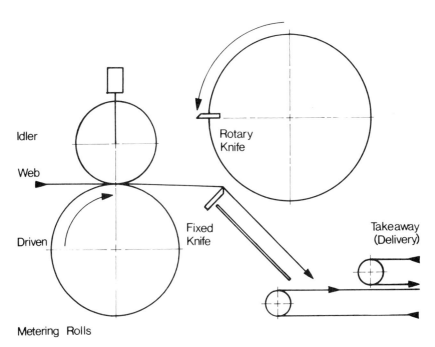

Fig. 21-1. Rotary sheeting.

to large, relatively slow-speed manufacturing processes, such as a cylinder board machine where the flexibility provided by the duplex is required.

In either case, the sheets thus made are received by a conveying system, slowed to a safe speed, and dropped into a collection system where they are stacked, either in a pile for discharge by a conveyor, or on a skid or pallet for the next operation.

Because accuracy requirements are not as stringent, envelope papers, wrapping papers and paperboard are generally considered as finished immediately after leaving the sheeter. On printing papers it was once common to guillotine-trim the paper to the exact size, also often splitting the sheets into smaller sizes. The opportunity for waste and labor reduction provided by this dual process led to the development of more precise sheeting equipment giving accuracies sufficient to allow sheeting directly into the final required size.

The essentials of a precision sheeting system consist of improved control and treatment of the web during the unwinding process, including automatic tension control, improvements to the metering section to minimize staircut; improved accuracy in the ratio of knife speed to web speed, particularly during acceleration and deceleration of the machine; and improved pile quality through more effective collection and jogging systems. Also, to improve

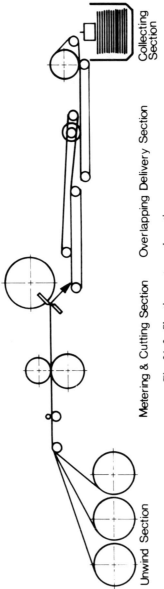

Fig. 21-2. Sheeting system schematic.

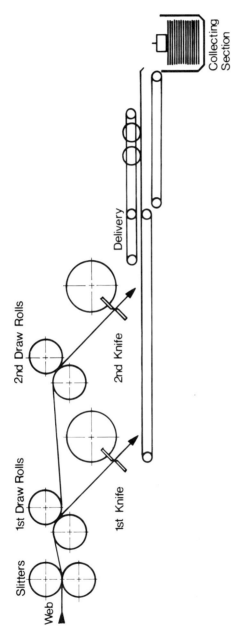

Fig. 21-3. Duplex sheeter.

quality of cut and quantity of rolls being cut, the double rotary knife design has become very common on most precision sheeters.

EQUIPMENT DESIGN AND SELECTION

The factors described below have a significant role in designing and selecting a sheeting system.

Sheet size, as has already been discussed, will change the equipment configuration significantly. The cut-size sheeter for small sizes, and the general duty sheeter, for a range of 20-80" in sheet length, are two extreme examples. Order size also has an effect. Large orders with few size changes make feasible the utilization of more expensive high-speed sheeting equipment with large-diameter roll unwinds and automatic splicing. Small orders require a smaller, more flexible machine, with simple rollstands of one or two-roll capacity. A machine like this is economically feasible for orders of 2000-4000 lb, with a maximum tonnage/day of 18-25. The larger, high-speed equipment, however, requires a production figure of 80-120 tons/day to be feasible.

The grades of paper to be sheeted will determine accuracy requirements and, in all probability, type of cut-off. For example, 20-24-point board can be sheeted accurately at high speeds with minimum dust on a fixed-bed knife skid sheeter with single rollstand. On the other hand, lightweight board and coated papers require double rotary knife sections (Fig. 21-4) and specially designed slitter sections to handle 300-400 lb under the knife. This is because of dust-free cutting capacity of the double rotary knife sheeter is generally higher than that of the fixed-bed knife sheeter.

Similarly, paperboard is generally sheeted to size minus zero, plus 1/16" without sophisticated unwind equipment. The multiple roll sheeting required for coated and uncoated printing papers makes it more difficult to achieve this accuracy, without special designs to minimize staircut and sheet length variations.

Even though the same grade of paper is used in all cases, the final use of the product after sheeting affects the sheeter design. For example, envelope stock has a very generous tolerance for inaccuracy of ± 1/8", while usage for stationery requires precision sheeting. The same paper used in school supplies would involve the addition of punching, ruling and binding equipment to the sheeting system.

Configuration of the sheeter is also tailored to the location. A large mill with a high volume of sheet orders can use a single-purpose high-speed design, while a converter will require a multi-purpose machine capable of filling smaller orders, and a printer can choose a machine that will produce skids only of the sizes that are necessary to fill his orders.

SHEETING 439

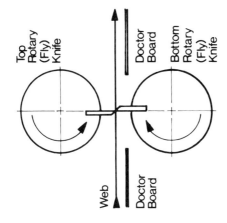

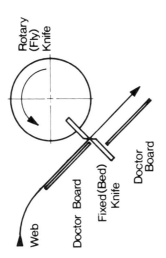

Fig. 21-4. Single rotary versus double rotary knife sections.

ROLL UNWIND EQUIPMENT

Unwind stands for sheeters may be classified as shaftless where moving supports with locking chucks enter the core from each end of the roll and secure the core to the arbor containing a brake, and shaft type, where a through shaft is used to support the roll on arbors.

Shaftless stands (Fig. 21-5) require a means of moving both arbors inward to chuck up the roll and shifting both arbors in either direction to align it properly with the web travel. Several types of shaftless stands are in use, individual stands, multiple stands with several arbors mounted on movable pylons, self-loading and turret.

Individual stands each have their own adjustment mechanisms, braking mechanisms and skewing adjustment. Mutliple shaftless stands are designed so that the pedestals holding the individual arbors move in and out for roll width adjustment, but chucking is done by an additional inward motion built into the arbor, normally pneumatic or manual.

Self-loading stands have a means of lowering the arbors to align with the core of the roll sitting on the floor. Another variety of these stands also has a horizontal drive mechanism, so that the entire stand may travel out into the aisle and then self-load a roll. Turret stands are usually of the two-position variety and are normally self-loading. Each set of arbors are movable in and out independently to allow the loading of one set, while the other holds the unwinding roll. A quick roll change is made possible by rotating the unwinding roll to the rear position, while bringing the spent core to the forward position for removal and reloading.

Shaft type stands (Fig. 21-6) are generally of three varieties: a tubular steel shaft with brake drum or disk at one end and two core chucks locked to the shaft by set screw to secure the core; a pneumatic shaft with expansion or protruding parts operated pneumatically to secure the core to the shaft; and

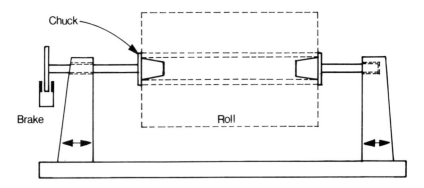

Fig. 21-5. Shaftless unwind stand.

SHEETING 441

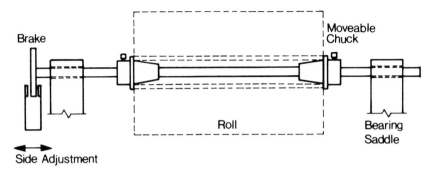

Fig. 21-6. Shaft type unwind stand.

stub shafts, which fit into either end of the core. The stub shafts are long enough to handle the range in width variations, but short enough to be lighter than the full shaft. They have not always been wholly successful and are in limited use at present.

Chucks in use on shaft type stands are generally tapered and fluted or straight and notched for use with notched cores. The size of the notch is not uniform in the industry, so chucks must be selected to match the size and number of notches in use.

Chucks used for shaftless stands may be straight and notched, tapered and fluted for fiber cores or expandable, either through pneumatic or cam action. In the cam action type, rotation of the chuck by the core moves the plates outward until the core is gripped securely, at which time the arbor begins to turn. The pneumatic variety has a bladder which expands the plates. Both types require regular maintenance.

Maintaining accurate sheet length means maintaining uniform tension in the web. A tension system to do this consists of a measuring device, which also may be a compensating device such as a dancer roll, and one or more brakes. The output of the sensing device is changed into a brake supply pressure which provides a braking action. Regenerative braking is not in use on sheeters at the present time, although some hydraulic braking systems are found.

On shaft type stands, a single brake per stand is used. On shaftless stands, a single brake per stand may be used successfully for notched cores, or for single-use fiber cores. Reuse of the cores generally makes two holding chucks and two brakes desirable, as the fiber core bears the entire braking load and consequently tends to lose the inside plies after a few uses. This is particularly true of 75 mm and 100 mm I.D. cores. Both drum brakes and disk brakes are in common use, with single or multiple bands or pucks.

Tension control systems (Fig. 21-7) must face a roll width range of 3 to 1, a roll diameter variation of as much as 8 to 1 and a tension adjustment range

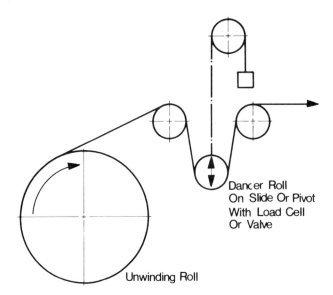

Fig. 21-7. Dancer roll tension control.

of approximately 6 to 1. It is therefore difficult, using 80-psi air, to design a system controlling accurately through a range of 144 to 1. The use of two brakes, or multiple bands or puck sets, with automatic or manual selection, can bring this ratio down to a more practical 15 to 1 or 25 to 1.

Since the major variation involved in range of pressure required comes from diameter change during unwind, a diameter sensing or diameter calculating device is occasionally used in place of an automatic tension sensing system. This achieves the majority of control automatically, but minimizes cost. This is practical when roll diameters are reasonably uniform within the set of rolls.

Heat dissipation is difficult in sheeter unwind stands, since rpm is low and air cooling inefficient. As a result, water cooled brakes are often applied, particularly where a single brake is involved.

Both shaft type and shaftless stands are also mounted in multiples on either rotating or shuttling cars (Fig. 21-8), in order to minimize downtime for roll set changes. These are desirable for roll diameters less than 45″. Shaft type stands are generally loaded by an overhead monorail; shaftless stands may be loaded by overhead monorail or by self-contained post or scissor lift. If roll-out stands are not used, a shuttle car and truck mounted in the floor are used to move the rolls into position in front of the stands. To simplify the installation, stands are usually located so that one shuttle track serves two stands.

Stand spacing is generally maximum roll diameter plus 6″. If shuttle track

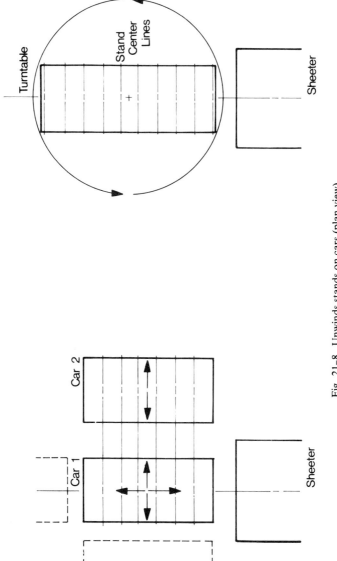

Fig. 21-8. Unwinds stands on cars (plan view).

is used, an additional space between the stands bordering the shuttle track should be allowed at maximum roll plus 12″. In a conventional sheeting system, where accuracy is not critical, overhead web support is limited to lead-in rolls immediately before the slitting station. In modern high-speed precision sheeters, each web is run to an overhead carrier roll, where the sheet run is such that a maximum of 5° wrap by multiple webs is maintained (Fig. 21-9).

Most sheeters either have skew adjustment in the backstand or skew rolls for each web in the carrier system. These can be moved longitudinally at one end to stretch a baggy edge. Bow rolls can be used to treat baggy edges or centers, and are frequently seen before the lead-in roll leading to the slitters. It has been found through experience, however, that multiple webs do not work well on a bow roll. Treatment of the paper with a bow roll is only of limited success, and static is generated by the movement of webs against each other. Bow rolls are most effective where each web can be run over a roll and run immediately to a straight carrier roll close to the draw rolls to capture the effect.

Web alignment is built into each unwind stand. In addition, edge guiding systems may be useful in certain circumstances. Where folding carton board is sheeted, slitters are not generally used. However, a master size roll can be slit with a single slitter successfully, if the unwind stand is guided automatically. The edge of the web is sensed, and the roll-stand shifted sideways to maintain alignment within ± 1/32″.

Edge guiding is also desirable for systems using automatic splicing of new rolls, since, due to the change from roll to roll of web characteristics, the web will run to a different position, even though the unwinding roll is in line. For such an application on lightweight webs, a web steering system consisting of two rolls in a framework which is rotated in a plane perpendicular to web travel is generally preferred.

Curl in the unwinding web is either roll set, which is in the direction of the web center line and downward, or manufactured curl which may be upward in the machine direction if a felt side curl, or downward in the cross direction if a wire side curl. Both roll set and machine direction curl can be treated successfully by running the web over a smooth bar or small-diameter roller (Fig. 21-10). The amount of curl removal is determined by tension in the web, penetration of the web by the decurler, and the diameter of the bar corner or small roller. For coated papers, 3/8-3/4″ diameter rollers are successful; for heavy paperboard, a 2″ diameter roller is sufficiently small; and for uncoated papers, a bar type decurler is generally used. It should also be noted that decurling adds friction and, therefore, tension to the web. Placement of the tension-sensing device after decurl will provide a measure of the total tension and assure that sheet length is not affected. Treating a cross-direction

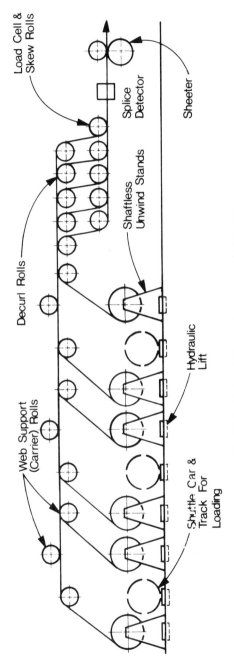

Fig. 21-9. Modern high-speed sheeter unwind system.

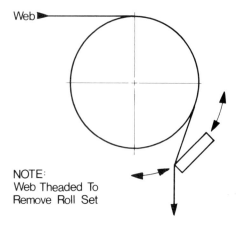

Fig. 21-10. Bar type decurl.

curl with present equipment has only limited effect. Residual curl or roll set can cause hangup of the sheet after the knife at the knife board, delivery takeaway, overlap or collecting station.

After the multiple webs are combined into a single web for slitting, a splice detector may be used to identify splices made in the rolls either at winding or at unwinding. This detection system can be one of measuring bulk variation, or measuring capacitance. Microprocessor logic is also supplied to calculate the travel of the splice into a cut sheet, and actuate a reject gate in the delivery system to prevent the splice from entering the collected stack.

SLITTING

Sheeting systems use shear type slitters (Fig. 21-11), generally consisting of a hardened narrow drum or anvil, and a smaller blade on top sharpened to one side. The drum is usually harder than the blade and may, in some designs, be cupped inward at the slitting surface. The bottom slitters are mounted with set screws on a keyed, driven shaft and are moved sideways for size adjustment. For sharpening, the slitters and shaft are removed, or the anvils may be split type, removable from the shaft directly.

The top slitter blades are mounted so that they can be adjusted for rake, toe-in, and penetration. Pneumatic side loading and lifting devices are supplied, or spring loading and manual lift may be used. Outside or trim slitters are opposite hand from each other to support the remaining web while slitting. This is critical to a clean slit. Some slitting sections have been built with spacers between the anvils, as it has been found that web support on both sides of the blades will provide improved slitting.

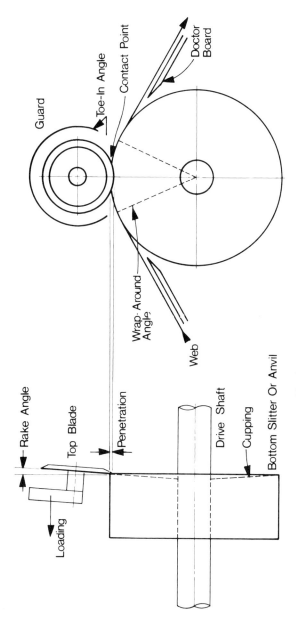

Fig. 21-11. Slitter nomenclature.

Critical adjustments of the slitter are depth of penetration, rake and toe-in. Although manufacturers' recommendations vary, typical settings for these adjustments are 0.05-1.0 mm for penetration, 1° of toe-in and 0°-1° of rake (see Fig. 21-11).

As the slitters wear, a wear ring develops from the outside edge inward, and the point of contact moves backward. As this wear ring increases, slitting quality eventually becomes marginal. Because of the difference in diameter and hardness, top slitter blades wear out approximately four to eight times as fast as the bottom anvils or drums. A carbide coating is often used to increase bottom slitter life.

The tip angle of the slitter blade is generally 30°-45°. Care must be taken to avoid runout and to provide a smooth finish with no wire edge. A final touch with a honing stone generally is used to remove this. Slitters also may be honed in use to touch them up and extend their life. The bottom slitters are usually driven approximately 5-15% faster than web speed. The bottom slitters drive the top slitters.

With increased emphasis on cut quality, and the introduction of double rotary knife sections with increased web thickness capability, dual slitter sections (in which the web is split into upper and lower sections, run over separate slitters and recombined before entering the metering section) have been used to increase slitting capacity. (Fig. 21-12). Another approach to increasing slitter capacity is the use of driven bottom blade slitters with top blade slitters, so that the work is shared more equitably by the top and bottom slitters.

Web support into and out of the slitter and absence of wrinkles or bubbles are essential to quality slitting. Slitter weave, or movement of the web in a sideways direction through the slitter, can be caused by inadequate tension, web oscillation or vibration, a bag in the web, slitter runout or dullness.

METERING SECTION

The metering section generally consists of two rolls, one driven and one idler roll. The bottom roll is generally larger in diameter, metallic and is the driven roll. The top roll is generally rubber- or felt-covered with a medium softness and is pneumatically loaded at both sides by air cylinders at the journals. The air cylinders have individual air loading adjustment to compensate for differences in web structure from side to side. The more typical rubber covering is designed with a chevron or spiral grooving which is intended both to spread the web and to allow the relief of tensions within the webs created by the nip.

To avoid the web sailing or looping between the knife section in the metering rolls, a small brush roll, often segmented, over small-diameter solid roll may be used on precision sheeters. Driven at a slight overspeed, this nip

SHEETING 449

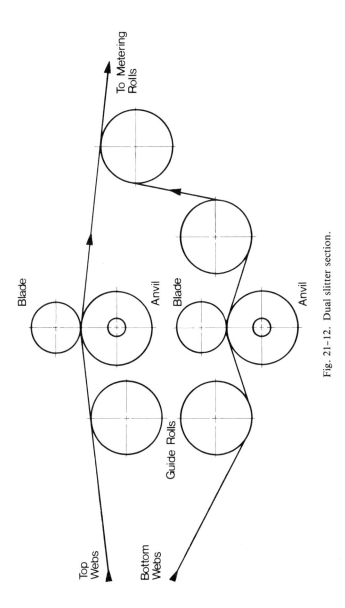

Fig. 21-12. Dual slitter section.

keeps the web smooth and flat. The small diameters allow the brush roll assemblies to be placed close to the knife section.

In older, conventional sheeters the web travel from the draw roll to the knife is often downward at a slope of approximately 45%. This drop is a hindrance to high-speed processing, so most modern precision sheeters have an approximately horizontal web flow from metering rolls to the knife section.

Staircut in multiple webs is caused by each succeeding web away from the bottom-driven metering roll having a larger driving radius than the previous layer by the thickness of the web. The more tightly the webs are held by the top metering roll, the more pronounced this staircut becomes. Staircut, therefore, is affected by the diameter of the driving metering roll. The percentage change becomes smaller as the diameter increases. The amount of wrap of the bottom draw roll, and the diameter and covering of the top roll, have also been shown to affect staircut. In some instances a slight wrap of the top metering roll is used.[1] Another factor in staircut is stretching of the web in the nip, which tends to increase in the middle plies.[2] With heavy webs such as paperboard, increased brake pressure on the outer webs may be used to reduce staircut.

The top metering roll is often ground with a slight crown and dubbing of the edges to allow for evenness of pressure across the web under compression.

The metering section is driven by the main drive of the sheeter. Power may also be transmitted to the knife by a variable speed transmission, and to the delivery system and the slitting system by belt drives.

Both dc drives and ac eddy current drives are in use. The drive system should include an adjustable thread or jog speed, speed regulation, controlled acceleration and deceleration and regenerative or mechanical braking. Ventilation should not be neglected, particularly on dc drives, as sheeters often operate for extended periods of time at full torque requirement, but below 50% design speed.

In sizing the drive, theoretical horsepower consumed by the brakes can be calculated by the formula below. To this result must be added horsepower consumed in the slitters, internal friction, delivery system drive and a contingency. These factors vary depending on machine configuration.

$$\text{HP} = \frac{T \cdot W \cdot R \cdot S}{33,000} \qquad (1)$$

where:

T = Max. tension (lb/linear in.)
W = Max. web width (in.)
R = Max. no. of webs
S = Max. speed (fpm)

Threading up the web is accomplished by raising the top slitters and the top metering roll and moving the first web through the roll. The roll is then lowered, and put into thread speed while additional webs are fed through. They continue to feed through the brush roll and the knife section as well. A threading convenience is a pneumatic web clamp, which can be used to accumulate the webs and then feed them through the metering section.

CUTTING SECTION

Simplex cutting sections are either of single rotary or double rotary design, as shown in Fig. 21-4. In single rotary the paper passes between a fixed bed knife and a rotating knife. Its advantages are simplicity, lower cost and wider speed range. Its disadvantage is primarily one of cut quality as the cut is not a true shear, but a compression cut as one edge of the knife passes the other. The knife does not contact the bed knife, but it is generally set close enough to cut a single sheet of newsprint cleanly when rotated by hand.

A knife section cuts on a helix, whether true or approximated, as the simultaneous cutting of the entire web would be a severe impact. Therefore, the cut starts at one edge of the paper and travels across to the other edge. The distance traveled by the web during the period of cutoff depends on sheet length and resultant knife speed. In a fixed-bed knife sheeter with constant rotating knife speed, the entire knife section is skewed in the horizontal plane to compensate for this and is adjusted every time sheet length is changed. The amount of shear or helix varies, with 1-2 mm/m of knife width being typical. The lower the shear, the louder the noise level and the less distortion is done to the edge of the cut.

In a single rotary with constant knife speed, knife path diameter is selected based on the sheet length range on the sheeter, with large diameters being used for longer sheets. A 4 to 1 sheet length range is available, but the speeds in excess of the 2 to 1 range are generally slower. Too slow a knife speed relating to sheet speed inhibits sheet flow, while too high a relative speed causes turbulence and kicking of the trailing edge.

Double rotary knives are synchronized to sheet speed by either a mechanical cyclic linkage or by an independent computerized digitally-controlled dc cyclic drive. On sheets longer than synchronous length (equal to knife path circumference), the knife slows down after cutting and speeds up to sheet speed just before beginning the next cut; on sheets shorter than synchronous length the mechanical linkage or cyclic drive accelerates the knives beyond sheet speeds between cuts. No squaring adjustment is required on the double rotary, as the knife is always at approximate sheet speed during the cut, and therefore the squaring is built in, depending on the shear rate.

The cut produced by double rotary knives is generally found to be less dusty than single rotary, as the knives actually pass each other and interfere in a shear type cut. While contact is not made, the end of the web, examined by magnifier, shows a two-sided cut like that produced by a scissors. The main advantages of the double rotary design are cleaner cut under equivalent knife loads, longer knife life, and lower noise level. A disadvantage is that the double rotary is generally speed limited by the forces of acceleration and deceleration required between cuts; synchronous speed is the maximum speed of the sheeter.

Intermediate transmissions (which vary knife speed in relation to web speed) are typically:

- Direct gear change for finite sizes
- Full range variable speed transmission
- Differential with separate driven input
- Combination gear and variable speed transmission
- Dual motor with digital control.

Advantages of a variable speed transmission are ease of adjustment and coverage of the full range. The major disadvantage is limitation in accuracy, particularly during acceleration and deceleration. Direct gear drives are limited in selection of sizes, unless combined with a variable speed transmission. They are also time-consuming to change, although gear-shift transmissions are now available at added cost.

Dual motor drives may be higher in cost and may require more sophisticated electrical maintenance. Their most significant advantage is simplicity and flexibility, with dial-in sheet length control, along with excellent reliability.

Rotary knives may be mounted tangentially, or radially. Cutting is generally theorized to involve primarily a radial load, and therefore radial designs tend to be more rigid and long-lasting, while tangential designs are usually easier to set, and are more useful for light cutting. The wedge type mounting of a radial design requires knife butt contact with the cylinder. In this design, after each knife grind, the knife must be shimmed out to compensate for the amount ground off. In a double rotary design, adjustment is by moving the knive interference path closer together with a tangential adjustment. To adjust a tangential, fixed-knife/fly-knife arrangement, the rotating knife is bent in a radial direction to match up with the fixed knife. On some fixed-bed knife sheeters a micrometer adjustment is available, which moves the bedknife closer to the rotary knife uniformly, once the rotary knife has been set for uniform cutting. Thus, as wear takes place, an adjustment can be made.

Knives are usually removed in pairs and reground to the original sharpness. Honing of the wire edge from the knife, and sometimes honing a lead into the leading edge of the knife is done. In addition, nicks and worn spots leading to spotty, poor or dusty cutting can be removed by use of a hone, with that portion of the knife then adjusted.

Typical knives are constructed of a hard material for the cutting edge which is puddled into a softer material. Approximately 1-1.25 mm can be ground off before approaching the end of the puddled material. In high-wearing applications, carbide coatings and cryogenic treatment have both been utilized. They are not widespread because of their expense, and because of the difficulty of regrinding and honing to compensate for wear and accidental damage.

Both electronic and mechanical sheet counters/multipliers have been used, taking one pulse per knife revolution to count the number of sheets made and collected. The multiplier must be changed for the number of webs. The output of the counting system may simply drive a dial indicator, an automatic discharge system and/or marking or tabbing equipment in the collection system which inserts marks at the 500 sheet count.

DELIVERY SYSTEM

The delivery section of the sheeter may consist of a high-speed takeaway section, a reject gate and an overlapping, low-speed delivery system.

The high-speed section is generally two sets of belts or tapes, 1-6" wide, which are arranged so that the cut edge of the web leaving the knife is guided onto the moving bottom belts and nipped by the throat of the downcoming top belts. This may be a single roll or two-roll nip. The advantage of the two-roll nip is that it draws the sheet with a softer nip.

In precision sheeters, a reject gate is normally used so that the delivered sheets are without defects from starts and ends of rolls or splices. Gates may also be connected to an electronic or mechanical inspection system for automatic rejection of defective sheets by mechanical lump/void detectors and/or optical spot detectors. A reject gate may in addition be used for sampling sheets for quality inspection.

Gate design is either interfering or non-interfering (Fig. 21-13). An example of an interfering gate is a solid or segmented metal foil which raises or lowers to direct the sheet from one path into another. Non-interfering gates use either an air blast to direct the leading edge of the sheet or the movement of a roller nip to change direction. While manual gate operation is acceptable at low speed, high-speed sheeters usually require electronic timing to assure gating of the sheet without a jam. A further problem is caused by the elimination of a sheet in the high-speed section, leaving a partial opening in the low-speed overlap section. Thus, if one or two sheets are to be gated out, either the previous cut must be held in the overlap position, or the overlap

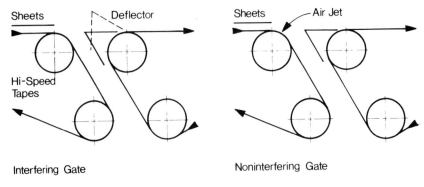

Fig. 21-13. Gates

section must be emptied. This requires, usually, rejection of around four cuts.

Overlapping

The overlapping system (Fig. 21-14) is the most common method of slowing down the cut sheets to prevent edge damage in collection. This is thought to occur at the range of a speed change of 200-300 fpm, depending on grade and basis weight. Usually, the sheet is corrugated with the use of corner lifters or varying diameter "wave" rolls, so that is can be projected out over the trailing edge of the previous sheet, falling on it and sliding over it to the overlap position. Overlap may range from 30-90%, depending on the speed of the sheeter.

Stop wheels or belts that are adjustable for sheet length are used to catch the leading edge of the sheet and align it on the slow-speed belt section. The sheets may also be guided by the passage of the high-speed tapes over the slow speed tapes into the stop wheels.

As sheeter speed increases, this method of overlap becomes unsatisfactory. Damage occurs at the stop wheels, and because the trailing edge of the sheet does not fall rapidly enough, the next leading edge collides with it. Solutions to this problem in use are vacuum overlaps and knockdown overlaps. A timed or constant vacuum snaps the trailing edge of the cut sheets downward out of the way, and then releases it to meet the stop wheels. An alternative is a mechanical knockdown, also timed, that must push the trailing edge downward out of the way, and then raise up before the leading edge of the next sheet.

Most of the foldovers or turndown corners occurring in sheeters, occur in the overlap section. At higher speeds, jam-ups are most likely to occur here because of front and tail collisions. Careful adjustment of the position and

SHEETING 455

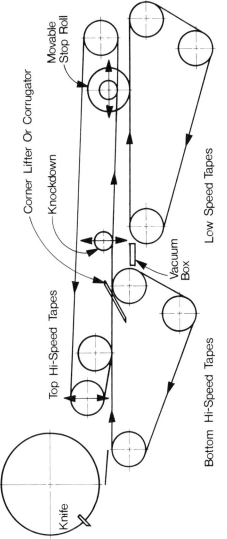

Fig. 21-14. Overlapping delivery.

tension of the belts and position of the stop wheels is also required to prevent skewing of the sheets on the low-speed belts.

COLLECTION SECTION

Collection sections are categorized as single and multiple point and skid, ream or carton load.

The most common sheeter collection system (Fig. 21-15) is simply a mechanical or hydraulic lift for building a skid of material. A movable front guide and side guides (dividers) are set to collect the sheets in the appropriate number of piles. The lift is lowered automatically, or manually, as the pile builds up, and the sheeter is stopped when the load is completed.

Carton load or ream collection systems (Fig. 21-16) require continuous operation in order to maintain productivity. Earlier designs consisted of dual or triple station systems, using directing gates and individual overlapping systems to collect the piles and discharge them on takeaway conveyors. With this design high-speeds are obtained, but set-up times become considerably longer than single point systems and accessibility is limited.

Single point continuous systems use a method of inserting a grid system to hold the sheets being collected, while the previously collected drop, or ream, is being discharged. Timing mechanisms and methods of making an opening for these grids are used in order to assure jam-free insertion, especially with lighter weights of paper.

Uniform piling of the sheets is essential to successful transportation and usage in printing presses or copiers. Uniform drop-off height, solid reference

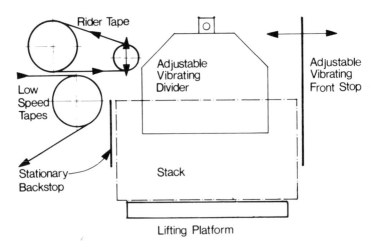

Fig. 21-15. Collection station (layboy).

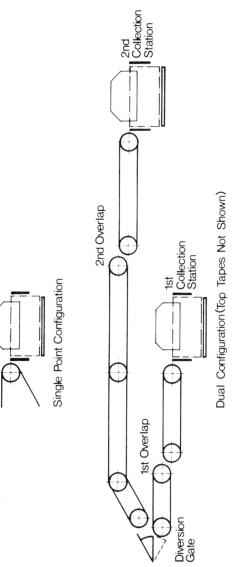

Fig. 21-16. Collection configuration.

guides, and vibration or oscillation to jog sheets against the reference guides are required to make a good pile. Low-speed, high-amplitude oscillation has most commonly been replaced with high-speed, low-amplitude vibration as sheeter speed and accuracy requirements have gone upward.[3]

Edge lift or corrugation, along with ionized air injection, are used to assure large sheets reaching the end of the pile successfully. Additional overhead and bottom drive rollers at the end of the low-speed belts assure that the trailing edges will move into the pile. Overhead belts or guide rods may extend over the top of the pile to prevent flying of lightweight sheets during the piling process. On small sheets the piling process is much easier to accomplish. If the dividers and frontstop are set too loosely, inaccurate jogging will result; if set too tightly, a non-flat pile and resulting jams will occur.

The trailing edge of the pile is usually the location of the ream markers or tabbers. The placement of ream markers from the side of the pile is simple, but often impractical due to multiple sheet piles or geometry of the discharge system. Insertion of the ream markers at the trailing edge of the pile requires an accurate timing system.

STATIC REMOVAL

Generation of static in web processing is a common problem. It is accentuated in sheeting by the processing of multiple webs and overlapping. Also, cut sheets cannot be controlled by tension or guiding completely and static-charged corners are attracted to each other and to pieces of the equipment, causing erratic piling and jam ups.

Most commonly applied over the top metering roll are ac or dc static bars, over and under the web between the metering section and the knife section, and at the high-speed, overlap section.

The overlapping process itself generates considerable static, which inhibits good piling. Static bars, while often used before the collection station, have limited effect because of the overlapped position of the sheets. A more effective method is to use ionized air at the trailing edge of the collection station to float the sheets into the pile. This air moves between cuts as they fall into the piling station. In order to maintain effectiveness, static systems must be kept clean and in good repair. The accumulation of dust can cause shorting and result in failure to operate.

Safety

In addition to design of catwalks and ladders, and belt and chain guards according to accepted practice, several specific points of protection are required by published standards.

Slitters must be guarded from accidental contact. A blade guard is the common application. The guard must prevent insertion of the hands during threadup. A nip guard must be in place ahead of the metering roll, also to prevent insertion of the hand during threadup. The knife section must also be guarded from accidental contact. Generally, a guard which encloses the entire knife section is utilized. It is attached to the frame and is electrically interlocked so that, when opened, the machine is inoperative. This guard may also cover the metering rolls.

A braking system on the main drive may also be specified, such that the sheeter is brought to a stop as quickly as feasible upon operation of an emergency stop button. These stop buttons should be placed at all usual operator's stations about the sheeter.

Discharge System

Automatic or semi-automatic discharge systems are often included as part of the sheeting system. In the case of the skid sheeter, these may consist of steel plates and conveyors which discharge the loads out quickly so that the sheeter may start up again without waiting for the use of hand trucks or power trucks. Automatic skid or pallet insertion systems are also available.

For carton or ream delivery sheeter, conveyor belts or flight conveyors are most commonly used. Where the reams are large and may have to be reoriented manually, pushers or belts used in connection with air-film tables are applied.

Factors Affecting Sheet Quality

Sheet quality (the suitability for further processing or end use) is affected by sheet length and width accuracy, squareness, dust, wrinkles and folds and quality of jog, the accuracy with which one sheet is piled on top of the previous.

In addition to the basic design accuracy, sheet length is affected by wear and backlash developing in the gear train or variable speed transmission, a change in tension, feeding of the web smoothly through the metering section without slippage and staircut.

Sheet width is affected by slitter sharpness and setting, sag in the web at a slitter, causing wander and changes in tension, also causing the web to move about at the slit.

Squareness is affected by fluctuation in web feed and tension as well as by adjustment of the knife rig. A web with one baggy edge will relax after the metering rolls in greater degree from one side to the other, causing a squareness problem. Both squareness and sheet length are affected by the flatness of the web in the metering and knife sections. The brush roll, if applied, will help push the web forward and keep it flat.

Dust, created by the sheeting process, is affected by knife sharpness and adjustment, slitter sharpness and adjustment, quality of the web, and speed and tightness of the take-away nip after the knife. Too fast and too tight a nip will cause a tearing action at the cut, leaving it ragged and dusty.

Wrinkles can come from a previous operation or be caused by a non-flatness in the web, which causes baggy wrinkles in the unwind to the draw (metering section). These can be ironed into hard wrinkles at the decurl, going over a carrier roll, or in the metering rolls.

Folded corners generally occur from impact on a doctor board after cutting or impingement against another sheet or guiding device in the overlap or in the collection station. Static increases their occurrence.

Poor jog is not generally a problem to the printer or to copy machines. In large size sheets, however, poor jog will lead to damaged edges of the sheet in transportation and, therefore, to feeding problems in the press.

Maintenance And Training

The modern high-speed precision sheeter is a far more complex piece of equipment than is normally realized. The most successful operations have a regularly scheduled preventative program, by maintenance personnel especially trained in the requirements of the sheeter. Similarly, operator training, adjustment and minor maintenance are essential. On lightweight paper, particularly, experienced operating personnel are essential to high production.

Critical maintenance areas are maintenance of the roll covering and slitters in the metering section, bearing and brake maintenance in the unwind stands, and bearing and knife maintenance in the knife section. Maintenance of the intermediate transmission in first-class condition is essential to continue accuracy. Worn and sagging tapes should be replaced in the delivery system. All roll bearings should be checked regularly, and the collection station should be checked thoroughly for loose bolts and bent dividers plates. Critical areas of operator adjustment are the braking system, slitters, knives, overlap and positioning of dividers and stops.

REFERENCES

1. Schulze, Peter. The Multi-ply effect (staircut): A sheeting problem. *TAPPI* **54**(8):1305 (1971).
2. Pfeiffer, J. David. The mechanics of a rolling nip on paper webs. *1967 Annual Meeting Proceedings, Technical Association of the Pulp and Paper Industry,* Atlanta, Georgia.
3. Fitzpatrick, D. Jog quality from paper sheeters. *1979 Paper Finishing and Converting Conference Proceedings, Technical Association of the Paper Industry,* Atlanta, Georgia.

22
Die Cutting

E. Raymond Schaffer
Bobst Champlain, Div.
Bobst Group, Inc.
Roseland, New Jersey

Die cutting may be defined as a process by which an object is cut or formed by use of a die. Webster defines a die as "any of various tools used to shape or impress an object or material."

The forerunner of all types of dies in use today is known as the mallet handle die. They were originally fashioned by a blacksmith to the desired configuration of the product to be produced. After the cutting edges were hand-sharpened and heat treated, a handle was attached to the back of the die which could be struck with a mallet to force the die through the material to be cut. Although crude by present-day standards, the "handle-mallet" approach is still used for the limited production of samples and custom products generally made of rubber, plastics, fabrics or leather. At about the turn of the century, a mechanical press was developed by the United Shoe Machinery Corporation to provide the pressure necessary to push or force the die through the material. This was known as the "ideal clicking machine" from which a second generation of dies takes its name—the clicker die. The clicker die is similar in construction to the mallet handle die in that the cutting blade is bent or fabricated to conform to the desired configuration. The structural stability which is built into this type of die is such that no external support is required to prevent the cutting knives from bending or otherwise distorting under the cutting load. For general-purpose applications these dies are manufactured to the commercial tolerance of ±0.38 mm. When closer tolerances are required, the die is machine tooled from a solid block of steel in which case accuracies to within ±.025 mm may be held.

The steel rule die is probably the most commonly used in conjunction with web and sheet fed equipment for the manufacture of folding cartons from either corrugated or solid paperboard stocks. Unlike the dies formerly discussed, the cutting knives are flexible and relatively thin. They must therefore be imbedded in or otherwise supported by a thicker, solid material such as plywood.

The male-female die, as its name implies, is made of two distinct (upper and lower) sections. Each section is made of hardened tool steel blocks which must be mated with extreme accuracy in order to accomplish the scissor type cutting action inherent to this form of die cutting. One important advantage of this die is that the waste can be extruded and disposed of through the female section. It is particularly useful when intricate stripping is required and is most applicable to the production of tags and labels, which are delivered in roll form.

In addition to the flat bed type dies as described above, rotary dies are also used. Because of their curved configuration, however, they are more difficult and time consuming to build and therefore more expensive.

An exception to this has been the development of a wraparound type die which consists of two light caliper plates on which the die image is produced respectively in male and female form by employment of photochemical etching techniques. As compared to the older die types, the wraparound concept is relatively new and has gained limited acceptance for the production of folding cartons from a continuous web.

The function of any die cutter is to operate (i.e., open and close, or rotate) the die, and to control the flow of material as it enters and exits from the die area.

In the process of rotary die cutting, the velocity of the web must be equal to the circumferential speed of the die. Flat dies require that there be no relative motion between the die and the web during the actual cutting portion of the cycle. There are two methods employed to accomplish this.

The first method utilizes eccentrically mounted rolls or cam controlled looper mechanisms to intermittently stop the forward motion of the web each time as the die closes. The units operating on this principle are known as reciprocating or stationary die cutters, of which there are two basic types, to be explained later on.

The second method permits the continuous flow of the web over which the die and its associated members pass back and forth in a swinging motion. The mechanism is so timed that the forward speed of the die coincides with the speed of the web during the instant of die penetration. This is known as the swing type punch. It is capable of handling light materials for the production of labels or tags of various types, as well as paperboard up to 25 points (0.6 mm) in thickness for production of small, specialty type cartons.

When producing tags or labels in roll form a rewind unit is provided. If the die is of multiple configuration across the web, such as in the production of tea tags, a slitting unit is provided to create ribbons of tags which are then separately wound on spools. In the production of pressure-sensitive labels, the shut height, or pressure exerted by the die, is adjusted to cut through the label material but not through the supportive backing. The operation is similar to that when producing tags except an additional rewind system is required to separate and eliminate the resulting scrap from around the labels. This continuous type scrap is referred to as a "ladder."

When the swing type punch is used for the production of folding cartons, male-female dies are preferred to facilitate the removal of scrap through the female section. If a steel rule type die is used, a separate pin type scrap removal system is required. In either case a carton delivery system is required to systematically remove the cartons in either shingled or stacked form.

As formerly noted, there are two basic types of reciprocating die cutters which can be distinguished primarily by their size, the type of material which they handle and the type of end product which they produce.

The smaller type handles paper, paper backed foil, pressure-sensitive and other materials up to 10 points (0.25 mm) in thickness for the production of labels and tags in roll form. This unit may also be used for perforating, notching, form punching and similar operations on films such as glassine and cellophane.

The larger type is of much heavier construction capable of processing materials in the 10 (0.25 mm) to 40 (1 mm) point caliper range. This unit is designed specifically for the mass production of folding cartons, which is the largest single use of die cutting equipment. As such, the remainder of this chapter will deal with the development and accomplishments in that particular field of die cutting.

FOLDING CARTONS

Web-fed cutting and creasing, today an established and accepted method of producing folding cartons in large volume, has a history dating back to the end of the first World War. Although it has come into common usage only since the early 1950's, there were installations of web-fed press equipment as early as 1918.

As the demand for larger quantities and faster production of packaging materials grew, it became evident that the small, slow equipment then in use was a critical limitation both to volume and production costs. Many users of ever-larger quantities of packaging found that they could no longer depend upon their customary sources of supply for their increased needs. A new design concept was clearly indicated.

Two European package machinery firms accepted this challenge and in-

464 WEB PROCESSING AND CONVERTING TECHNOLOGY AND EQUIPMENT

itiated engineering programs aimed at solving the speed, volume and cost problems associated with the production of high-quality folding cartons. The designers recognized that since the paperboard used for producing folding cartons was manufactured in continuous web form, it would be consistent to maintain the continuous web principle for the printing and die cutting equipment as well.

The earlier installations, therefore, generally included an unwind stand, a web feed unit, a number of rotary letterpress or dry offset printing stations, and fabricating equipment suitable to the product being manufactured. In the case of folding cartons, this was usually either a reciprocal or a rotary die cutter.

The rotary die carton cutter afforded a marked increase in production speed over the relatively slow reciprocal cutting and creasing equipment. This advantage was eclipsed, however, by the relative costs involved in the construction of the cutting and creasing dies. Because of this necessary cylindrical construction of the rotary die, it was far more costly and less flexible than the relatively simple, flat, reciprocal type die.

From a practical point of view, the use of rotary die cutting was therefore limited to those products which were produced in large quantities with little or no expectations of design or size change.

RECIPROCAL DIE CUTTERS

The reciprocal die cutting approach, working from a continuous web, has developed into the present-day "work horse" of the folding carton industry, despite its rather humble beginnings.

Since the production of most folding cartons necessarily includes printing, most cutting and creasing units are presently placed in-line with a number of printing stations—generally rotogravure and some flexographic. Figure 22-1 outlines a typical press line which is representative of the present state

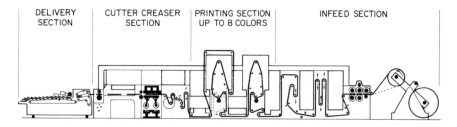

Fig. 22-1. Cutter creaser press line.

of the art for the high-speed production of folding cartons. It is made of four major sections, which are briefly explained as follows:

The press infeed section accepts the boxboard stock in roll form and, by virtue of automatic splicing equipment, presents a continuous flow of web to the printing stations under a controlled level of tension.

The printing section may consist of up to eight color units. Any number of those units may be equipped so as to permit printing on either side of the web. Each unit has its own drying system capable of completely drying the printed matter prior to its arrival at the next printing unit. Web length compensators are located between printing units to control the printed matter from the first unit so that it arrives at each succeeding unit in proper registration. These compensators are under the automatic command of a photoelectronic system which constantly scans and corrects the position of the printed matter as it appears in the vicinity of each printing location.

The function of the cutter-creaser section is to cut and crease the cartons in proper register with the printed matter and to remove all internal and external waste within the die form.

The carton delivery section accepts the cartons in their die cut form from the cutter creaser, breaks the tabs that are required to hold the carton form together and deflects the individual cartons to a slow-moving conveyor belt, thus causing the cartons to be formed into individual, closely shingled streams. At this point the cartons may be manually lifted from the conveyor belt and stacked on pallets. However, the high production capabilities of modern-day equipment (in instances up to seven thousand cartons per minute) practically demands that this stacking process be automated as well. A stacking unit is therefore generally provided to receive the cartons in shingled rows, stack them to a predetermined count and place the stacks on a delivery belt for removal by the stock handlers.

As stated earlier, the function of the cutting and creasing section is to perform this function in precise register and remove all scrap within and around the cartons. Figure 22-2 depicts the three subsections provided to accomplish this:

- The infeed section, which interrupts the continuous flow of web from the printing section and intermittently feeds it to the cutter creaser.
- The cutter creaser section, in which the cutting and creasing die is mounted, converts the web into carton form each time the web is momentarily stopped by the action of the infeed section.
- The stripper and transfer section, which transports the cartons in died-out form from the cutter creaser to a pin cylinder which extracts the waste.

466 WEB PROCESSING AND CONVERTING TECHNOLOGY AND EQUIPMENT

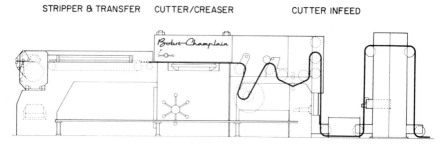

Fig. 22-2. Cutter creaser sections—modern.

It is of interest to note that the basic concepts for production of folding cartons has not changed to a significant degree over the past two decades. Rather, the improvements gained during this period are the result of refinements made in the execution of those concepts. As an example, let us review a typical cutter creaser system as depicted in Fig. 22-3 as it existed in the early 1960's.

Note that the same subsections exist—namely, the feed section, which creates the stop and go motion to the web, the reciprocating motion of the cutter section for stamping out cartons at the instant the web is stopped by the feed sections, and the rotary type stripper section for the removal of waste.

To gain appreciation of where the refinements are made, let us make a section-by-section comparison between the former and present-day equipment.

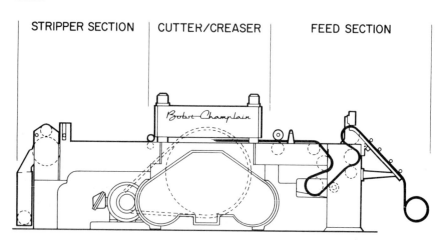

Fig. 22-3. Cutter creaser sections—earlier model.

DIE CUTTING 467

FEEDER

Starting with the feed section of the earlier model, we shall, with the aid of Fig. 22-4, follow the web as it passes through each unit as shown within this figure.

The Angular Side Guide

Its purpose is to laterally align the web relative to the position of the die. A series of steel rollers, angularly mounted on a steel support table, forces the web against a straight edge which is attached to the support table. The built-in side motion of the entire structure permits alignment corrections to be made while the press is running.

The Metering Unit

Its function is to meter out the exact amount of web required for each feed-up or cutting cycle. It consists of a steel driven roll in contact with a rubber

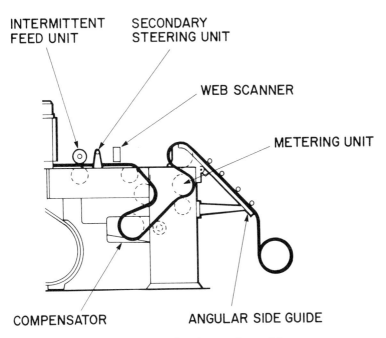

Fig. 22-4. Feed section—earlier model.

covered pressure roll. Since the steel driven roll is of fixed diameter, proper selection and installation of four change gears is required to satisfy the feed-up length for a given job.

The Compensator Roller

As the web leaves the metering roll nip point, it wraps the compensator roll. This is adjustable in that it regulates the physical web distance between the metering roll nip point and the cutting and creasing die. In this manner it serves as the registering device for maintaining proper position of the printed matter with respect to the died-out cartons. The unit is motorized and can be controlled manually by pushbuttons or automatically when used in conjunction with the web scanner.

The Web Scanner

This optical electronic device constantly monitors the arrival of a printed reference mark within each printed repeat length. The distance this reference mark must travel to arrive in proper position under the die is then calculated electronically. The compensator motor is energized accordingly, thus positioning the compensator for automatic and continuous control of cut-to-print register.

The Secondary Web Steering Unit

This unit controls the final side positioning of the printed web immediately before entering the die area. It can be manually adjusted while the equipment is running. In addition to adjusting the sidelay it can, by virtue of its pivoting motion, adjust the angularity of the web in the event that the printed copy is not square with respect to the die.

The Intermittent Feed Unit

This unit intermittently feeds web to the cutter creaser. It consists of a series of free turning upper wheels which nip the web against a mating set of lower wheels. The lower wheels are commonly driven by an intermittent drive mechanism. Since the nip pressure on the web remains constant throughout the entire cycle, the forward motion of the web follows the stop-and-go speed profile of the intermittent drive mechanism. The intermittent drive is timed to the motion of the cutting head so that the web can be fed into the cutter creaser only when the die is opened sufficiently to provide unobstructed passage. A slight overspeed is built into the mechanism to cause the web to be pulled tight for an instant around the compensator roll. This serves to re-establish correct cut-to-print relationship during each feed-up cycle. Any repeat length within the size range of the press can be fed to the cutter creaser by changing two gears.

The limitations with respect to speed and accuracy of this earlier equipment may be summarized as follows:

- The absence of automatic web side-guiding makes it necessary for the operator to constantly monitor and make manual adjustments to hold cut-to-print register within acceptable limits.
- No means to control the natural curl of the stock was provided within close range of the cutter creaser. The resulting inability to present a flat web under the die at all times caused the web to jam up in the die, resulting in undue material waste and press stoppages.
- The free loop condition under which the web was fed under the die permitted the web to wander sideways, causing erratic cut-to-print behavior.
- The vibration set up in the web at the instant it was pulled tight around the compensator roll caused an erratic slipping condition as the web passed through the intermittent nip point. This resulted in erratic cut-to-print behavior in the running direction.
- The initial forces resulting from the stop-and-go motion of the intermittent drive mechanism resulted in high maintenance costs and reduction of press speed.

Present-day performance, in terms of higher permissible running speeds with greatly improved performance reliability plus an end product of superior quality, testifies to what extent these problems have been resolved. With the aid of Fig. 22-5, let us review the units or subsections contained within the cutter creaser in-feed section as it exists today.

The Web Steering Unit

Although free standing, it is considered a part of the feed section since its function is to maintain proper side register of the printed material as it is cut and creased. The two upper rolls are commonly mounted in a pivoting structure which causes the web to be steered into the desired web path. A photoelectric sensing head, capable of identifying either the edge of the web or a printed mark, senses any deviation in the web path (or printed mark) and signals the steering rolls to pivot accordingly. This continuous monitoring and automatic control relieves the operator from the need to make manual corrections after an error is detected.

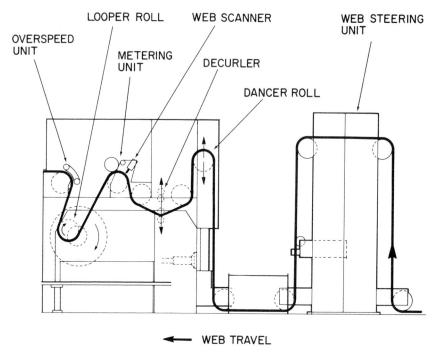

Fig. 22-5. Feed section—modern.

The Dancer Roll

This roll, which is free floating, serves to maintain a constant tension on the web as it enters the metering roll nip point. It is mounted on air cylinders which, when air is applied, cause the roll to move in the upward direction. Since under normal running conditions it is pushing against the web, a balanced condition occurs when the upward force of the roll is equal to the downward pull of the web. In this manner, it becomes a take-up device which compensates for any differential in speed or change in web length between the last printing station and the metering roll nip point.

The Decurl Unit

As noted earlier, the ability to eliminate or control the natural curl in the web is of vital importance for consistent high-speed performance. The addition of this unit in close proximity to the cutting head has significantly reduced the number of press stops formerly required to clear web "ball ups" under the die or carton jams throughout the stripping and delivery sections.

Control of the unit is semiautomatic. At the time a machine splice passes through, the decurl rolls automatically return to a present position which has been found satisfactory to accommodate the outer wraps of each new roll of stock.

Note that movement of the decurler would have resulted in a change of web length had it not been compensated for by the action of the dancer roll.

The Metering Unit

Similar to the older equipment, its basic function is to meter out the precise amount of web required for each feed-up cycle. Its basic speed for a given repeat length is governed by the proper selection of the two change gears. A secondary drive input is initiated by an electric motor. Through planetary gearing contained within a feathering drive box, the electric motor, by virtue of its speed and direction of rotation, can add or subtract from the basic speed of the unit.

The Web Scanner

Its purpose is identical to that formerly explained. It does not, however, apply its intelligence to a compensator unit, which is no longer required. As it is presently coupled to the feathering drive electric motor, automatic print-to-cut running register is maintained by minute speed regulation of the metering roll. This more direct approach reduces response time, making it possible to maintain register within closer tolerances than formerly experienced.

The Looper Roll

This free turning roll is mounted on an eccentric mechanism which makes one revolution for each stroke of the cutter creaser. In this way it alternately collects and pays out web as required by the closing and opening of the die.

The Overspeed Unit

This consists of a series of free turning belt assemblies which force the web against a lower constantly driven roller. The surface speed of this roll is at all times greater than the speed of the web passing over it. This assures that the web is under tension at all times as it passes around the looper roll. In this respect, the overspeed unit may be considered a slave which is constantly pulling at the web and feeding it under the cutting head as dictated by the motion of the looper roll.

472 WEB PROCESSING AND CONVERTING TECHNOLOGY AND EQUIPMENT

The advantages of this present arrangement for intermittent web feeding are twofold:

1. Since the web is being held under constant tension at all times, it no longer tends to weave as it enters the die area.
2. The elimination of the intermittent drive with its objectionable initial forces permits higher production speeds along with reduced maintenance costs.

CUTTER CREASER

Figure 22-6 represents a typical die layout. Recognizing that at the moment of cutting, the die makes instantaneous impression contact with the full area of material to be cut and scored, a considerable impression pressure is required to cut cleanly and to score uniformly.

As shown in the illustration, there are 39 cartons which are produced simultaneously during each stroke of the cutter creaser. Each inch of cutting knife requires a force of approximately 300 lb pressure to cut through the stock, and each inch creasing rule requires approximately 225 lb to properly

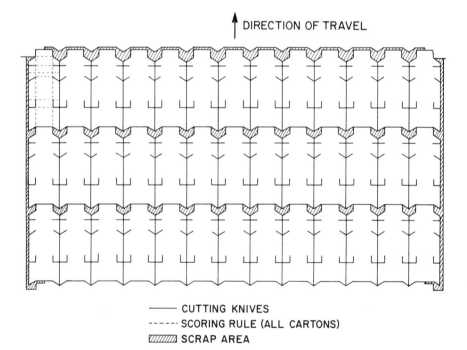

Fig. 22-6. Typical die layout.

form the scores. Since there are 988″ of cutting knives and 1,116″ of scoring rule involved, a total of 274 tons of pressure is required for each impression.

This represents approximately 68% of the 400-ton rated capacity of the modern equipment. Recognizing that these forces can be generated and controlled at a repetitive and sustained rate in excess of five times a second, one may begin to appreciate the precise, rugged dependability which must be designed into equipment of this type.

For comparative purposes let us first examine the older type cutter creaser shown in Fig. 22-3. More detail of this unit is shown in Fig. 22-7, which will serve to explain its function.

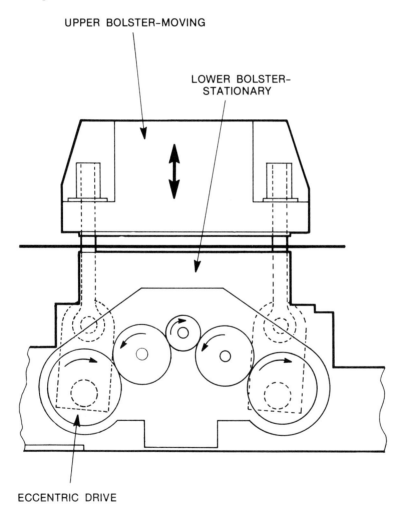

Fig. 22-7. Cutter creaser proper—earlier model.

The Upper Bolster

Also referred to as the cutting head, it is made of a single slab of steel 16″ thick, which weighs approximately 15,000 lb. The lower surface is precisely machined and supports the upper chase in which the cutting and creasing die is mounted.

The Lower Bolster

As a part of the base structure, it contains the lower chase assembly which supports the stock at the instant of cutting and creasing.

The Drive Components

The upper bolster being the moving or reciprocating member is secured to four driving posts which pass through the stationary lower bolster and are connected through linkage to eccentric shafts contained within the base structure of the unit. In order to withstand the forces with which we are dealing, the posts are constructed of forged and tempered alloy steel 4½″ in diameter. The two crank shafts which support two posts each are 15″ in diameter. The inertial forces resulting from the speed and mass of the moving members, particularly the upper bolster, limited the speed to 275 strokes/min even though the entire unit was mounted on shock absorbing material to minimize objectionable floor vibration in the surrounding area.

In the pursuit of higher speeds in conjunction with wider equipment in order to keep abreast of the ever continuing demands for greater production, it became obvious that the mass and weight of the reciprocating members must be appreciably reduced.

The manner in which this has been resolved is outlined in Fig. 22–8, which is representative of present day design. The obvious difference is that the upper bolster now remains stationary and the lower bolster becomes the moving member.

The total weight of the reciprocating members has been reduced to less than one-half as compared to the former design. Inertial loading is therefore no longer the speed-limiting factor. Rather, speeds are presently limited by the rate at which the stock can be introduced to and removed from the die area.

There are other features of note built into the cutter creaser which improve its efficiency.

- Recognizing that the required cutting force changes relative to the amount of cutting and creasing rule within the die and the density of the

DIE CUTTING 475

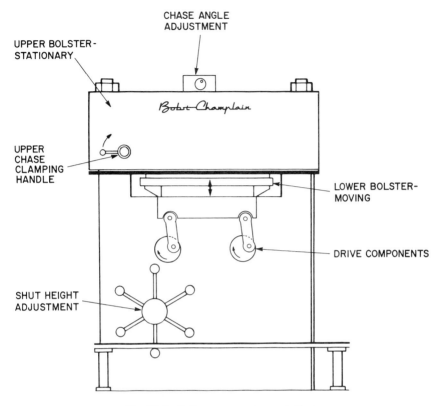

Fig. 22-8. Cutter creaser proper—modern.

stock, a ready means is provided to compensate for this by turning the shut height adjusting headwheel. This causes the lower bolster to be minutely raised or lowered with respect to the base of the unit, thereby regulating the pressure between the cutting plate and the die when the lower bolster is at the uppermost limits of its travel. This is commonly referred to as "top dead center." The control is also useful during a given job to slightly increase the cutting load to compensate for the normal wear of the cutting knives as the job progresses.
- By turning the chase pivoting handwheel, both the upper and lower chases can be simultaneously pivoted or "skewed" for purposes of squaring the printed matter to the cut. This is a more positive and direct approach than attempting to "skew" or deflect the web from its normal path as earlier described.
- The clamping handle provides for quick and positive means for clamping the die chase securely to the upper bolster.

STRIPPER

The stripper or waste removal section is made up of three major subsections, as shown in Fig. 22-9.

The Kicker Section

Immediately after the cutting and creasing cycle is completed, the new web entering the cutter creaser pushes the die cut portion of the web between the upper and lower kicker rollers. The nip created by these rolls is cam operated. As the nip closes, the die cut form is positively engaged and broken away from the oncoming web. Now in sheet form, the cartons, which are held together by tabs, are transported to the stripping cylinder.

The Transfer Section

Made up of a series of belts, wheels and brushes, this section positions and propels the die cut sheets to the stripping cylinder in proper register.

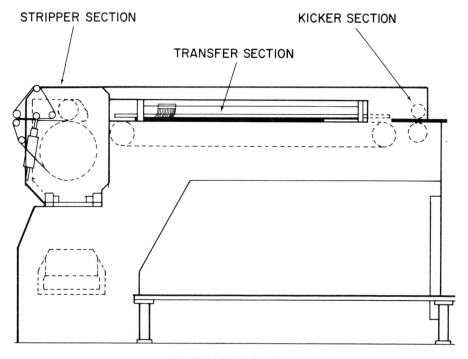

Fig. 22-9. Outfeed section.

DIE CUTTING

The Stripping Section

The manner in which the waste is extracted may be more readily understood by referring to Fig. 22-10. As the die cut sheets leave the transfer section, they enter the nip point created by the upper and lower stripping rolls. The upper, commonly referred to as the anvil roll, is rubber covered and supports the stock as the stripping pins penetrate the scrap. The lower cylinder, or stripping shell is pre-drilled and tapped to receive the stripping pins. A die template is used in conjunction with the drilling and tapping operation to assure that the pin locations conform to scrap areas of a given die form. The scrap portions are impaled by the pins and are pulled down around the stripping shell, while the cartons, supported by the upper and lower belts, continue to pass straight through to the delivery sections. Removal of the scrap from the pins may best be described with the aid of the enlarged view of the pin assembly shown in Fig. 22-10. The pin itself is secured in a fixed position to the threaded body. A free sliding sleeve located between these two members, when pushed toward the head of the pin, surrounds the barbed head, thereby extracting any impaled waste. This sleeve is actuated by its proximity

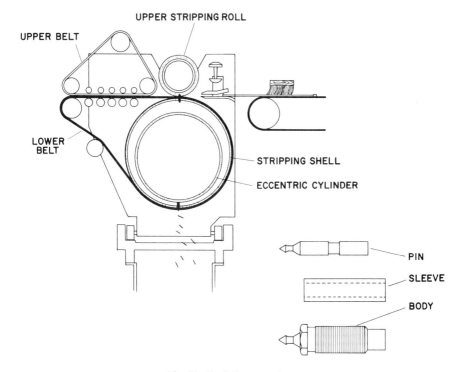

Fig. 22-10. Stripper section.

to the eccentrically mounted inner cylinder. Thus, at the impingement, or 12 o'clock position, the sleeve is fully retracted and the pin head is exposed to penetrate the scrap. As the rotation continues, the base of the sleeve contacts the eccentrically mounted inner roller and is forced to extend, causing the scrap to be pushed free from the pin. The scrap then falls onto a conveyor belt for removal from the press area.

The cartons, now free of scrap but still attached or tabbed together, proceed to the delivery section where they are rearranged and placed on a final delivery table as individual precounted stacks. With the aid of Fig. 22-11, let us examine the subsections provided to accomplish this task. While doing so, keep in mind the quantity of cartons which must be handled during a given time period. As an example, consider a job utilizing the die form shown in Fig. 22-6. The cutter creaser running at 180 strokes produces a total of seven thousand cartons/min. This requires that the delivery during each minute of operation must separate 180 die cut sheets into seven thousand individual cartons; then count and stack them into 14 piles consisting of 500 cartons each.

DELIVERY

Again referring to Fig. 22-11, we shall describe the function of each of the four major sections which make up a typical delivery system.

The Collector Section

The sole purpose of this section is to reduce the speed of the die cut sheets by a factor of two. A series of tapes running one-half the speed of the stripping cylinder provide two separate paths. A cam operated deflector blade deflects the die cut sheets alternately into each path. Due to the difference in the length of each path, two sheets emerge simultaneously at the exit point. From this point on, the superimposed sheets (which eventually become cartons) are treated as a single form.

The Tab Breaker Section

As its name implies, the function of this section is to receive the die cut sheets from the collector, break the tabs that were holding the sheets together and convey the now individual cartons to the belt delivery section. The hardware provided to do this consists of individual upper and lower tape sections which are driven at a slightly higher speed than the exit nip of the collector. This causes the leading row of cartons to accelerate as it enters the tab breaker nip, thereby breaking it free from the rows following. One tape sec-

DIE CUTTING 479

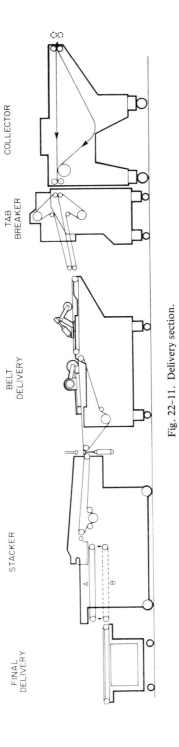

Fig. 22-11. Delivery section.

tion is required for each row of cartons in the across direction. The angular placement of each section with respect to the centerline causes the cartons passing through to be deflected sideways, thus breaking the tabs and creating separation between the rows of cartons as they arrive at the belt delivery section. The top and side view schematically presented in Fig. 22-12 will serve to clarify these functions.

The Belt Delivery Section

Consisting of two independently driven belt sections, its purpose is to accept the cartons from the belt tab breaker section and deliver them in separately shingled streams to the stacker section. Each belt section follows press speed, but may be trimmed to change the rate of shingle. Recognizing that the cartons could be arriving on the delivery at speeds in excess of 600 ft/min, it is desirable to run the first belt section as fast as possible to minimize the shock of the fast deceleration, but still maintain the integrity of the shingle configuration. The slower speed setting of the second belt section creates the tighter shingled pattern which is desirable during the stacking operation.

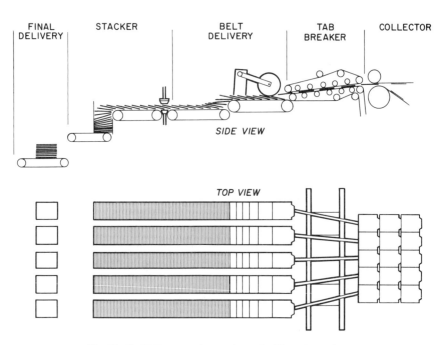

Fig. 22-12. Delivery section—schematic side and top views.

The Stacker Section

The cartons as they proceed through the stacker are dropped into individual pockets where they are side and end jogged into precise stacks. Upon command of a counting device which has been preset to the desired number of cartons per stack, the streams are interrupted and the pockets emptied in preparation for the next sequence. The mechanisms involved to accommodate this action are outlined in Fig. 22-12. The indexing sequence is initiated by the clamp which stops the oncoming streams. At the instant the clamp comes on, the transfer belts accelerate, causing all unclamped cartons to move quickly into the pockets. The elevator table, which forms the bottom of the pockets, then moves down to clear the completed stacks from the side joggers and end plate. The belts, which are a part of the elevator system, move the stack on to the final delivery table, after which the elevator table raises to receive the oncoming streams of cartons which have already been freed of the clamp. This entire sequence takes place within five seconds.

DIES

Let us now look closely at those components which do the actual cutting and creasing of the cartons: namely the die and its counterpart, the cutting or counter plate. As depicted in Fig. 22-13, the die consists of a number of thin, flexible steel cutting and creasing rules which are cut and formed to make up the desired die configuration. These are held in place by a base material which has been prepared or channeled to conform to the outline of the die form. Cork or rubber strips are glued along both sides of the cutting rule for

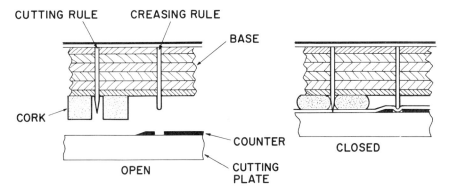

Fig. 22-13. Steel rule die sections.

the purpose of ejecting or quickly releasing the cartons from the cutting rule immediately after the cut has been made.

The cutting plate, which is attached to the lower chase, supports the stock as it is cut and creased. A counter material, after being glued to the cutting plate, is channeled to conform to the placement of the die creasing rule. This serves to crease or score the cartons for proper folding. There are other ways of forming these channels which will be discussed later.

Steel-rule die preparation and construction, which began prior to the turn of the century, has of necessity developed into a highly sophisticated art, in order to produce the more complex and dimensionally stable cartons required for the modern high-speed folding, gluing and filling operations.

There are several basic types of steel-rule dies which are generally used in conjunction with reciprocating cutter creasers. These may be explained as follows.

The block die is the first known type of steel-rule die. It is made up of individual plywood blocks, commonly referred to as furniture, which are cut into shapes representing the panels and flaps of the cartons to be produced. The precut and formed cutting and creasing rule is then inserted between the blocks, after which the entire assembly is locked up in a die chase. Dimensional accuracy of this type of die is largely dependent upon the individual skill of the die maker and the manner in which it is locked into the chase. The block die has played a major role in the industry over the years and will probably continue to do so in the future. However, its popularity has diminished somewhat in favor of one-piece die construction which, due to modern technology, can be prepared with extreme accuracy.

The jig die is made from a single piece of plywood on which an outline of the die is drawn or scribed. A jigsaw is used to cut the slots in accordance with the outline. Precut and formed cutting and creasing rules are then inserted into the slots. The slots are interrupted in various places, thus forming bridges which hold the plywood together. Before insertion, the cutting and creasing rule must be undercut to straddle these bridges. Originally, the accuracy of the jig die was entirely dependent upon the expertise of the individual who must first draw the die image on the board by hand and then manually follow those lines with a jigsaw. Modern techniques, such as X and Y systems, step and repeat processes, and numerically controlled computer systems have automated the production of the scribed image, thereby removing the human error in the preparation of the die board prior to cutting. Together with improved jigsaws designed or equipped specifically for die making, jig dies are capable of producing cartons with an acceptable level of accuracy.

Laser dies are virtually identical to jig dies in their structural form. The difference lies in that the slots for receiving the cutting and creasing rule are cut by a laser beam. With this system, automation is carried to the final step in that the die board is automatically moved under the laser beam in accordance with the die image. This advanced method provides for excellent slot quality and extreme dimensional accuracy since the human error has been entirely removed.

The bonded steel die is unique in that the base is made up of three distinct sections. Two thin steel plates form the top and bottom surfaces. By use of computer-controlled photographic techniques, the die image is identically reproduced on each plate which has formerly been treated with a photoresist material. The slots are then chemically milled into the plates to receive the cutting and creasing rule. After being properly aligned and spaced by use of studs to create the desired die thickness, they are welded together. The cutting and creasing rule is then inserted, after which the voids between the plates are filled with an epoxy bonding material. The dimensional stability of this type of die is superior since the materials of which it is constructed are virtually unaffected by temperature and humidity changes.

As noted earlier, the cutting plate, being the supportive member of the die, must be prepared to include channels for forming the creases in the cartons. They must be very closely aligned to the creasing rule in the die to assure that the cartons fold squarely and easily without rupturing the stock. The depth and width of the channels is determined primarily by the caliper and type of stock to be processed. The several ways of forming the channels are described as follows with the aid of Fig. 22-14.

The original method, still in extensive use today, utilizes a sheet of pressed board of the correct thickness which is securely glued to the cutting plate. A sheet of carbon paper, carbon side down, is placed over the pressed board. The cutting plate and die are inserted and locked in their operational position within the cutter creaser. By jogging the cutter creaser through its cutting cycle, the die image is transferred to the pressed board, which then serves as a guide for manually cutting the channels. The pressed board must be cut and removed from all cutting locations and the edges of the remaining pressed board sections must be tapered or skived to permit smooth passage of the die cut material.

An alternate method utilizes commercially available metal strips in which the channel has been formed. The strips are attached to the cutting plate by an adhesive backing. Proper alignment with the mating cutting rule is assured by a slotted plastic locating member which is supplied as a part of the channeled strips. The slot in the plastic locator serves to attach the channeled strips to the creasing rule of the die. After this is done, the die and cutting plate are positioned and locked in the cutter creaser. Jogging the cutter creaser through its impression cycle causes the channeled strips to firmly adhere to the cutting plate. The plastic locators are then stripped away and

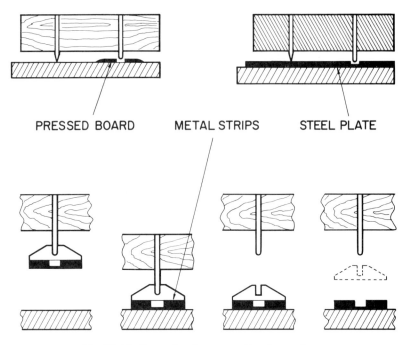

Fig. 22-14. Crease forming preparational procedures.

discarded. This system, compared to the hand-cutting method, is superior in that it reduces preparation time and affords greater accuracy. Also, the creasing channels are less subject to wear by virtue of their steel construction.

The most advanced method of crease forming utilizes a solid cutting plate in which the creasing channels are precut, generally by a chemical milling process. It can only be used, however, in conjunction with dies that remain dimensionally stable (not affected by temperature and humidity changes), and which have been developed by these more sophisticated computer controlled techniques as noted above. These same techniques are then employed to properly place the channels in register with the creasing rule throughout the entire die area. Both the die and cutting plate contain registering holes into which closely-fitted pins are inserted. The die is then locked into operating position with the cutting plate suspended by the pins. Jogging the cutter creaser through impression lowers the cutting plate and bonds it to the original cutting plate on a filler plate which has been previously positioned and secured within the lower chase. The advantages of this system are quick and positive setup, uniform scoring, and greater production speeds. The latter is due to the smoother flow of the stock through the cutter creaser because of the one-level cutting and creasing profile.

23
Embossing and Related Processes

John A. Pasquale III

Liberty Machine Co., Inc.
Paterson, New Jersey

Embossing is a method by which a web is textured or decorated by the use of a pattern roll pressing against a backup roll under controlled conditions. One can emboss both thermoplastic and non-thermoplastic webs by choosing the proper roll arrangement to deform the web.

In embossing thermoplastic materials, the web is deformed by preheating and pressing it with a cooled embossing roll to set in the pattern and cooling the web to retain the pattern. The degree of preheating to soften the web must be carefully controlled so that no melting or degrading of the web will take place. Only the amount of heat needed to satisfactorily emboss the product should be applied so that the heat removal process is made as efficient as possible.

Embossing a non-thermoplastic web, such as paper, textile, foil and the like, is accomplished by applying pressure which exceeds the elastic limit of the substrate and imparts a texture. This type of texturing often involves the use of male and female rolls, either two rolls with matched patterns made of steel or other metal, or the use of a steel pattern roll in contact with a filled backup roll which takes a permanent deformation for a given pattern by running the steel embossing roll in contact with the backup roll during a "running in" process. In some cases, special rubber-covered rolls can also be used. They do not require "running in" due to their ability to deform.

THERMOPLASTIC WEBS

Embossing of thermoplastic webs is usually achieved by using an engraved metal roll pressing against a rubber-covered backup roll. The metal roll is cooled with a refrigerated solution to remove heat from the product and to set in the embossed pattern. The backup roll is internally cooled, mainly to increase the life of the rubber covering. The roll may also be cooled externally by a water bath and squeeze roll system, especially if the web is an unsupported thermoplastic film having a tendency to adhere to the hot rubber surface.

Embossing of a web is dependent upon many variables, such as:

- Degree of preheating/rheological properties of the product
- Sheet thickness
- Hardness of the rubber backup roll
- Embossing roll pattern and its cooling capability
- Post-emboss cooling.

A fine balance exists between the preheating and the removal of heat to set in the texture. Applying the appropriate amount of preheat but insufficient cooling results in the inability to retain the embossing. The best embossing system is one that optimizes heat input and removal for a given product thickness and speed.

When the preheat temperature is too high, a thermoplastic web which has a specific melt temperature will break down and melt as that temperature is attained. Therefore, it is important to operate just below melting temperature to impart the best possible retained texture. A thermoplastic product which has no specific melt temperature (such as polyvinylchloride) will gradually soften and ultimately break down with increasing temperature. The optimum temperature at which to emboss this type of product is the temperature at which appropriate flow takes place without thermal degradation of the product.

The hardness of the backup roll often plays a role in the finished product's texture. If a roll is too hard, a good definition of fine surface patterns might be achieved. However, it may not allow the displacement of material within the product for deep embossings. If the roll is too soft, it will allow deep embossings to show through the back of the product, making it objectionable in some applications.

The displacement process, which takes place in the web when texturing, is important to consider. When light texturing, or embossing, is performed, the amount of displaced material is small with respect to the total volume of the sheet. In this case, a thermoplastic material can flow sufficiently to allow the total displaced volume to be redistributed within the total sheet volume, or the face, without showing through the back of the web. When doing deep texturing which involves significant volumetric displacement of the thermoplastic material, a point will be reached where the product itself cannot locally deform to allow the displaced volume of material to be absorbed through the full sheet. When this occurs, the embossing pattern must be allowed to show through the back of the sheet, or the limiting compressability of the sheet will prevent the depth and final shape of the embossed texture from taking place.

The need to cool the product after it has left the embossing roll is another important factor. Appropriate post-cooling facilities, usually cooling drums or rolls, are used to bring the sheet temperature as close to ambient conditions as possible before it is wound up in the roll. When cooling a thin sheet, the problem of retained heat is minor because a thin sheet releases heat easily. Most of the cooling takes place at the embossing roll and the stripper roll which follows it. In heavy sheet texturing, although the surface of the web might feel cool, the heat is retained in the body of the sheet. If this remains unchecked, the heat will cause a loss of embossed grain later.

Embossing units can be placed in various geometric positions. They are usually either vertical, where the web path enters in a horizontal manner, as shown in Fig. 23-1a, or they can be placed in a horizontal fashion, as shown in Fig. 23-1b. Under special conditions and for certain applications, it might also be advantageous to find them disposed at a particular angle, as shown in Fig. 23-1c.

Position is not only a matter of choice. The preheated web should enter the embossing nip perpendicularly to the line of action of the embossing and backup rolls so the web is not prematurely cooled by striking either the embossing roll or backup roll first. It is acceptable and sometimes desirable that the backup roll be contacted first. Many webs are unstable in their preheated condition and will be easily creased if they enter the nip unsupported. A short arc of contact before the embossing action of the nip allows the web to be flattened. It is also preferable to contact the backup roll first because the surface temperature of the backup roll is higher than that of the cool embossing roll, thereby removing the least amount of heat from the preheated web. Furthermore, the heat is removed from the back side rather than the face which is to be embossed.

PRODUCTS

Embossing is used not only for decorative effects or to imitate products such as leather or fabrics, but also to enhance the physical properties of a product to increase its surface area by imposing various geometric designs, thus improving the functionality of the web.

The automotive and upholstery industries use embossing to achieve natural looks from a synthetic product. Their use of leather-type grains helps to impart a softness and warmth to a purely synthetic product. Wall-

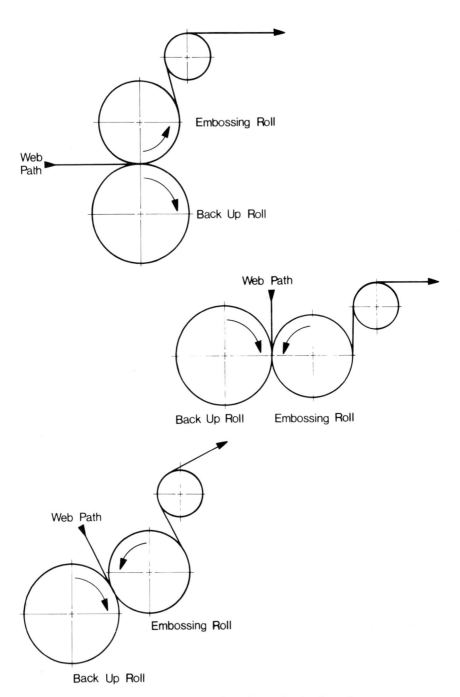

Fig. 23-1. Various geometric positions of embossing units.
    a—Vertical
    b—Horizontal
    c—Inclined.

covering, flooring and garment industries use textured products for aesthetic or functional purposes. While wall-coverings are textured to improve the appearance, flooring products such as tile or sheet goods are textured to give a functional surface which accepts wear, does not show dirt as readily and may be printed in register with the texture.

Disposable products such as diapers, garbage bags, etc., are examples of the embossing strengthening a sheet with the imposition of regular geometric patterns to control stress distribution.

EMBOSSING MACHINES FOR THERMOPLASTIC WEBS

Figure 23-2 shows a piece of equipment which includes a preheating drum with the possible use of additional radiant heat, an embossing section consisting of a metal embossing roll and a rubber-covered backup roll and appropriate cooling facilities.

Polyvinylchloride film of moderate thickness of 0.2-0.3 mm is unwound from a tension controlled unwind to assure that it is not stretched during the process. It enters the preheating area by passing over a spreader roll and is applied to the heat drum by use of a lay-on roll. The lay-on roll is a rubber-covered, pneumatically operated roll, whose hardness is usually in the range of 50-55 durometer, Shore A and whose function is to lay the web onto the heated drum surface, allowing for intimate contact and preheating.

The pneumatic operation of the lay-on roll is achieved through the use of double-acting air cylinders attached to the moving bearing housing. The air cylinders are provided with air regulating equipment and a four-way directional valve which is either manually or electrically operated. A regulator and gauge is provided to control the force which the lay-on roll exerts on the web and heated drum. The drum is steam or hot oil heated and is usually 1-1.5 m in diameter. Steam heat is preferred as it responds more readily to temperature changes than hot oil. In the case of polyvinylchloride film, the steam is more than adequate, since a drum temperature of 150°-175°C is used at a typical speed of 20-30 m/min. For this type of web processing, the arc-shaped radiant heat unit around the periphery of the drum is not required, as it would be used mainly for heavier webs.

The steam heated drum is usually of a double-shell construction, as shown in Fig. 23-3. Saturated steam passes through the center shaft of the drum entering from one end and is fed to the annular section formed by the inner and outer shells of the drum through passages. As the steam fills the annular chamber and performs its work, condensation takes place and is removed through similar passages at the opposite end of the drum by a syphon tube and exits through the shaft opposite the steam inlet. For uniformity, the steam from the center shaft usually enters the annular chamber at both ends

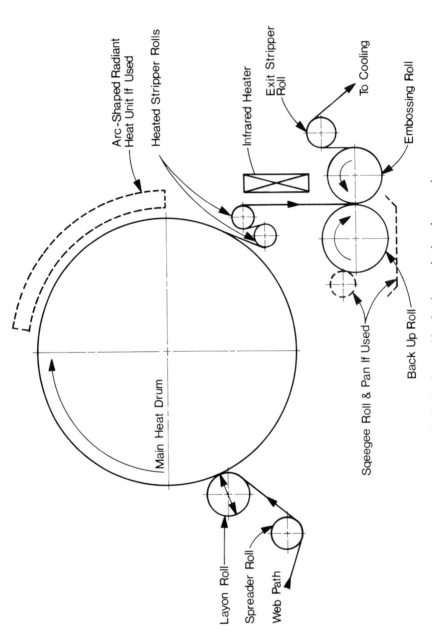

Fig. 23-2. Embossing machine for thermoplastic web processing.

EMBOSSING AND RELATED PROCESSES 491

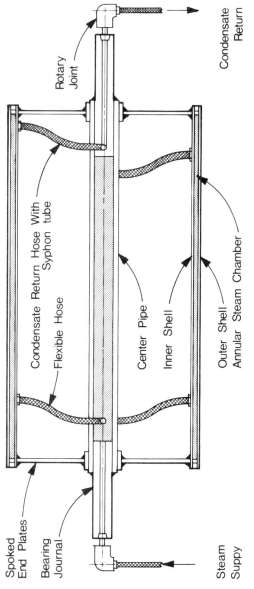

Fig. 23-3. Steam heated drum.

of the drum. Both the steam and the condensate enter and leave the drum through rotary unions furnished with bronze hoses to withstand the temperature of the steam. The rotary unions are pipe connections which allow the drum to rotate freely while they remain stationary to provide a solid connection to the steam pipe and condensate return system. The inner section of the rotary union is usually equipped with graphite bearings and seals, permitting rotation of the inner member with respect to the outer casting.

For hot oil heated drums the construction is similar, but spiral windings forming passageways or channels are provided in the annular space between the inner and outer shells to allow the oil to flow in a prescribed path and in a most efficient manner in order to promote good heat transfer. The hot oil is pumped through the center shaft of the drum and enters through conduits similar to the steam heated drum design, and after it has done its work, it leaves by similar passages at the opposite end of the drum. It is appropriate to pump the hot oil at rates which will cause it to flow through the drum annulus passageways in a turbulent manner to provide the best heat transfer coefficient.

The preheated web is stripped away from the main heating drum by use of the heated stripper rolls. These rolls help to maintain the temperature of the web and allow it to be passed into the embossing section.

The stripper rolls are usually single shell construction and vary in size from 100–150 mm for steam heated systems. Steam is fed through a syphon-type rotary union with bronze hoses, allowing the steam to enter through one hose into the shaft of the roll and the condensate to exit through the syphon tube placed in the roll and connected to the rotary union. The condensate will exit from a separate bronze hose on the same side as the steam entry. The infrared heater shown on the drop into the embossing section is primarily to maintain the ambient temperature around the web and to assure that the surface to be embossed will be entering the nip at approximately 160°C.

The embossing section consists of a 250–300-mm diameter, double-shelled embossing roll which is designed to allow a cooling solution to pass through it in an efficient manner to remove heat as rapidly as possible, setting in the texture.

Embossing rolls are usually of a double-shell design with a spiral wrap for the most efficient passage of the cooling medium. The internal construction would therefore be very similar to Fig. 23-3. Embossing rolls are designed either by mill engraving, which is a machinery operation in which the roll is mechanically formed using a die, or by electroforming, whereby the roll is manufactured by using an electrical charge to remove and displace the metal.

In the mechanical process a tool is manufactured from artwork presented to the engraver. The tool is often formed both by machine and hand-forming processes. It is hardened and used to manufacture a die which is the piece ultimately used to engrave the roll. The die is an exact opposite of the roll, meaning that if there is to be a raised male portion on the roll, the die will have a recessed female portion. Once the die is manufactured, it also is hardened and is placed in an engraving machine which is usually a lathe where pressure is exerted between the die and the roll body. The roll is engraved using a very fine speed control, and pressure is constantly controlled between the die and the roll body. It takes numerous passes of the die over the roll body to bring the roll to its finished design. It is inspected and then hardened and chrome plated to give a surface which will withstand wear. The chrome plating may be either a polished or dull chrome, depending upon the finish that is to be imparted by the roll.

The electroformed roll is also hardened and plated, but the formation process uses electrical charge for the removal of the metal to get a desired pattern.

The design of the embossing rolls is also important since pressure is exerted between the embossing roll and the backup roll. This is especially true in embossing units that deal with non-thermoplastic materials, resulting in embossing forces of 100–150 kg/linear cm. The double shell construction, along with the spiral winding, produces a roll which is structurally strong while still having excellent heat transfer capability. The roll body acts as a composite section giving minimal bending under load. To compensate for the bending that does occur, backup rolls are usually crowned with the degree of crown depending upon the embossing load. The crown can be a straight taper from the center out to the two ends or a very subtle parabolic taper. The center of the roll is larger than the two ends.

The cooling solution is applied to the roll through the use of rotary joints and rubber hoses, with the solution passing in at one end, flowing through the spiraled annular construction of the roll, and being discharged at the other end. The cooling medium used may vary from plain water to a low-temperature cooling solution such as a mixture of water and ethylene glycol or isopropyl alcohol. The solutions have freezing temperatures well beneath that of water and are used when sub-freezing temperatures are required to give the greatest amount of cooling capability.

However, it is not always desirable or possible to use temperature differential as the driving force to maximize cooling. In instances where low temperatures result in surface condensation on the embossing roll, one must resort to using higher flow rates of a warmer solution, thereby depending upon increased heat transfer coefficient rather than on temperature difference. Roll design becomes very important from the standpoint of heat transfer efficiency, and care must be taken to properly size the entrance and exit ports, to permit the required flow rate, and the annulus, to assure that the cooling solution is passing through it in the turbulent region of flow, thereby promoting the highest heat transfer coefficient and maximizing the cooling effect.

In some applications, separate pumps have been used to pass a cooling solution through an embossing roll at unusually high flow rates to promote maximum cooling instead of using a lower temperature cooling medium and risking condensation on the roll surface.

The backup roll is a 65–70 durometer, Shore A, hardness for a medium depth pattern. This roll is also cored for internal cooling, but primarily to keep the interface of the steel and rubber covering cool and avoid premature bond failure. The covering used on the roll should be abrasion resistant and able to withstand heat. Typically, ethylene propylene terpolymer (EPT) rubber and carbon-loaded neoprene coverings are used.

The two rolls are brought together hydraulically or pneumatically with the embossing roll usually being the non-driven, freely movable roll, while the backup roll is motor driven. It is important that synchronization between the main heating drum and the embossing section is accurately maintained to prevent stretching the product as it passes from one section to the other.

Embossing pressure might range anywhere from 45–60 kg/linear cm. Certain textures will require more, while others less, but this is a typical range for a product 0.2–0.3 mm thick.

It is important that the embossing roll is wrapped by the thermoplastic web. This enables cooling to take place while the web is still in intimate contact with the textured embossing roll. The web is then removed by use of an exit stripper roll which is also cooled, passing the web to the post-cooling section.

Post-cooling is more important on heavier gauge materials, where residual heat in the web would later result in a temperature rise in the rewound roll, causing a loss of embossing. In a sheet of this thickness, most of the cooling is achieved by the embossing roll and exit stripper roll. For the heavier sheets, it is recommended that dual shell cooling drums averaging 460–610 mm diameter and cooled through rotary joints be used to complete the cooling system.

Post-cooling is best achieved by the use of conductive heat transfer rather than forced air cooling tunnels. By placing the embossed side of the web against the cooling drums, the embossed surface is maintained at a low temperature level, forcing heat out the back side of the product. Drum cooling also provides more uniform cooling. In cooling tunnels, any air imbalance or temperature gradient from one side to the other results in uneven cooling and a difference in the retained embossed texture of the final product.

In-line Embossing

Other embossing unit designs take advantage of web processing at elevated temperature. Examples are embossing after forming the film by calendering or extrusion and the embossing of coated products that have been processed in a heating oven prior to the texturing.

In calendering, the web comes from the calender system pickoff roll set and enters the embossing nip, which is either mounted on the calender frame face or is separately supported, and additional preheating using infrared heaters might be necessary, depending upon the distance from the pickoff rolls to the embossing section. The use of an embosser in line with a calender also provides the most stress-free sheet, since no stresses are introduced prior to embossing and cooling.

In an extrusion embossing operation, the extruded sheet comes directly from the die into the embossing nip. Most extruded flat sheets are formed by a chill-casting method. This employs a smooth chilled roll opposed either by a second roll of the same design or a rubber-covered roll. In extrusion embossing, the sheet passes into the nip of the embossing section and the residual heat of the process is used.

In embossing in-line with a coating process, a substrate is coated with a thermoplastic material and later fused in an oven. The product can be embossed directly without additional preheating. Thermoplastic coating has been applied to a substrate which may have a memory of its own, making this type of embossing different from that of calendered or extruded film embossing.

In these instances, the embossing unit operates in the same manner, except that preheating is not required. This type of embossing is often referred to as in-line, distinguishing it from self-contained embossing units, as described earlier.

NON-THERMOPLASTIC EMBOSSING

Non-thermoplastic materials are embossed in a manner similar to the thermoplastic materials except that the force between the embossing roll and the backup roll is greater, so that the elastic limit of the material being deformed can be reached. In such instances, a heated embossing roll is used, since it eases the load required to emboss the product.

The predominant factor in embossing non-thermoplastic materials is the use of male/female rolls. This roll combination includes either metal male/female rolls, a metal embossing roll operating against a filled roll, or a metal embossing roll acting against a rubber-covered roll.

The use of metal male and female embossing rolls gives the most concise and definitive texture possible. The engraved patterns on both rolls are carefully matched to allow for insertion of the respective male parts into their counterpart. The problems with this type of embossing are the costs of the rolls, since both are precision manufactured metal rolls, and also the danger that results if the two rolls come together with no material between them. This is usually controlled by the use of positive stops which prevent the rolls from crashing together in the event of a web break. The process is usually limited to the embossing of heavier materials than with the other methods.

The most commonly used combination of rolls is the metal embossing roll and the filled backup roll. This backup roll is either a saturated paper or a textile fabric composition, both of which have been densely packed to form a very solid homogeneous structure. The core is a large-diameter metal shaft over which the paper or fabric composition is placed and tightly packed. The roll is then finished to a specific diameter and when placed in the embossing machine is usually "run in" by bringing the metal embossing roll into contact with the filled roll under load. By continually running the two rolls in contact with each other, the embossing roll pattern is imparted to the surface of the backup roll, giving a male/female characteristic without the use of two mating metal rolls. This combination is excellent for thin materials as well as heavier materials, since embossing roll to backup roll contact is inconsequential with one metal roll and a composition backup roll.

The use of a rubber-covered backup roll for non-thermoplastic materials depends strictly upon the embossing load. There are certain rubber compositions, such as cast urethanes, which will act similar to a composition roll. They do not require "running in" but will deform under the load of the embossing roll, giving the resilience of a rubber covering while still allowing local deformation to occur similarly to a composition roll.

In most instances, the backup rolls are much larger in diameter for non-thermoplastic materials than those used for the thermoplastic webs. The reason is primarily because of embossing loads and wear. The larger surface gives less revolution of the roll; hence less wear of the surface as compared to a smaller roll. The larger diameter gives a better structural member to resist deflection. In some instances where embossing loads are extremely high, three-roll embossing units are used with the metal embossing roll sandwiched between two composition or rubber covered rolls so that the forces applied between rolls will cancel and allow for greater embossing force with minimal deflection. This type of embossing unit would have a fixed center roll (embossing roll) with both composition or rubber rolls moving by hydraulically operated cylinders working in opposition to each other.

VACUUM EMBOSSING

Another method of embossing, although limited in certain aspects by sheet thickness and embossing depth, is vacuum embossing.

In this technique, a heated web is placed against a roll which has female indentations and which is connected to a vacuum pump. The heated sheet is sucked into the interstices in the roll while simultaneously being cooled to give a retained pattern. The side of the film against the vacuum forming roll will receive the greatest amount of detail while the opposite side will be less defined.

Thin films are more readily embossed by vacuum method than those which are heavier. The pattern also plays a role in the finished product. The thinner film will deform more readily than a heavier film and more easily in a larger depression than in one which is small.

A rubber covered backup roll can also be used to bring the web intimately into contact with the vacuum embossing roll. Embossing force, however, is virtually nonexistent since all the action for the web deforming is in the vacuum system. The vacuum forming roll is a special design whereby positive displacement vacuum pumps are used to create the suction. The degree of vacuum is controlled by both the internal design of the roll and the piping circuitry. The roll surface has holes drilled into it to allow the air to be sucked in and ultimately to allow the web to be drawn into the depressions. Rotary joints and piping connected to the vacuum pump are used to complete this system.

If a film is overheated or too thin with regard to the amount of vacuum being drawn, it may develop pinholes or rupture. If a web is insufficiently heated or is too thick, it may not form to give the definition required in the finished product.

VALLEY PRINTING

Valley printing is a method of decorating which includes not only the embossing of a texture but also the transference of ink into the valleys of the embossed product. It is achieved by applying ink to the raised male surface of an embossing roll in either single or multiple color schemes, and the transference of that ink during the embossing or texturing process so that it is deposited into the valley of the product. It gives a printed and three-dimensional effect to the product.

Figure 23-4 shows a typical embosser/multi-color valley printer used in the decoration of a continuous web. This particular unit is used primarily in the manufacture of flooring products, especially embossed and valley printed tile.

Figure 23-5 shows a schematic diagram of how patterned applicator rolls apply ink of differing colors to select raised surfaces on the embossing roll. The construction of the embossing roll is similar to the standard roll, except that the raised male portions that are to be printed must be in the same radial plane so that ink can be appropriately applied. Raised male portions of the roll are textured with ink receptors such as small lines or cells, the same as

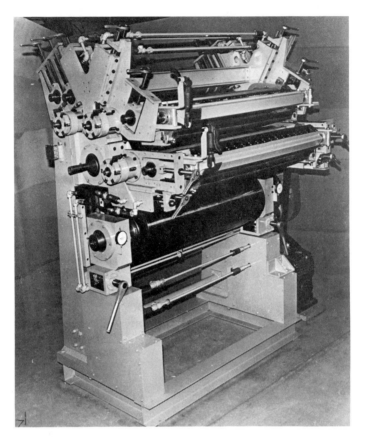

Fig. 23-4. Embosser/multi-color valley printer. (*Courtesy Liberty Machine Co., Inc., Paterson, New Jersey.*)

those used in rotogravure printing, which allows the ink to be retained and deposited in a uniform manner when the embossing pressure is exerted. If these receptors were not present, the ink would exude from the valleys and wash up onto the surface of the web.

Valley printing allows printing in register with a texture. In the case of single color valley printing, no registration is required since ink is applied to the complete raised surface of the embossing roll and later transferred into the valley of the product. In multiple color valley printing, the applicator rolls are patterned, or sculptured, so that only segments that will be printing with a particular color are left intact, while the other segments of the roll surface are cut away manually or by laser beam so as to selectively print in particular areas. This also requires that the pattern rolls be kept in register as in any printing operation. This is usually achieved by using a central gear system with registration gears mounted on each of the pattern rolls with ap-

EMBOSSING AND RELATED PROCESSES 499

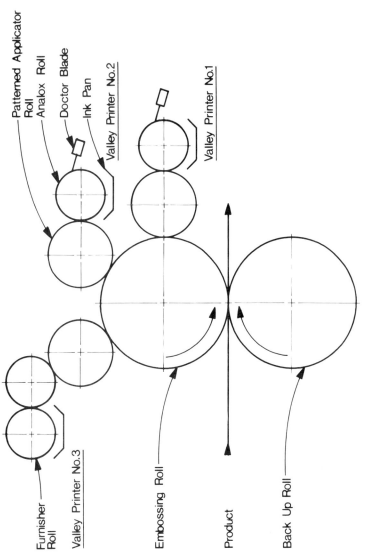

Fig. 23-5. Embosser/multi-color valley printer (schematic).

propriate registration control in both the machine and cross-machine directions.

The valley print unit features a rubber-covered applicator roll which, as noted above, is either full-faced (for single color applications) or sculptured (for selective or multi-color printing) and an engraved and chrome plated analox, or transfer, roll, which provides the ink in a metered amount to the applicator roll. Both rolls are mounted in a frame assembly which is actuated to contact the embossing roll either pneumatically or hydraulically. There is also a mechanical adjustment between the analox roll and sculptured roll to control the ink transfer process. A doctor blade system and ink pan assembly complete the printer system. Applicator roll durometer and analox roll engraving are the determining factors in the amount of ink that is applied to the embossing roll.

The embossing process for a thermoplastic web still involves the use of a preheated product and a relatively cool embossing roll. The embossing roll temperature must also be more accurately controlled since it involves the application of ink where temperatures beyond 50°C can often be detrimental and may cause drying of the ink or may change its flow properties.

Conversely, if the roll is cooled too much, moisture may condense on its surface, especially in humid conditions. Where the surface condensation might not have been detrimental in plain embossing, it will mix with the ink and cause application and ink adhesion problems.

The use of high-velocity, mass flow-type, heat transfer rather than low-temperature media is most appropriate in this situation. Embossing roll diameters as large as 600 mm have been used to further assist in the cooling process.

Floor tile or similar flooring materials require a smooth back so that contact can be made with the surface to which it is applied. If a rubber-covered backup roll is used, it will cause displacement through the back of the product, sometimes making it unsatisfactory for its end use. When embossers for the flooring industry were first provided, very hard rubber backup rolls were used (90 durometer, Shore A, or higher). This was satisfactory in many of the early textured pattern applications, but as patterns involving reproduction of slate, stone, brick, etc., came into the marketplace, a need arose for a different backup roll.

When using the rubber-covered backup rolls, the embossing roll is allowed to operate at its full depth. This means that the male portions of the embossing roll penetrate the product to the full extent of their formation. If the amount of material displaced is too great to be distributed within the sheet thickness, it will cause material to show through the back side of the product. When steel backup rolls came into use, it became necessary to provide an adjustable gapset mechanism which included a fail-safe stop system to prevent the embossing roll and steel backup roll from crashing together in the event of a web break. Since the use of the steel backup roll requires that all of the material being displaced in the embossing process be distributed through the thickness of the sheet, embossing roll design has changed to allow clearances between the full depth of the roll engraving (total depth) and the depth at which the male portions of the embossing roll penetrate the product (working depth). The clearance that results between the total depth and working depth is designed to allow for the material displacement. If the displacement is too great for the clearance, then the desired embossing depth may not be achieved. In a valley printed product, this is especially detrimental since the ink that would deposit in the valley would begin to move up the side walls, giving an undesirable shadowing, or ghosting, effect.

POLISHING

Polishing is accomplished by using a smooth, chrome plated, highly polished roll working against a rubber-covered backup roll. The same type of embossing equipment as described earlier is used with some variations, along with the appropriate conditions of preheat and cooling and with the proper formulation of material, to give a polished or patent leather look.

A good grade thermoplastic sheet should be used in polishing to give a patent leather look. This sheet is heated to the appropriate preheat temperature, activating the plasticizer so that it tends to flow to the surface, and if that surface is then contacted by a highly polished, chrome plated, smooth roll which is cooled and opposed by a relatively hard (perhaps 80 durometer, Shore A) rubber backup roll, a finish will be imparted which would give a patent leather look.

It is important that all the heat is removed from this product, since, if it were to be rewound hot, it would lose some of its polished look. It is best to cool by contacting the web on the polished, or face, side.

Likewise, if a matted embossing roll is used with a rubber covered backup roll, a dull finish will be achieved which is often required on a web which is later to be printed, since the matte surface is more receptive to the printing ink.

In the matted, or dull-look finish, uniformity of surface texture is most important. Most matte roll surfaces are either electrically or mechanically etched and are usually chrome plated to protect their surfaces.

Preheating is again a requisite, as is cooling of the matted roll. The use of conventional embossing equipment with the opposed rubber backup roll will give the desired effect. In the event the matte roll is not uniform, the finished sheet will mirror the nonuniformity.

A small decrease in thickness results during web polishing.

BUFFING AND SANDING

Many other techniques, such as buffing, sueding, brushing and sanding, are used for finishing of thermoplastic and non-thermoplastic webs. Many natural and synthetic fiber fabrics, including knitted materials, woven fabrics and non-woven materials, as well as the full complement of thermoplastic materials, including vinyls, urethanes, rubber, cork and other flexible materials, can be processed in particular machinery which imparts a surface finish to the product.

In buffing, a smooth, uniform finish to a product is imparted by using the action of a buffing cylinder and a rubber covered backup roll. Pressure is exerted between the two rolls with the web of material passing between them and with the use of differential speed between the web and the buffing roll.

Shown in Fig. 23-6 is a schematic representation of a typical unit that is used for sueding or sanding. It features an unwinding section for the web of material, a sanding cylinder and rubber pressure roll, a cleaning brush for the surface of the sanded web, and a rewind. Careful tension control is required as is extreme accuracy of the sanding and pressure roll. The sanding roll is usually a steel roll with coated abrasive material applied to it. It is internally cooled to remove the heat generated during the friction process and to prevent distortion of the highly accurate roll. The rubber pressure roll is position controlled through the use of micrometer handwheels whose adjustment is very accurate, allowing for reproducible settings during a given process.

The material is unwound by a pinch roll which rests against the rubber pressure roll, forming a drive nip. The sheet then passes between the rubber pressure roll and the sanding cylinder which is operated at a much faster speed than the surface speed of the web. The sanding effect causes the release of fibers and dust which are exhausted through a hood enclosure system. The final web cleaning is achieved by a cleaning brush, which is also placed in a hood. The material is then rewound into finished rolls.

If sueding is required, then an appropriate roll in the sanding position is used, the finish of which will determine the final web surface effect. The sueding process requires a final brushing before the desired finish is imparted.

PERFORATION

Perforating is another web texturing process which is performed to render various webs porous. Materials that can be perforated are not limited to plastic sheets but also involve thin metals, cardboard, tapes, etc.

Several perforating methods are used for web materials:

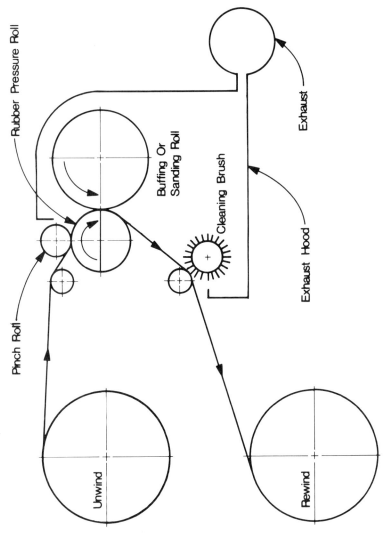

Fig. 23-6. Buffing/sanding machine.

- Die punching
- Piercing
- Melting.

Perforating of continuous webs by die punching is usually done on rotary perforators. The minimum hole size is limited, since die punching requires removal of material, which becomes exceedingly difficult with decreasing hole size. If the operation is carried out correctly, die punching give openings with clean edges. Die cutting is discussed in detail in Chapter 22.

Figure 23-7 shows the machinery for perforation by piercing which consists of opposing rolls. The perforating roll is a pinlike roll with protrusions which pierce the web as it passes between the nip of the perforating roll and a rubber backup roll.

In processing both thermoplastic and non-thermoplastic webs, the material is unwound from its roll, laid flat by the use of nip and spreader rolls, and applied to the surface of the perforating roll by using pinning wheels. It is perforated by pressure exerted from the backup roll through the web, causing the perforations to result. The web is finally stripped away from the perforating roll and is rewound.

The perforator itself is essentially an embosser which, instead of making depressions in the web, perforates the material. In the processing of non-thermoplastic materials, the shape of the perforating protrusions, the pressure exerted between the two rolls, and the durometer of the backup roll dictate the size and final shape of the perforation. There is no need to use any heat in this application.

Perforating by piercing is also described by Chandler.[1] It is possible to obtain as many as 75 punctures per square centimeter by repeated passes

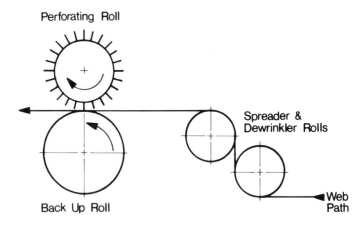

Fig. 23-7. Perforating machine.

through the perforator. Pressure-sensitive tape is also perforated by a piercing method.[2]

Many different perforating methods involving melting and a combination of melting and piercing have been developed and are used for specialized applications. The great interest in perforated products has been in pressure-sensitive adhesive tapes and bandages for applications against the skin, sweat bands, upholstery materials and similar products where the porosity improves the product utility.

Perforating of pressure-sensitive tapes by use of a heated pattern roll and a smooth anvil roll has been used in various combinations.[3,4,5,6,7] Porous pressure-sensitive tape has been prepared by perforating through the adhesive only.[8,9]

Thermoplastic sheet has been perforated by contacting the sheet with a perforated cylinder and deforming the film into the perforations by fluid pressure. The film ruptures when the deformation exceeds the ultimate elongation.[10] A similar method is used by preheating the sheet.[11]

Polyethylene terephthalate and other thermoplastic films can be perforated by placing the film over a steel roll with indentations on the surface. Heat is applied to the film by a gas burner, causing the film to melt in the area of indentations, but not in the area where the film is in contact with metal surface because the heat is conducted away by the steel.[12,13]

TRANSFER PRINTING

Heat transfer printing consists of several different processes. These processes are:

- Hot stamping
- Hot transfer of preprinted pattern
- Hot transfer of designs printed with sublimable inks.

Hot stamping or hot transfer of preprinted decals is usually done on an individual basis rather than continuously on a moving web. The method usually requires a time/temperature relationship where the heat and pressure of the application causes the transfer, or stamping, process.

The hot transfer of designs printed with sublimable inks was originally developed to apply printed dyes onto fabrics. Later, the process developed into the application of printed patterns onto thermoplastic and other types of webs.

In the heat transfer printing of dyes onto fabrics, the key feature is the use of sublimable dyes which turn directly from the solid state to the gaseous state when heated above a certain temperature. This temperature is the same temperature at which the dye will fix itself permanently onto the fiber structure of the fabric.

Typical temperatures for fabric processing may range from 200°–220°C, depending upon the type of dyes being used. Dwell times vary from approximately one-half minute to one minute, depending again upon fabric weight, speed and the type of dyes.

The dyes are printed onto a release paper, usually by a rotogravure printing method. The paper is rewound and disposed of after it has been used. Transfer printing is an easier method to print a product than the direct printing.

In using sublimable dyes, the complete fiber is colored rather than surface printed as you would in conventional rotogravure processing. This makes for rich, deep tones. It enables complete coverage of the fabric web with no distortion of the pattern, since the transfer was made by intimate contact of the transfer paper and the product.

Figure 23-8 shows a typical transfer unit used to apply a sublimable dye to a fabric. It generally consists of a large drum which can be heated either electrically or by hot circulating oil (the latter method is preferred).

Usually an oil heated, main center drum of approximately 180–250 cm diameter, double-shell constructed, is used. Circulating hot oil is pumped into the center shaft of the drum through stainless steel hosed rotary joints. The rotary joints are of a high-temperature type since the circulating oil may be at temperatures approaching 245°–260°C. The oil passes from the center shaft to the annulus of the double shell construction through spokes or hoses. The annulus is spiralled so that the oil flows through passages in a prescribed manner. Pumping rate and flow velocity are important since turbulence is desirable for maximizing heat transfer. The oil then exits the opposite end of the drum and returns to the heating and pumping system for reprocessing as in any ordinary closed-loop system. The surface of the drum is smooth, and although not necessary, may be chrome plated for prevention of rust. The bearings which support the drum should be of a high-temperature design using either special high-temperature greases, or, preferably, a circulating oil lubrication system which includes a heat exchanger and cooling section.

The drum is wrapped with a continuous, high-temperature blanket which may be made of material such as Nomex® so that it can withstand the temperature required of the process. The blanket is tensioned and guided through appropriate hydraulically controlled tensioning devices and automatic guiding equipment. The unit also includes an unwind system for the fabric, appropriate guiding and tension control equipment to deliver the fabric to the transfer heating drum, and a rewind for the finished product. The transfer paper has an unwind section with tension control and guiding

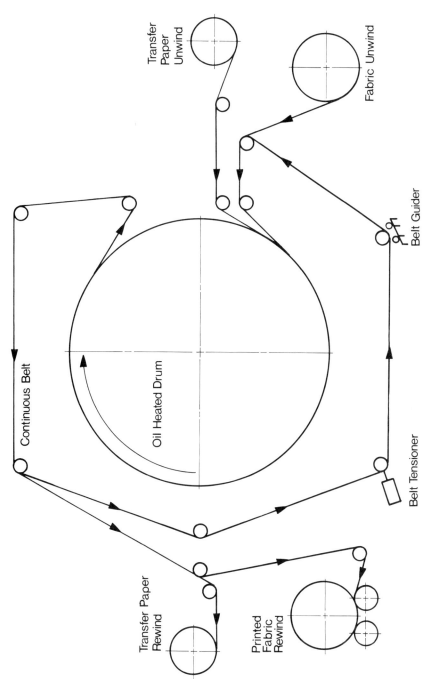

Fig. 23-8. Fabric heat transfer unit.

equipment as well as appropriate lay-on and combining roll system to bring the web to be printed and the transfer paper into intimate contact. Also included is a rewind section for the exhausted paper. All the components are driven through an electrical drive system with appropriate controls.

The key factor is the dwell time at temperature to effect the transfer. One would want to use the minimum temperature and dwell time at which the dyes will sublimate and penetrate the fiber.

The importance of tension control and guiding of both the product to be printed and the heat transfer paper prior to their combining on the center drum cannot be overemphasized. Once the two items have been brought together at the drum, they are held in intimate contact by the continuous blanket, enabling the process to take place.

Once the product has been printed, it can be textured in any typical embossing machine.

Dry printing by the heat transfer method on fabrics gave rise to a similar process for thermoplastic webs.

In this type of process, paper is coated with a heat-sensitive coating which is later printed, usually by the use of rotogravure printing equipment. Since the printing is done on a non-extensible paper web and under controlled temperature and humidity conditions, it allows for the finest in color separation and accuracy that gives a higher quality print than possible by the wet printing process. It is particularly desirable to use the transfer printing process when dealing with stretchy thermoplastic webs that are dimensionally unstable.

Figure 23-9 shows a typical method by which a heat transfer paper can be used to transfer print to a thermoplastic substrate.

The complete machine consists of an appropriate substrate unwind complete with tension and guiding controls to deliver the web to the transfer section; a preheating section which can be either a radiant type, convection type or a drum type heater; the transfer nip assembly, which includes a heated steel roll that can be either steam or hot oil heated, opposed by a rubber backup roll which can be loaded pneumatically or hydraulically against the steel roll to effect the transfer. The product is then cooled and rewound into a finished roll. The machine also includes the transfer paper unwind with its tension and guiding controls, as well as a scrap paper rewind. It may also require additonal infrared heating to preheat the transfer paper, activating the release coating and facilitating the transfer at the point of contact with the substrate.

In transferring to a thermoplastic sheet, temperatures for the transfer may vary from 105°-180°C. It is especially important to note that some thermoplastic films exhibit a very low distortion temperature. Therefore, it is necessary to heat that substrate slightly to 65°-80°C, usually by contact

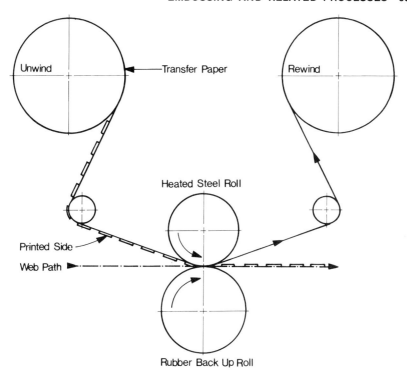

Fig. 23-9. Thermoplastic heat transfer unit. (*Courtesy Sublistatic Corp. of America, New York.*)

heating, while applying the heat to release the print from the transfer paper by using the infrared heating system for that section. Formulation of the release coating on the transfer paper plays an important role when processing materials requiring low-temperature application. When transfer printing a polyvinylchloride film, it can be heated to temperature of 135°–150°C without distortion problems. In this case, preheat of the heat transfer paper is not needed, since the temperature of the film contacting the heat transfer paper may be enough to cause the transfer of the print to the substrate.

The temperature of the steel roll and the amount of wrap of the transfer print paper to that roll is also important. The roll is heated; therefore, it brings the heat transfer paper up to temperature to cause the release of the print. When this print is released, it is actually a film of ink so that the transfer is similar to that of transferring a decal onto a surface. The amount of dwell time of the heat transfer paper on that steel roll and the angle at which the exhausted paper comes away from the print may be critical in certain applications.

If the substrate is to be subsequently embossed, the retained heat from the process, along with additional radiant heat prior to entry into an embossing nip, is all that will be necessary to conclude the operation.

Cooling the substrate prior to winding is again an important factor, and the use of cooling drums or rolls is recommended.

Transfer printing often covers many surface profile defects in thermoplastic sheets which would not be covered if conventional web decoration techniques involving wet printing were used.

Those parameters that are most important in the transfer of print from a release paper to a thermoplastic substrate include the temperature of the substrate and its preheating drum or medium, the temperature of the roll around which the transfer print paper is wrapped for its preheat and transfer purposes, the amount of dwell time between the heat transfer paper and the transfer roll, the transfer nip pressure where the print separates from the heat transfer paper and is applied to the substrate, the tension control between the substrate and the heat transfer paper and, finally, the speed of the web being processed.

Temperature at the transfer point nip should be as close as possible to the softening temperature of the substrate when it is most receptive to the print. The force, or transfer nip load, most suited to effect the transfer from the release paper to the substrate may vary between 10 and 20 kg/linear cm. The speed of the process will depend in part on the thickness of the substrate and its heat load characteristics, and also on the ability to heat the transfer paper so that the proper release can be affected.

REFERENCES

1. Chandler, F. J. U.S. Patent 2,081,219 (to Perfotex Co.) (May 25, 1937).
2. Scholl, W. M. U.S. Patent 3,170,354 (February 23, 1965).
3. Hannauer, G. Jr., Grimes, E. M. and Schaar, C. H. U.S. Patent 3,214,795 (to Kendall Co.) (November 2, 1965).
4. Schaar, C. H. U.S. Patent 3,073,304 (to Kendall Co.) (January 15, 1963).
5. Schaar, C. H. U.S. Patent 3,088,843 (to Kendall Co.) (May 7, 1963).
6. Hannauer, G., Jr., Grimes, E. M. and Schaar, C. H. U.S. Patent 3,243,488 (to Kendall Co.) (March 29, 1966).
7. Schaar, C. H. U.S. Patent 3,214,502 (to Kendall Co.) (October 26, 1965).
8. Blackford, B. B. U.S. Patent 3,085,572 (to Johnson & Johnson) (April 16, 1963).
9. Blackford, B. B. U.S. Patent 3,161,554 (to Johnson & Johnson) (December 15, 1964).
10. Raley, G. E. and Adams, J. M. U.S. Patent 4,252,516 (to Ethyl Corp.) (February 24, 1981).
11. Zimmerli, W. F. U.S. Patent 3,054,148 (September 18, 1962).
12. Schaar, C. H. U.S. Patent 3,012,918 (to Kendall Co.) (December 1961).
13. MacDuff, R. U.S. Patent 3,394,211 (to Kendall Co.) (July 23, 1968).

24
Static Electricity

Joseph J. Keers
3M Company
St. Paul, Minnesota

Static electricity can be a major problem in the web processing industry. Solvent coating, laminating and extruding processes are all affected by static electricity.

There are five major static related problems associated with web processing. These are:

- Dust and lint
- Fire and explosion
- Degraded product
- Handling problems
- Personnel shocks.

Static electricity on a web of paper or film will attract dust and lint to the web and cause defects in coating or contaminate the film. Arcing or sparking of static electricity has enough energy to ignite solvent vapors resulting in an explosion and fire. Destroyed or degraded products due to static most often occur in the electronics industry when working with transistors, but it can also cause defects in release paper or puddling of an adhesive coating. Handling and personnel shock problems are not as common in web processing, but can be very serious when they do occur. If a machine operator gets zapped while working on the equipment the involuntary reaction to the shock could cause a serious accident.

Although static electricity is a very common phenomenon, until recently, little research has been devoted to it when compared to other fields of electricity. The reason for this is that static charge very often causes problems, and we want to eliminate it. Flowing electric currents, on the other hand, are very useful and have been studied extensively.

A static charge is generated whenever two materials come together and then separate. This is called tribo-electric charging or contact electrification. The word *tribo* is from the Greek word for rubbing; hence, we have electricity generated by rubbing or the friction associated with many contacts and separations.

Although the actual mechanism is not fully understood, it is generally assumed that when two materials contact, and separate, one material has a greater affinity for electrons and the electrons transfer to that surface. The surface that acquired the electrons assumes a negative charge, and the other surface has a positive charge because it lost electrons. Figure 24-1 illustrates this principle of tribo-electric charging.

This phenomenon occurs in all material to some degree, depending on several factors such as the intimacy of contact, speed of separation, and conductivity of the material. If two very smooth surfaces come together, the total number of points of contact are greater than when rough surfaces are in contact. This means more charges can be generated because of the greater intimacy of contact. The faster the speed of separation, the greater the generation of charge since a slow separation allows the electrons a chance to leak back to the original surface.

Conductivity plays a dual role in static charging. Since the charges are free to move on a conductive surface, they tend to distribute themselves over the entire surface rather than stay localized where they were generated. This spreading of the charge over the surface results in a lower voltage because the charge density is lower. The second factor is that because it is conductive,

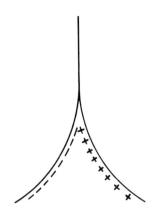

Fig. 24-1. Static charge generation.

any part of the material that comes in contact with an earth ground will immediately allow all the charge on the material to drain to ground.

Over the years scientists have developed tests to determine how certain materials charge relative to other materials. A list called the tribo-electric series shown in Table 24-1 gives some of the more commonly used materials in web processing. Any material above another in the list takes on a positive charge, relative to that material, and the other a negative charge. Therefore, depending on which material on the list it comes in contact with, it can become either positively or negatively charged. The farther apart the materials are on the list, the higher the level of charge. This is an empirical list that can be used as a guideline in determining how susceptible two materials will be to static charging.

CONDUCTIVE VERSUS NONCONDUCTIVE

Whether a material accumulates a charge depends on its conductivity. Conductive materials lose their charge rapidly when they come in contact with a ground. Non-conductive materials retain their charge because even though they might come in contact with a ground, the charge cannot move across the surface to the grounded point.

In practice we rarely measure the conductivity of a material, but rather its resistance. Resistance is a measure of the difficulty that electrons have in flowing through a material. It is measured in ohms and a highly conductive material such as a piece of copper wire may have a resistance of the order of one-millionth of an ohm. A non-conductive material such as a sheet of polyethylene would have a resistance across its surface of over one billion ohms. Based on its resistance, a material can be classified as either a conduc-

Table 24-1. Triboelectric Series

Asbestos	Steel
Glass	Sealing wax
Human hair	Hard rubber
Nylon	Acetate rayon
Wool	Nickel-copper
Fur	Brass-silver
Lead	Synthetic rubber
Silk	Orlon
Aluminum	Saran
Paper	Polyethylene
Cotton	Teflon
	Silicone rubber

tor, semi-conductor or non-conductor. Static problems are usually associated with materials classified as non-conductors.

There is a class of materials called antistatic because they do not tend to accumulate static charge. They are usually non-conductive materials that have been treated with a chemical to make them slightly conductive. This chemical is most often a quaternary salt that is applied to the surface. The salt will absorb moisture from the air, even at relatively low humidity, and decreases the resistance across the surface of the material. Other chemicals are added to the bulk of the material and bloom to the surface, which also absorb moisture, allowing the static charge to drain away.

STATIC ELIMINATION

It is very easy to generate a static charge, but eliminating the charge can be difficult. There are three basic methods for neutralizing a static charge. If the material is a conductor, simply connect it to ground and drain the charge away. An example would be the grounding of a person in a solvent coating operation by the use of conductive flooring and conductive footwear or grounding straps.

If the material is a non-conductor, one solution to the problem is to make the material conductive by adding an antistat to it. As mentioned before, it is not always possible to make a non-conductor conductive, since it may have an adverse effect on the material, such as reducing the adhesion of a coating to a web of plastic film. A second method is to add to the charged surface a charge of the opposite polarity. This can be readily accomplished by using ionized air.

Ionized air is normal room air that has been broken up into positive and negative ions. Since positive and negative charges attract each other, ionized air has a very short life, lasting only fractions of a second, so a supply must be continuously generated. The appropriate ions are attracted by the charge on the surface and neutralize that charge. The generation of ionized air requires energy. The source of this energy is the chief difference between the various methods used to ionize air.

There are three types of air ionizers:

- Induction
- Electrically powered
- Nuclear powered.

INDUCTION NEUTRALIZERS

Induction type static eliminators obtain the energy required to ionize air from the charge on the material itself. The device typically consists of a series of sharp needle-like points, as shown in Fig. 24-2. The charge on the surface of the material induces a high charge onto the sharp point. This high concentration of charge on such a small surface exerts a force on the air surrounding the point. The intensity of this electrical field causes the air to ionize, and the strength of the field required to start the induction ionization process is quite high. A minimum of 4000-5000 volts is necessary for a well-designed induction needle system. On a poorly designed unit as much as 10,000 volts may be needed to start the process. Below the threshold (starting voltage) no ionization occurs. As the voltage increases above the threshold, the amount of ionized air increases. If the voltage on a moving web is 30,000 volts, more than enough ions are generated to neutralize the charge on the web. This type of static eliminator is independent of the speed of the material passing by the device; only the voltage on the charged surface determines the level of ionization.

Induction type static eliminators come in a variety of sizes and shapes. The most common is a series of needle points mounted on a bar connected to ground. Tinsel, another form of an induction type neutralizer, is metallic and can be grounded. Tinsel requires as much as 10,000 volts on the web before it becomes effective. As a general rule, induction type devices are most effective in eliminating the problem of shocks to personnel but cannot be depended upon to solve other kinds of problems.

ELECTRICALLY-POWERED NEUTRALIZERS

Electrically powered static elimination equipment is made up of two basic components: the static eliminator or ion-producing source and its high-

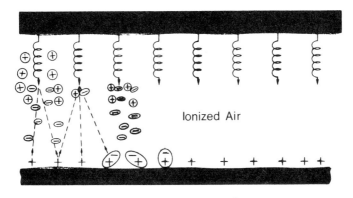

Fig. 24-2. Induction device.

voltage power supply (Fig. 24-3). The neutralizing portion generally consists of one or more electrified needles that are rigidly held less than an inch from a grounded metal housing or proximity ground rod. High-voltage supplies of 5000-8000 volts ac are used to power units of this type. On neutralizers where no housing or proximity grounding is used, power supplies of as high as 15,000 volts are necessary. Even with increased supply voltage, the ionizing capabilities of these devices are only a fraction of those having proximity grounds. Ion generation from electrical static eliminators occurs in the air space surrounding the highly charged needle points. The mechanism of ion production is just the reverse of that of the induction type neutralizer. Instead of the required charge concentration being induced in the needle tips, it is continuously supplied by the high-voltage ac power supply. The electrical field surrounding the body being neutralized then attracts the charge-balancing ions. The ion supply from an electrically powered static eliminator, unlike the induction device, is limited. As higher web speeds are reached, the device's neutralizing efficiency drops off.

The neutralizing needles may be connected to their power supplies either capacitively or directly. Direct coupling, in which the ionizing needles are attached directly to the high-voltage supply, results in what is termed a hot bar (Fig. 24-4a and b). If any of the needles on a hot bar are accidentally arcing from one of the needle tips to ground, the supply voltage will drop instantly, and the bar can become ineffective. Hot bars are not desirable in areas where personnel are apt to touch the needles or electrodes. This is especially true when they must be used in the vicinity of dangerous machinery. Involuntary muscular responses resulting from accidental shocks could result in serious injury.

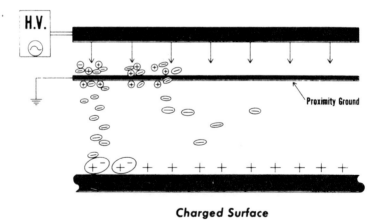

Fig. 24-3. Electrically powered device.

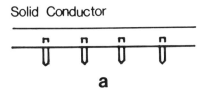

Fig. 24-4a. Direct coupled high voltage.

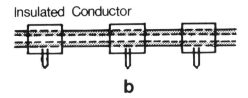

Fig. 24-4b. Capacitively coupled high voltage.

By capacitively coupling the needles and power supply, a shockproof bar is produced. In this case the needles are each imbedded in an individual conductive sleeve surrounding the high-voltage transmission cable. The cable conductor acts as one plate of a capacitor, its insulator as the dielectric and the outer sleeve to which the needle is attached is the other plate (Fig. 24-4b). The shorting or arcing to ground of a needle on a shockproof unit will not appreciably affect the efficiency of the rest of the bar. If someone accidentally touches a needle, they will receive at most a sort of pinprick discharge to the skin. The use of electrical static eliminators of either the hot or shockproof variety in atmospheres containing explosive vapor concentrations or dust clouds should be avoided.

NUCLEAR-POWERED NEUTRALIZERS

Nuclear-powered static eliminators create ions by the irradiation of the air molecules in the immediate vicinity of the object being neutralized. These devices are generally composed of a nuclear source secured inside a metal housing and covered with a protective grid or screen (Fig. 24-5). Their ionizing zone, or cloud, unlike those of induction or electrical type devices, is spread out over a very large area.

Alpha-emitting isotopes are commonly used as the nuclear source. A high-speed alpha particle can influence a nearby air molecule with sufficient energy to actually strip off one of its outer electrons. This forms a positive ion. The loose electron can then attach itself to a second molecule creating a corresponding negative ion.

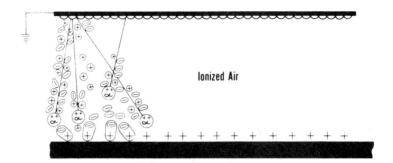

Fig. 24-5. Nuclear powered device.

Three different isotopes have been used to make industrial static elimination equipment: americium-241, radium-226 and polonium-210. Americium-241 has a relatively long half life of 458 years. It is an alpha emitter; however, it also gives off a significant amount of gamma radiation. The latter can be deleterious to human tissue over extended periods of time. Radium-226 has an extremely long half life of almost 1,700 years. Like americium-241, its alpha emissions are accompanied by harmful gamma rays. Polonium-210, however, is virtually a 100% alpha emitter and as such is not externally hazardous. It has a radioactive half life of 138 days and decays directly to safe, non-radioactive lead. One year is considered to be the useful life of a static eliminator containing polonium-210. By the end of this time the isotope has decayed to about 16% of its original strength. These devices are renewed annually by their manufacturers according to a lease arrangement.

There are two methods of source fabrication for static neutralizers on the market today: the rolled foil type and the microsphere type. In the rolled foil source the isotope is sealed between a thin layer of gold and a relatively thick silver backing. The isotope in the microsphere type is chemically bonded and physically sealed into small ceramic beads which are then nickel plated. The finished microspheres are then bonded securely onto an aluminum substrate.

COMBINATION BAR

The combining of both the nuclear and induction types of static eliminators into one housing results in an all-purpose device—a combination bar. The different methods of ionization complement each other in such a way as to eliminate the drawbacks of each other. A unique advantage of this type of eliminator is that there is no limit to the number of ions produced. The more charge the device sees, the more ions it will produce. That is, the higher the voltage on the surface or the faster the material goes past the device, the greater the number of ion produced.

STATIC-CAUSED PROBLEMS

Using ionized air to neutralize static charge requires an understanding of electrostatic fields and how they are affected by the surrounding environment. The capacity of a web to hold charge varies with the type of material and its electrical capacitance. The strength of an electrostatic field is dependent on the amount of charge and the capacitance of the material. For the same amount of charge, a low-capacitance material exhibits a strong field. If the capacitance is increased, the field strength decreases. It is the strength of the electrostatic field that determines the severity of the static-caused problems. The strength of a field is usually measured in thousands of volts per meter of distance from the charged surface. Many instruments that measure static charge are calibrated for a fixed distance from the surface and indicate the field strength in thousands of volts.

Some of the static caused problems and the associated voltage level are given below.

- Fire and explosion—1500 V
- Dust attraction—2000 V
- Handling problem—4000 V
- Discharge/arcing—7000 V

Because capacitance has an effect on the electrostatic field, the placement of a neutralizing device is very critical. If a charged web comes near or in contact with another surface such as a roller or guide, the electrostatic field collapses in that area, and the field strength is greatly reduced. The reason is that the effective capacitance of the web increases to the point where the voltage on the charged surface is practically zero. Since an electrostatic field is necessary to attract ionized air to the charged surface, a collapsed field will not attract any ions. The collapsing of the field does not affect the quantity of charge on the web. As soon as the web leaves the area of increased capacitance, the field will rise again to its previous strength. Figure 24-6 illustrates the positioning of a bar-type device near a roller or other surface that will effectively increase the capacitance of the web. The optimum position of a bar-type device is 2.5 cm above or below the web at a point that is 15 cm away from rollers, guides, plates, machine framework, etc.

Neutralizing static charge with ionized air is not a permanent solution. It

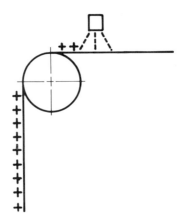

Fig. 24-6. Position of device near a roller.

merely provides a positive or negative charge to offset the charge on the web. Should the neutralized web come in contact with and separate from another surface, a charge can be regenerated. This means that static should be eliminated where it causes a specific problem and ignored if it is not causing a problem. It might require the application of several bars on one piece of equipment to solve all the problems or sources of static.

It is necessary to neutralize the charge on only one side of the web to control it on both sides. Generally the charge is generated on only one side, but its effects appear to be on both sides. By providing an oppositely charged ion to either side will neutralize the charge on both sides. There are some instances, because of multiple layers or coatings, where the charge must be removed from both sides. These situations can be identified by using a static meter to insure proper treatment.

FIRE AND EXPLOSION PROBLEMS

Of the five major problems caused by static electricity, fire and explosion are the most critical for the web processor. Many coating and laminating operations use flammable solvents as part of a coating or adhesive. A static discharge in a solvent-laden environment can cause an explosion and fire.

Three conditions are necessary to start a fire. A source of ignition, oxygen and a flammable material. In coating operations electrostatic discharge is the source of ignition. Discharge occurs whenever the electric field associated with the level of charge on the web is strong enough to overcome the electrical resistance of the surrounding air. This usually happens when the web comes close to a grounded surface such as a roller or the frame of the machine. If the difference of potential between the charged web and the grounded surface is sufficient, the air will break down—forming an arc or discharge. The arc can have sufficient energy to ignite a solvent vapor, and this will cause a fire if the other conditions of the right amount of solvent and oxygen are also present.

Having the right amount of oxygen to support combustion usually isn't a problem, but if the amount of solvent vapor is either too high or too low, a fire will not occur. The lower explosive limit is approximately 7% by volume for most common solvents. Less than this is too low a level of solvent to ignite. Under most conditions, the upper explosive limit is not reached.

Eliminating the static charge eliminates the source of ignition and reduces the possibility of fire. A good ventilation system to reduce the concentration of solvent vapor is also recommended. The combination nuclear and induction static eliminator is recommended because it can handle both high-voltages and high-speed webs, and it doesn't have the limitations of the induction or the electrically powered devices. Electrically powered devices can cause an arc due to the high potential applied to the needle points.

It has been demonstrated that voltage levels as low as 1500 volts on a web can ignite certain vapors. This is well below the threshold voltage of even the best induction neutralizers.

Also, the combination nuclear and induction type handles all voltage levels and cannot generate an arc itself when properly grounded. The device should be placed at a point just prior to where the web enters the area of solvent laden air. This is usually after the last roller before the web enters the coating station.

DUST AND LINT PROBLEMS

Dust and lint problems are also of great concern to web processors. Products such as magnetic tape and photographic film are particularly susceptible to dust and lint particles on the base material. A strong electrostatic field can attract airborne dust particles as much as 15-25 cm away from the surface.

Removing static charge eliminates the attractive force and prevents dust and lint from being attracted to the web. The neutralization of the charge does not remove the dust that has already been deposited on the web. This must be removed by other means such as ionized air nozzles or a wiping system, depending on the size of the particle to be removed.

Products such as magnetic recording tape for computer or video applications are seriously affected by dust particles as small as $1-2\mu$ in size. When coated over by a layer of magnetic oxide, the particle size is magnified, causing a raised spot in the coating that results in a lost or imperfect electronic signal in the final product.

A similar problem occurs in the coating of photographic emulsions. Extremely small particles of metal can be attracted to the base material. These particles can react chemically with the photosensitive coating, leaving marks and spots much larger than the original particle size.

Another area where dust and dirt are critical is in the production of plastic bags for intravenous solutions. These bags hold pharmaceutical solutions that are injected into the human body. It is absolutely critical that the plastic webs be clean before the two layers are laminated together to form a bag.

In all the above applications, a static eliminator should be used on the base material as it unwinds from the roll. By neutralizing this charge one cause of contamination is eliminated. Dust that is on the web before the process is started or dust that falls on the web due to gravity should be removed by a suitable web cleaning system. However, if the method of cleaning involves contacting the web, a static eliminator must be used after the cleaning.

A safe voltage level for the prevention of dust attraction would be less than 2000 volts. Electrostatic fields below this level generally do not have enough force of attraction to act on particles that are an inch or more away from the web.

DEGRADED-DESTROYED PRODUCTS

Electrostatic discharges can burn pinholes in the release coating of a carrier web. When the product is peeled from the release paper, parts will stick to the paper because of the static-caused pinholes. The carrier web should be neutralized after the release coating is applied and also when the web is unwound during the product coating operation.

Light-sensitive coatings are affected by the blue arc of a static discharge. This can occur almost anywhere the unprocessed photographic film is handled—in manufacturing, in the camera, and during processing. Neutralization is most practical during manufacturing and processing. Electrically powered devices usually cannot be used for this application because the corona discharge, associated with them, can fog the film. Nuclear devices are very effective, but care must be taken to place the device facing the base side of the film and the film should not be allowed to stand still in the presence of the device. Most films have about a 5-sec safe period. As long as the film is moving as little as 5 ft/min, no noticeable fogging will occur.

HANDLING PROBLEMS

Handling problems usually imply a loss of control over the movement of the product during processing. Generally speaking, this is not a problem with web processing because the web is under tension and the force of attraction caused by the static charge is not sufficient to overcome the tension on the web. However, when the web is converted into sheets and the material is being pushed rather than pulled through the process, control of the product can be a serious problem. Sheeting, jogging, stacking, collating and a variety of other operations where the material is not under positive control are affected by static. The neutralization of the charge, at a point just prior to where the problem occurs, will solve these problems.

Ballooning is a handling problem that can happen when webs are under tension. It occurs when two webs are brought together in a laminating process. If charges on the webs have the same polarity, they repel each other, causing the web to move away from each other. Again, a static eliminating bar placed across each web eliminates the charge and stops them from repelling each other.

PERSONNEL SHOCKS

If a static charge discharges to or from a person, the individual feels an electrical shock. These shock can be mild like the tingle you feel when touching a doorknob in the winter, or very strong such as a 2" arc from a roll of film charged to 50,000 volts.

Personnel shocks are usually considered a nuisance; however, in some instances, they can become a real safety problem in web processing. Quite often, a person can build up an induced charge from a web which will arc from the individual to any grounded area. If this occurs in the presence of a flammable solvent, it can cause an explosion and fire. In the manufacture of rubber products, such as V-belts, static discharge from the machine operator can be a serious problem. If an operator receives a shock from a web or roll of material while working around the machinery, injury can happen because of the involuntary reaction to the shock.

To remove the charge from the web or roll of material, an induction needle bar is all that is needed if the only problem is personnel shocks. If an explosion hazard exists, a nuclear or combination device is required since arcing can occur at voltages below the threshold of an induction static eliminator.

Ionized air blowers or conductive floor mats are recommended for the neutralization of charge on a person. The mat will ground the body through most footwear, except heavily insulated soles such as rubber or cork. In this instance, a specially designed shoe strap can be used to connect the person to ground. A shoe strap is a small ribbon of conductive plastic that is placed inside the shoe and runs around the outside the shoe and is adhered to the bottom of the heel. It will make electrical contact through the person's stocking, and doesn't have to be in contact with the skin.

An ionized air blower floods the area with ionized air to neutralize the charge on the person's body before it builds up to a level that can discharge.

Subject Index

Italics indicate a major source of information

ABS, 107
Absorptance, infrared radiation, 274
Absorption, infrared radiation, 274-275
Accelerated electrons, *See* Electron beam
Acceleration
  electron current, 342, 345
  electron voltage, 342, 345
Accelerator, electron beam, 333, 336-340, 346
Accumulators, 391, 394, *425-429*
Acid etching engraving, 31
Activated carbon bed, 308
Adapters, extruder, *129-140*
Adhesion promoters, 124, 125, 235
Adhesives, radiation curable, 29
Adsorption, gas, 308-309
Air
  blowers, 305-307, 356
  dampers, 307
  dryers, *288-302*
  drying, *250-271*, 281
  ductwork, 307
  entrapment, holt melt, 171
  flow, 256, 259, 301, 305
  greased bar, 144
  heating, 302-303
  horsepower, 261
  impingement, *255-262*, 359
  ionizers, 514, 515
  moving, 301, 305
  nozzles, *255-261*, 264-267, 296-297, 356
  recirculation, 301, 308
  velocity, *262-264*
Air-knife coaters, 34, *54-59*, 103, 219
Airless spraying, 97, 100

Allowable flammable solvent concentration, 310
Aluminum coating, *200-204*, 209, 211
Analox roll, 500
Analyzer, combustible gas, 308-310
Annealing rolls, 354, 355, 359
Applicator roll, 35-37, 57, 75, 500
Arch dryer, 140, 288, 294
Artblade, 44-45, 47
Atactic polypropylene, 147, 184
Atomization, *97-98*
Atomizers
  air, 97-100
  bell, 103
  electrostatic, 98, 100, 102
  hydraulic, 100
  impingement, 98, 100, 101
  rotary disk, 102-103
  swirling nozzle, 100, 101

Back ionization, 179
Backlash elimination, 222
Back scatter sensors, 430
Back-up roll, 224, 226, 485, 486, 488, 489, 490, 494, 496, 499, 500
Bar coaters, *68-72*
Bar, Mayer, 68, *70-71*, 219
Bar scrapers, 219, 220
Bar, smoothing, 22, 71-73
Bar spreaders, 381, 414
Barrier coatings, 161, 168, 200-204, 209, 210
Bearings, 115, 116, 295
Belt delivery section, diecutting, 480

SUBJECT INDEX

BEMA coater, 167
Bending, roll, 116–118, 493
Bent bars, 414
Bernoulli principle, 266
Beta gauge, 428
Bevelled blade coater, 35, *38–41*, 48
Blade coaters
  applicators for, 35–38
  Artblade, 44, 45, 47
  Billblade, 51, 52, 53
  bevelled blade, 35, *38–41*, 48
  blade-roll, 51
  Chilply, 52, 54
  Combiblade, 44, 46
  Constacoat, 44, 45, 46
  constant blade tip angle, 44, 45
  inverted, 35
  low angle, 35, 38, 41
  puddle, 48, 49
  rod-blade, 45, 48
  S-matic, 44, 45
  Sym-Lam, 50
  two-blade, 50, 51
  Twinblade, 50
  Vacply, 48, 49
Blade coating, 10, *34–54*, 55, 66
  angle, blade, 39, 40, 41, 43
  design, blade, 65–66
  pneumatic loading, 43
  pressure, 40, 42–45, 47
  theory, *38–42*
Blank coaters, 169, 170
Blistering delamination, 231–233
Block copolymers, 147, 152, 169
Block die, 482
Blowers, 305–307, 356
Bonding, 29–31, 224–225, 228–235, 237, 238–240
Bowed rolls, 381, *414–417*, 444
Brakes, 368, 395, 417, 440–441
  disk, 374–375, 441
  drag belt, 395, 397
  drum, 373–375, 395, 397
  magnetic, 395
  mechanical, 373
  pneumatic, 140
  regenerative, 140, 373, 441
  unwind, 373–375
Brush coater, 34, 78
Brushing, 502
Bubbles, adhesive coating, 232, 236

Buffing, 502–503
Bull nose knife, 65–66
Burner, catalytic, 282, 284
Burners, gas. *See* gas burners
Burst cut, 380
Butt splice, 390
Butyl rubber, 111, 169

Calender drive, 119
Calendering, 2, 77, *105–121*, 165, 213, 495
Calendering, friction, 112
Calender lamination, *107–110*
Can dryer, 268
Caratsch process, 175–176
Carton blanks, 169, 170
Cast calendering, 77
Cast coaters, 76–78
Cast embossing, 77
Catalytic burner, 282, 284
Catenary oven, 266, 288, 293
Cathode ray tube, 336
Cellophane, 124, 184, 209
Center winders, 389, 422
Champion-Hamilton coater, 12
Charge neutralizers, *514–519*
  electrical, 515
  induction, 514
  nuclear, 514
Chilled iron rolls, 114
Chill rolls, 140–142, 144, 195–196, 238
Chilply coater, 52, 54
Choker bars, 136, 139
Chucks, 369, 370, 371, 388, 440, 441
Clam shell oven, 301
Clay coating, 76, 94, 95, 103
Clicker die, 461
Clutch, 374, 422
Coathanger die, 131
Coaters
  air knife, 34, *54–59*, 103, 219
  bar, *68–72*
  BEMA, 167
  blade, *See* Blade coaters
  blank, 169, 170
  brush, 34, 78
  cast, 76–78
  Champion-Hamilton, 122
  Combined Locks, 11
  Consolidated-Massey, 10–12
  Contracoater, 9

SUBJECT INDEX 527

curtain, 148, *150-156*
direct roll, 162-164
extrusion, 107, *122-146*
floating knife, 64-65
fountain, 156
Genpac, 157
gravure, 4, *15-33,* 140, 161, 169, 173, 175
gravure-reverse roll, 4
hot melt, *147-172*
hybrid gravure, 24
inverted blade, 35
inverted knife, 65
Kimberly-Clark-Mead, 11
kiss roll, 74, 103, 160, 169
knife, *60-67,* 91, 161, 181
knife-over-blanket, 63-64
knife-over-roll, 8, 60, 61, 64, 94
Levelon, 67-69
meniscus, 78-79
Microtransfer, 12
offset gravure, 4, *24-27,* 140, 161
Park, 157-159
pressure roll, 169
puddle, 35, 48-50
reverse offset gravure, 26-27
reverse roll, *3-10,* 158, 160, 162-165, 169
roll, *3-33,* 60, *67-78,* 148, 158, 162, 169
roll flinger, 103
rotary screen, *81-96,* 173, 176, 179, 329
St. Regis-Faeber, 11
screen, *81-96*
sequential knife, 64-65
slot orifice, 35, 37, 148, *156-158,* 159, 168
spray. *See* Spray coaters
squeeze, *72-74,* 125, 140, 169
transfer roll, *10-14*
wax, 147, 168-171
Westvaco, 11
wire wound rod, 68, 70-71, 219
Zimmer, 163, 166
Coating
barrier, 161, 168, 200-204, 209-210
blade, mechanism of, 66-67
coil, 293
dams, 4, 120
extrusion, 107, *122-146*
film, 125, 209-211
knife, mechanism of, 66-67
metallized, 200-204, 209-211
metering systems, 213, *218-222*
pan feed, 3-4, 6-9, 22
paper, 125
pattern, 169
powder. *See* Powder coating
pressure, 66
primer, 125, 140
release, 509
roll, *3-33, 67-78*
simultaneous two side, 50
spray. *See* Spray coating
thickness measurement, 394, 428, 431
vacuum, 200-204, 209-211
wet waxed, 169
Cockroft Walton power supply, 335-336
Coefficient of
absorption, IR, 274
heat transfer, 251, 254-255, 260-264, 267, 306
reflection, IR, 274
transmission, IR, 274
Coextruders, *131-139. See also* Extruders
Coextrusion
choker bars, 136, 139
coathanger die, 131
combining adapter, 133, 139
dual slot die, 133, *137-138*
flow restrictor vanes, 136
multi-manifold die, 138
Cogging, 222
Coil coating, 293
Colburn equation, 254
Collection section, sheeting, 456
Collection section, diecutting, 478
Combiblade coater, 44, 46
Combined Locks coater, 11
Combining adapter, 133, 139
Combustible gas analyzer, 308-310
Compensator roller, diecutting, 468
Concentricity, 227
Conduction drying, 252, 254, *268-271*
Consolidated-Massey coater, 10-12
Constacoat, 44-46
Constant rate drying period, *251-252,* 257, 270, 289, 301, 308
Constant torque, 395
Contracoater, 9
Control
edge guide, 407-408
infrared emitters, 277
temperature, 142
tension, 140

528 SUBJECT INDEX

Control (*cont.*)
   thickness, 432
   thyristor, 196
   web guiding, 403
   winding, 195-197
Convection dryers, 140, *255-267*
Convective heat transfer, 173, 254
Conveyor dryers, 288, 295-296
Cooling
   cans, 120
   rolls, 163, 225, 240, 355, 486-487, 493-494
   UV irradiation unit, 326
Cooling down period, 252
Core chucking, 388, 440, 441
Core cutting, 421, 422, 424
Core shafts, 369
Corona discharge, 235, *241-247*, 249
Counterflow air dryer, 257, 356
Counter roller, screen coating, 89
Crossing, roll, 116-118
Crowning, roll, 162, 493, 226, 414, 116-117
Crease forming, 484
Creasing rule, 472, 474, 482-484
Cross firing, electron beam, 343-345
Crosslinking, 250
Curing, 250
Curling, web, 228, 229, 293, 313, 315-316, 444, 446, 470
Curtain coaters, 148, *150-156*
Cutback, pressure roll, 19
Cutter creaser, 465-466, 468, 472, 474, 475, 481, 483
Cutting section, sheeting, 451

Dampers, air, 307
Dams, 4, 120
Dancer roll, 400, 401, 422, 470
Deadband, 406-408
Deaerator, 156
Decurling, 316, 444, 446, 470
Deflection, roll, 225-227
Delamination, blistering, 231-233
Delivery, die cutting, 478-479
Delivery, sheeting, 453, 454
Detector, splice, 222, 223
Die cutting, *461-484*
   belt delivery, 480
   collector, 478
   compensator roller, 468

   cutter creaser, 465-466, 468, 472, 474-475, 481, 483
   decurl, 470
   delivery, 478-479
   drive, 471-472, 475
   feed, 465, 467-468, 470
   folding cartons, 463, 465, 466, 468, 478, 480, 482
   kicker, 476
   looper roll, 471
   lower bolster, 474-475
   metering, 467, 471
   overspeed, 471
   reciprocal, 464
   stacker, 481
   stripper, 465-466, 476-477
   tab breaker, 478
   tension, 470
   transfer, 465-466, 476-477
   upper bolster, 474-475
   web scanner, 468, 471
Dies. *See also* Extrusion
   adjusting bolts, 133
   block, 482
   clocker, 461
   extruder, 129, 131, 133, 137-140
   flat, 462
   fountain feed, 4, 9-10
   jig, 482
   laser, 482
   punching, 504
   steel rule, 461, 463, 481-483
Differential winding, 421, 424
Diffusion, 251-252, 257, 301
Direct coaters, 3, 16, 125, 140, 158, 160-161, 329
Direct gravure coaters, 16-18, 125, 140, 161, 329
Direct heaters, 302
Disk brakes, 374-375, 441
Doctor blade, 20-21, 219, 221
Double roll dip saturation, 217
Double rotary sheeter, 438, 451-452
Double wall roll, 225
Drag belt, 395, 397
Draw rolls, 354, 359
Driers. *See* Dryers
Drilled roll, 114-115
Drives, 126-127, 140, 195, 197, 271, 422, 450-452
   calender, 119
   die cutting, 471-472, 475

SUBJECT INDEX

inertia, 222
regulator, 222
winder, 379, 386, 388
Drum
 brakes, 373-375, 395, 397, 441
 dryers, 270-271, 296, 489, 491-492, 506
 support, 140
 winder, 368-387
Dry bonding, 29-31, 224-225, 228-229, 231-232, 234, 238-240
Dryers, 225-271, 285-301, See also Nozzles
 air
  floatation, 296-297
  foil, 297
  handling, 301-302
  heating, 303
  moving, 305-307
  recirculation, 308
 arch, 140,
 can, 268-271
 catenary, 266, 288, 293
 clam shell, 301
 conduction, 268, 270
 construction of, 288-289
 convection, 140, 255-267
 conveyor, 288, 295-296
 counter current flow, 257, 356
 dielectric, 285-289
 drum, 268-271, 288, 296, 298, 489, 491-492, 506
 festoon, 288-289, 292
 floatation, 141, 256, 288, 293
 head of, 217
 heat recovery, 304-305
 high frequency, 252, 254, 290
 high velocity, 268
 idler roll supported, 293, 295
 impingement, 255-257, 270, 293, 298
 inert gas, 310, 312-314
 infrared, 271-284
 lamination, 268
 oven length, 300-301
 parallel, 255-257
 performance, 306
 roll, 268, 269
 tenterframe, 289, 298-299
 textile, 298-299
 through, 255, 268-269
 U-type, 293-294
 vertical, 289, 293, 296
 zoned, 308

Drying, 229, 250-318
 conduction, 252, 254, 268-271
 constant rate, 251-252, 257, 270, 289, 301, 308
 convection, 252, 254-267
 cooling down period, 252
 curves, 252-254, 270
 falling rate, 253, 263, 301
 heat transfer, 244-255
 infrared radiation, 252, 254, 256, 270, 271-284
 latex coating, 252
 process, 250-252
 rate, 251-253, 256, 257, 263, 268, 270, 271
 through, 268
Dry lamination. See Dry bonding
Dry printing, 508
Dual slot die, 133, 137-138
Ductwork, 307
Duplex
 sheeter, 434-435, 437
 winders, 389, 423-424
Dust, 521
Dynamitron accelerator, 333, 336, 348

Edge contamination, 231
Edge guide, 140, 403, 406-408, 444
Edge sensor
 mechanical, 404-405, 408
 photoelectric, 406-408
 pneumatic, 404-405, 408
 ultrasonic, 407
Elastomer rolls, 226-228
Electroformed roll, 493
Electron accelerators, 333, 336, 338-340, 346
 acceleration current, 342, 345
 acceleration voltage, 342, 345
 beam guns, vacuum metallizing, 185-187, 189-191, 206
 cathode ray tube, 336-338
 Cockroft Walton, 335-336
 Dynamitron, 333, 336, 348
 insulating core transformer, 334, 336
 linear, 336-337, 340, 346
 maintenance, 340
 power supply, 333
 shielding, 347
Electron irradiation, 331-351
 absorption, 343
 beam power, 345

530 SUBJECT INDEX

Electron irradiation (*cont.*)
  crossfire, 343–345
  curable adhesives, 29
  dosage, 342, 345, 347, 349
  festooning, 338, 339
  ionization intensity, 343
  penetration, 342–343
  point source, 336, 337
Electrostatically charged powder, 178–180
Electrostatic spraying, 98, 100, 102
Embossing, *485–510*
  cast, 77
  nonthermoplastic, 495
  rolls, 488–496, 499–500
  vacuum, 496
  vinyl, 120
Emitters, infrared radiation, 277
Engraving
  acid etching, 31
  cylinders, 16–18, *31–33*, 492–493
  electromechanical, 33
  mechanical, 31
  pyramid, 31–32
  quadrangular, 31–32
  tri-helical, 31–32
Etching, plasma, 241
Ethylene-acrylic acid copolymer, 132–133
Ethylene, vinyl acetate copolymer, 147, 233
Expander rolls, 415, 417, 418
Expanding chucks, 369–371, 388
Explosion, 520
Extruders, *122–146*, 246, 247
  adapters, 129, 133, 139
  choker bars, 136, 139
  coathanger die, 131
  coextruders, 131–139
  combining adapter, *133–137*, 139
  dies, 129, 131, 133, *137–140*
  die adjusting bolts, 133
  dual slot die, 133, *137–140*
  feed-block, *133–137*, 139
  flow restrictor vanes, 136
  melter, 150
  multi manifold die, 133, *138–140*
Extrusion, *122–146*
  coating, 107, 122–146
  laminating, 122–146
  neck-in, 131
  safety, 145–146
  web handling, *140–145*

Falling rate drying period, 253, 263, 301
Fans, *305–307*, 356
Festoon accumulator, 426–427
Festoon dryers, 288–289, 292
Festooning, irradiation, 338–339
Filled roll, 496
Film
  orientation, *353–363*
    double bubble, 360
    draw stage, 354, 359
    flat table, 363
    machine direction, 353–354
    sequential, 352
    simultaneous sheet, 352, 358, 361
    simultaneous tubular, 352, 358
    spindle unit, 361
    umbrella type, 363
  shrink, 349
  splitting, 1–3, 18, 162
  treatment, corona discharge, 235, 241–242, 244, 246–247, 249
  wiping, 1
Finishing, hot melt, 170–171
Fire, 520
FLAKT floatation nozzle, 267
Flame polishing, 170
Flame treatment, 124, 241
Flammable mixture, 308, 313
Flexible packaging, 145, 209
Flexography, 246, 464
Floatation
  dryers, 141, 256, 288, 293
  FLAKT nozzle, 267
  nozzles, 264–265, 267, 296–297
Floating knife coater, 64
Fluidized bed coating, 173, *178–179*
Fluoropolymer coated roll, 233
Flying splice, 193, 368, 372, 390–391
Foam skimmer, 218
Foaming, hot melt, *171–172*
Foils, 124
Folding cartons
  crease forming, 472, 474, 482, 484
  diecutting, 464, 465–468, 478, 480, 482
Follower arm, 397, 398
Force transducer, 401–403
Fountain coaters, 156
Fountain rolls, 213, 214, 216–218
Friction calendering, 112
Fusible interlinings, 173–176
Fusion, 250

Galvano screens, 81, 83, 85–87
Gap control, 116, 222
Gas adsorption, 308, 309
Gas burners
  direct refractory impingement, 282–283
  gun type, 302
  premix, 282
  radiant tube, 282–283
  surface conbustion, 282–283
Gas heated infrared, 281
Genpac coater, 157
Gravure coaters, *See* Coaters
Green bond strength, 232
Grooved rolls, 362, 414, 416, 419
Guiding, web, 229, *401–404,* 408, 410, 469
  alignment, 444
  controls, 403
  edge, 140, 403, 406–408, 444
  intermediate, 409
  positive displacement type, 410, 412–413
  rolls, 197, 412
  side, 218
  steering, 140, 410, 411, 444, 468, 469
Gun spraying, 98–100

Hardness, rubber roll, 226–228
Haze, adhesive coating, 232, 236
Heat exchangers. *See also* Dryers, Drying
  recuperative, 304
  refrigerant, 305
  wheel, 304
Heat recovery, 304, 311
Heat transfer, 141, 251, 254–255, 257, 270–272, 275, 353. *See also* Drying.
  coefficient, 251, 254–255, 260–264, 267, 306
  Colburn equation, 254
  conduction, 268–271
  convection, 173, 254–267
  Pohlhausen equation, 255
Heated drum dryers, 270, 271, 296. *See also* Dryers
Heaters. *See* also Dryers, Drying
  direct, 302
  induction, 119
  infrared, *271–284,* 492
  line, 302
  oil, 303
  steam, 303
Heating rolls, film, 353, 355

High-frequency drying, 252, 254, 290
High vacuum coating, *182–212*
High velocity dryer, 268
Holders, coating rod, 71
Hook knife, 65
Hot melt, 161, 162
  air entrapment, 171
  BEMA coater, 167
  blank coaters, 169–170
  coaters, *147–172*
  deaerator, 156
  EVA, 169
  finishing, 170–171
  flame polishing, 170
  foaming, 171–172
  Genpac coater, 157
  knife coater, 161
  polyethylene, 169
  roll coaters, 148, 158, 169
  slot orifice coater, *156–158*
Hot roll dryer, 268, 269
Hot stamping, 505
Hot surface drying, 270
Hydraulic
  brakes, 441
  forces, 215, 217
  spraying, 100
Hydrodynamic force, 40–42, 67
Hydrostatic pressure, 67

Idler rolls, 215, 217, 228, 378, 383, 401, 427
Idler roll bearings, 295
Immersion, *214–218.* *See also* Saturation
Impingement
  atomizers, 98, 100–101
  dryers, 255–257, 270, 293, 298
  nozzles, 256–257, 265, 267
Impregnation. *See* Saturation.
Impression roll, 21–22
Impulse force, 40–41
Incineration, 308, 310–311
Induction heating, *186–189*
Induction neutralizers, electric charge, 514
Inert gas system drying, 310, 312, 314
Inflatable shaft, 219–220
Infrared
  absorption, 274–275
  convection heater, 281
  drying, 252, 254, 256, *270–284*

Infrared (cont.)
  radiation, 271-276, 278, 280-281, 353, 354, 356, 359
  radiators, 170, 173, 277, 281
  reflection, 274
  reflectors, 276, 324-326
  thickness sensors, 430
Inks, printing, 94, 246, 247, 505-506
Inks, subliming, 505-506
Interferences, air flow, 259
Inverted blade coaters, 35-38
Inverted knife coaters, 65
Ionization gauge, 428
Ionized air, 458, 514-515
Ionomer, 133
Irradiation. See Radiation

J-box accumulator, 428, 429
Jet fountain, 35, 37-38
Jig die, 482
Joints, rotary, 493, 506

Kimberly-Clark-Mead coater, 11
Kicker, diecutting, 476
Kiss roll coaters, 74-76, 103, 160, 169
Knife
  bull nose, 65-66
  contours, 65
  hook, 65
  radius, 65, 66
  spanish, 65-66
Knife coaters, 60-67, 91, 181
  coating mechanism, 66-67
  floating knife, 64-65
  hot melt, 161
  inverted knife, 65
  knife-over-blanket, 63-64
  knife-over-roll, 8, 60, 61, 64, 94
Knives
  rotary, 420
  sheeting, 65, 435-460
  slitting, 377-381, 418-421

Label stock, 168, 463
Lacquer screens, 81-87
Lamination, 28-31, 107-110, 141-145, 162, 168, 224-240
  calendering, 107-110
  dry bond, 224-225, 228-229, 231-232, 238

drying, 268
extrusion, 122
moisture assist, 236, 239
nip, 108, 109
roll to roll, 225
wet, 28-29
Lap splices, 390, 391
Laser die, 482
Latex drying, 252
Layboy, 456
Lay-on roll, 489, 490
L-calender, 107-108, 119
LEL. See Lower explosive limit
Levelon coater, 67-69
Linear accelerator, 336-337, 340, 346
Lint, 521
Lithography, 329
Load cells, 401-403
Looper roll, diecutting, 471
Lower bolster, diecutting, 474, 475
Lower explosive limit, 308, 310, 313

Magnetic brakes, 395
Magnetic tape, 211
Magnetron cathodes, 211
Magnetron sputtering, 191
Mass transfer, 251-252, 255-257, 301
Mayer rod coaters, 68, 70-71, 219
Mechanical brakes, 373
Melt coaters, high viscosity, 162-168
Melters, 148-151
  continuous, 150
  double arm, 148
  extruder, 150
  grid, 148-150
  drum, 149, 151
  roll, 149, 151
  tank, 148-149
Meniscus coater, 78-79
Mercury lamps, 320-322, 324, 326, 327
Metal coil, 178, 293
Metal foil, 124
Metallized
  barrier coating, 200-204, 209, 210
  film, 209
  paper, 207, 209
  yarn, 197
Metallizing
  sputtering, 182, 185, 191, 198, 202, 204-206, 210, 211

transfer process, 208-209
vacuum, *182-212*
Metering, reverse roll, 219, 221
Metering, saturation, 220, 221
Metering section, sheeter, 448
Metering systems, coating, 213, *218-222*
Metering unit, diecutting, 467, 471
Microcomputers, 128, 129, 132
Microcrystalline wax, 169
Microtransfer coater, 12
Moisture addition, 315, 316
Moisture assist, laminating, 236, 239
Mount Hope roll, 381
Multi-manifold die, extrusion, 133, *138-140*

Neck-in, extrusion, 131
Neutraliziers, static charge, *514-519*
Nip
  contact area, 227-228, 230-231, 237
  feed, coating, 3-6
  lamination, 108-109
  pressure, 116-117
  rolls, 359, 399
Nozzles, air
  air-knife, 54, 56
  atomization, 98
  distance and width, 261
  FLAKT, 267
  floatation, 264-267, 287, 296-297
  flow interference, 259
  foil, 256, 266, 267, 296-297
  impingement, *255-261,* 264, 265, 267
  round, 256-259, 261, 356
  slotted, 257-259, 261
Nozzles, spray, 98, 100, 101
Nuclear gauge, 428
Nuclear static neutralizers, 517
Nylon, 124, 133, 184, 360

Offset blades, saturation, 220-221
Offset gravure, 4, *24-27,* 140, 161, 329
Oil heaters, 303, 492, 506
Orientation. See Film orientation
Ovens. See Dryers
Overspeed, diecutting, 471
Orifices. See Nozzles
Oxbow effect, 118

Packaging, flexible, 145, 209
Pan feed, coating, 3-4, 6-9, 22

Paper
  coating, 76, 125
  finishing, 105
Paraffin wax, 123
Parallel dryer, 225-257
Park coater, 157-159
Pattern coating, 169
Pattern rolls, 485, 498
Penta screen, 83-85, 91
Perforation, 502, 504
Photoinitiators, 319
Planck's law, 272, 273
Plasma etching, 241
Pneumatic blade, 43
Polishing. See Smoothing
Polyacrylate, 355
Polyamide, 124, 147, 209
Polycarbonate, 184
Polyester, 124, 125, 147, 174, 197, 204, 209, 210, 245, 355, 356
Polyethylene, 107, 111, 123, 125, 132, 133, 137, 141, 147, 169, 184, 209, 233, 236, 241, 245, 248, 347, 361
Polyisobutylene, 169
Polypropylene, 124, 147, 184, 236, 241, 245, 248, 354, 355, 359, 360
Polystyrene, 355
Polyurethane, 93, 95, 125
Polyvinylchloride, 105, 107, 110, 165, 176, 355, 486, 489
Powder coating, *173-181*
  Caratsch process, 175-176
  distribution, 173
  electrostatic charge, 178-180
  feeding, 179-180
  fluidized bed, 173, 178
  paste, 176
  Saladin process, 176-177
  scatter, 173-174
  sintering, 173
  spot, 175, 177, 178
Preheating, web, 233-235
Pressure head, 152-154
Pressure rolls, 19, 235, 236
  coater, 169
  footprint, 224-227
Pressure-sensitive adhesive
  hot melt, 150
  labels, 168, 463
  tapes, 94, 95, 110, 116, 236, 423, 424, 505
Pre-wet section, saturation, 213-214, 217

534 SUBJECT INDEX

Primary winders, 364, 365, 371
Primer coatings, 125, 140
Printing
   dry, 508
   flexographic, 246, 464, 465
   hot transfer, 505, 509
   inks, 94, 241, 246, 247, 505, 506
   lithographic, 329
   rotogravure, 141, 387, 498, 506
   subliming inks, 505, 506
   valley, 497
Proportional control, 406, 408
Puddle coaters, 35, 48-50
Pull rolls, 140, 398, 399
Punching dies, 504

Quartz electrode, 245, 246, 248

Radiation
   electron beam. *See* Electron irradiation
   infrared. See Infrared
   sources, 321
   ultra violet. *See* Ultra violet radiation
Razor comb, 420
Razor splitting, 380, 418, 420
Reactive coating, 182
Reciprocal diecutters, 464
Reciprocating spray coater, 99
Reel spools, 365-366, 371, 374, 383
Reflection, coefficient of, 274
Reflectors, infrared, 276, *324-326*
Regenerative brakes, 140, 373, 441
Reject gate, sheeting, 453, 454
Release coating, 509
Re-reeler, 369, 387, 388, 391
Resistance heated boat, *186-189,* 206
Retractors, web, 233
Reverse offset gravure, *26-27*
Reverse roll coating, *3-10,* 158, 160, 162-165, 169
Reverse roll metering, 219, 221
Rewinding, 369, 387, 388, 391, 395, 403, 409, 410, 423, 424, 427. *See also* Winding
Rider rolls, 371, 375, 383, 385, 388, 391
Rodholder, 71
Rods, wire wound, 68, 70-71, 169, 219, 329
Roll
   analox, 500
   annealing, 354, 355, 359

applicators, 35-37, 57, 75, 500
back-up, 224, 226, 485, 486, 488-490, 494, 496, 499, 500
bearings, 115, 116, 295
bending, 116-118, 162, 493
bowed, 381, *414-417,* 444
calender, *114-119*
chill, 140-142, 144, 195, 196, 238
chilled iron, 114
coaters, *See* Coaters
cooling, 140-142, 144, 163, 195, 196, 225, 238, 240, 355, 486, 487, 493, 494
crossing, 116-118, 162
crowning, 116, 117, 162, 226, 414, 493
cutter, 421, 422
dancer, 400, 401, 470
deflection, 225-227
draw, 354
drilled, 114-115
electroformed, 493
embossing, 488-490, 492-494, 496, 499, 500
engraved, 16-18, 31-33, 492, 493, 495
expander, 414, 415, *417-419*
fountain, 213, 214, 216-218
gap, 116
grooved, 362, 414, 416, 419
guide, 197, 412
dryer, 268, 269
filled, 496
fluorocarbon covered, 233
heated, 114, 115, 119, 224, 225, 237, 353, 355
hot melt, 148, 158, 169
idler, 215, 217, 228, 378, 383, 401, 427
impression, 21-22
induction heated, 119
lay-on, 489, 490
looper, diecutting, 471
Mount Hope, 381
nip, 359, 399
pattern, 485, 498
pressure, 19, 169, 224-227, 235, 236
pull, 140, 398, 399
rewind, 145, 371, 375, 383, 385, 388, 391
rider, 371, 375, 383, 385, 388, 391, 424
rubber covered, 142, 143, 226-228
skewing, 226
sleeve, 132
smoothing, 228
spreader, 228, *381-384, 414-417,* 444
squaring, 413, 414

squeeze, 219
steering, 140
stop, 222
stripper, 120, 487, 490, 492
tension, 388, 402
unwind, 440
web guiding, 197, 412
Rotary
  disk atomizer, 102, 103
  screen coating, 81, 173, 176, 179, 329
  joints, 493, 506
  knife, 420
  sheeter, 434, 435, 439, 451
  spreader, 382
  unions, 225, 492
Rotogravure, 464
Round nozzles, 256-259, 261, 356
Rubber calendering, 105-108

St. Regis-Farber coater, 11
Safety, *145-146*, 235, 458
Sanding, 502, 503
Saturators, 162, *213-223*
  doubleroll dip, 217
  immersion, 214-218
  metering, 220, 221
  pre-wet, 213, 217
  saw tooth, 215
  scrapers, 219, 220
  waterfall, 216
Screen coaters, *81-96*, 173, 176, 179, 329
Screens
  Galvano, 83, 85-87
  Galvano direct, 81
  lacquer, 81-87
  Penta, 83-85, 91
  silk, 329
Sensors
  back scatter, 430
  edge, *404-408*
  safety, 235
  tension, 399, 400
  thickness, 430, 432
Separator pan, 56, 57
Sequential knife coater, 64, 65
Shaft, inflatable, 219, 220
Shaftless winding, 367, 384, 385, 440-442
Sheeting, *434-460*
  collection section, 456
  counter, 453

cutting section, 451
delivery system, 453, 454
double rotary, 438, 451, 452
duplex, 434, 435, 437
knife, 65, 451
layboy, 456
metering section, 448
overlapping, 454, 455
reject gate, 453, 454
rotating knife, 451
rotary, 434, 435, 439, 451
safety, 458
sheet length, 441, 451
simplex, 434, 451
single rotary, 451
size, 435, 438, 441, 451
staircut, 450
Shielding
  electron, 347
  UV radiation, 328
Shrink film, 349
Shutter, UV radiation, 326
Silicone, 94, 109, 245
Skewing, roll, 226, 444
Skimmer, foam, 218
Slat expander, 414, 415, 419
Sleeve, roll, 132
Slicer, 420
Slip additive, 247
Slip clutch, 422
Slitting, 377, 384, 386-388, 394, 418, 420, 423, 424
  burst cut, 380
  knives, *377-381*, 418-421
  razor, 380, 418, 420
  score, 379
  shear, 376-381, 383, 425
  trim, 444, 446-449, 451
Slot orifice coater, 35, 37, 148, 156, 159, 168
Slot nozzles, 356
S-matic coater, 44, 45
Smoothing bars, 22, 71-73, 157, 228, 501
Solar control films, 210-211
Solvent incineration, 308, 310, 311
Spinning disk atomizer, 97, 98
Splice
  butt, 390
  detector, 222, 223, 444, 446
  flying, 193, 368, 372, 390, 391
  lap, 390, 391
  relief, 120

536 SUBJECT INDEX

Splicing, 140, 368, 390, 392, 418
Splitting, film, 1-3, 18, 162
Spools, 365, 366, 371, 374, 383, 425
Spray coating, *97-104,* 173, 179-180
  airless, 97, 100
  atomization, *97-98*
  atomizers, 97-98
  reciprocating, 99
Spray gun, 98-100
Spreader rolls, 228, *381-384,* 388, *414-417,* 444
Sputter coating, 182, 185-186, 191-192, 198-200, 202, *204-207,* 210, 211
Squaring rolls, 413, 414
Squeegee blade, 87-92, 95
Squeeze roll coaters, *72-74,* 125, 140, 169
Squeeze rolls, 219
Stacker section, diecutting, 481
Stand
  self loading, 440
  shaft type, 440-442
  turret, 440
  unwind, 442-445
Static aerate, 179
Static electricity, *511-523*
  bars, 458
  charge, 444
  elimination, 458, *514-519*
Steam heating, 303, 489, 491, 492
Steel coil, 180
Steel rule, die, 461, 463, 481-483
Steering. See Guiding, web
Stefan-Boltzmann law, 272
Strike-through, 63, 230
Striping, 75
Stripper, diecutting, 465, 466, 476, 477
Stripper rolls, 120, 487, 490, 492
Styrene butadiene rubber, 110, 147, 152, 169
Sueding, 502
Surface
  tension, 241, 242
  tension tests, 243
  treatment, *241-249*
    corona discharge, 235, *241-247,* 249
    flame, 124, 241
    oxidation, 124
    treat level measurement, 243
  winders, 382, 398
Sym-Lam coater, 50

Tab breaker section, diecutting, 478
Tachometer, 222
Tapes, pressure-sensitive, 94, 95, 110, 116, 236, 423, 424, 505
Temperature control, 142
Tendency drive, 295
Tension, web, 140, 195-197, 215, 217, 218, 228, 233, 240, 247, 368, 374, 376, 383, 394, 396, 399, 401, 441, 470
  control, 400-402, 470
  sensing, 140, 388-400, 444
  winding, 397
Tensioning, 395, 399
Tenter,
  clips, 357-359
  frame dryers, *289-299*
  frames, 359, 361, 413
  heating systems, 356, 357
  tracks, 358, 361
Textile dryers, 289, 299
Texturing, 485, 486
Thermal bonding, 224, 225, 228, 233-235, 237, 239, 240
Thermoplastic polymers, 110, 124
Thermoplastic webs, 485
Thickness measurement, 202, 203, 394, 428, 430-432
Thin film technology, 183
Through dryer, 196
Thyristor control, 196
TIR. See Total indicator runout
Torque, 140, 371, 375, 385, 390, 395, 397
Total indicator runout, 3, 12, 19
Tracking, 394
Transfer roll coaters, *10-14*
Transfer section, die cutting, 476, 477
Transmission, infra red, 274
Traversing frame, 432
Triboelectric series, 513
Trim slitter, 444, 446-449, 451
Turning bar, 144
Turret winder, 366, 368, 372, 440
Twinblade coater, 50

Ultra violet radiation, *319-330*
  cooling, 326
  curable coatings, 29, 94, 329
  mercury lamps, 320-332, 324, 326, 327
  photo initiators, 319

SUBJECT INDEX 537

shielding, 328
shutter, 326
Unwind, 140, 144, 367, 368, 371-374, 383, 390, 402, 409. *See also* Winding
    brakes, 373-375
    roll, *440-443*
    stand, 367
    tension control, 140
Upper bolster, diecutting, 474, 475
U-type dryer, 293, 294

Vacply coater, 48
Vacuum embossing, 496
Vacuum metallizing, *182-212*
    aluminum coating, 184, 200-204, 209-211
    barrier coating, 200-204, 209, 210
    capacitor roll coaters, 203
    chamber, 186, 187, 199
    coating plants, 206
    coating sources, *185-189,* 206
    continuous, 193, 195
    electron beam guns, 185-187, *189-191,* 206
    film, 209-211
    paper, 207, 209
    processing sequence, 185
    roll coaters, 203-205
    semi-continuous, 192, 195
    sputtering, 182, 185, 191, 198, 202, *204-206,* 210, 211
    thickness, 202-203
    transfer process, 208-209
    two-source system, 197-198, 200, 201
    vacuum pumps, 200-202, 206
    yarn, 197
Variable speed transmission, 452
Vertical dryers, 289, 293, 296
Vinyl
    calendering, 107, 116
    embossing, 120
    plastisols, 90, 93, 95, 107

Wallcoverings, 489
Wax, 168, 169
Wax coaters, 147, 168-171
Web
    alignment, 444
    curling, 228, 229, 293, 313, 315, 444, 446

    guiding. *See* Guiding, web
    handling, 140-145, *394-433*
    path control, 228
    retractors, 233
    scanner, 468, 471
    steering. See Guiding, web.
    tension. See Tension, web
    tracking, 394
    winding. *See* Winders, web
    wrinkling 226, 233, 235, 240
Wedge blocks, 222
Weir, curtain coater, 152, 153
Westvaco coater, 11
Wet bonding, 224, 225, 229, 230
Wet laminating, 28-29
Wet strength, 215
Wettability, 241, 242, 247
Wet waxed coatings, 169
Wien's displacement law
Winders, web, *364-394,* 397, 422
    center, 389, 422-424
    control, 195-197
    differential, 421, 424
    drive, 379, 386, 388
    drum, 386, 387, 391
    duplex, 389, 422-424
    flying splice, 193, 368, 372, 390-391
    primary, 364, 365, 371
    rewinding, 369, 387, 388, 391, 395, 403, 409, 410, 423, 424, 427
    secondary, 369, 371, 391
    shaftless, 367, 384, 385, 440-441
    stands, 442-445
    surface, 382, 389
    tapered tension, 140
    tertiary, 389
    vacuum metallizing, 195
Wire wound rod, 68, 70-71, 169, 219, 329
Wrap angle, 399
Wrinkles, 226, 233, 235, 240

Xenon lamp, 322, 324

Z-calender, 107, 109
Zinc coating, 203
Zinc evaporation, 186, 187
Zimmer coater, 163, 166
Zone dryers, 308